D0487709

Ajith Abraham, Crina Grosan, Vitorino Ramos (Eds.)

Stigmergic Optimization

Studies in Computational Intelligence, Volume 31

Editor-in-chief
Prof. Janusz Kacprzyk
Systems Research Institute
Polish Academy of Sciences
ul. Newelska 6
01-447 Warsaw
Poland
E-mail: kacprzyk@ibspan.waw.pl

Further volumes of this series can be found on our homepage: springer.com

Vol. 14. Spiros Sirmakessis (Ed.)
Adaptive and Personalized Semantic Web, 2006
ISBN 3-540-30605-6

Vol. 15. Lei Zhi Chen, Sing Kiong Nguang, Xiao Dong Chen
Modelling and Optimization of Biotechnological Processes, 2006
ISBN 3-540-30634-X

Vol. 16. Yaochu Jin (Ed.)
Multi-Objective Machine Learning, 2006
ISBN 3-540-30676-5

Vol. 17. Te-Ming Huang, Vojislav Kecman, Ivica Kopriva
Kernel Based Algorithms for Mining Huge Data Sets, 2006
ISBN 3-540-31681-7

Vol. 18. Chang Wook Ahn
Advances in Evolutionary Algorithms, 2006
ISBN 3-540-31758-9

Vol. 19. Ajita Ichalkaranje, Nikhil Ichalkaranje, Lakhmi C. Jain (Eds.)
Intelligent Paradigms for Assistive and Preventive Healthcare, 2006
ISBN 3-540-31762-7

Vol. 20. Wojciech Penczek, Agata Półrola
Advances in Verification of Time Petri Nets and Timed Automata, 2006
ISBN 3-540-32869-6

Vol. 21. Cândida Ferreira
Gene Expression on Programming: Mathematical Modeling by an Artificial Intelligence, 2006
ISBN 3-540-32796-7

Vol. 22. N. Nedjah, E. Alba, L. de Macedo Mourelle (Eds.)
Parallel Evolutionary Computations, 2006
ISBN 3-540-32837-8

Vol. 23. M. Last, Z. Volkovich, A. Kandel (Eds.)
Algorithmic Techniques for Data Mining, 2006
ISBN 3-540-33880-2

Vol. 24. Alakananda Bhattacharya, Amit Konar, Ajit K. Mandal
Parallel and Distributed Logic Programming, 2006
ISBN 3-540-33458-0

Vol. 25. Zoltán Ésik, Carlos Martín-Vide, Victor Mitrana (Eds.)
Recent Advances in Formal Languages and Applications, 2006
ISBN 3-540-33460-2

Vol. 26. Nadia Nedjah, Luiza de Macedo Mourelle (Eds.)
Swarm Intelligent Systems, 2006
ISBN 3-540-33868-3

Vol. 27. Vassilis G. Kaburlasos
Towards a Unified Modeling and Knowledge-Representation based on Lattice Theory, 2006
ISBN 3-540-34169-2

Vol. 28. Brahim Chaib-draa, Jörg P. Müller (Eds.)
Multiagent based Supply Chain Management, 2006
ISBN 3-540-33875-6

Vol. 29. Sai Sumathi, S.N. Sivanandam
Introduction to Data Mining and its Applications, 2006
ISBN 3-540-34689-9

Vol. 30. Yukio Ohsawa, Shusaku Tsumoto (Eds.)
Chance Discoveries in Real World Decision Making, 2006
ISBN 3-540-34352-0

Vol. 31. Ajith Abraham, Crina Grosan, Vitorino Ramos (Eds.)
Stigmergic Optimization, 2006
ISBN 3-540-34689-9

Ajith Abraham
Crina Grosan
Vitorino Ramos

Stigmergic Optimization

With 104 Figures and 27 Tables

Springer

Dr. Ajith Abraham
IITA Professorship Program
School of Computer Science
and Engineering
Chung-Ang University, 221
Heukseok-dong
Dongjak-gu Seoul 156-756
Republic of Korea
E-mail: ajith.abraham@ieee.org;
abraham.ajith@acm.org

Dr. Crina Grosan
Department of Computer Science
Faculty of Mathematics
and Computer Science
Babes-Bolyai University
Cluj-Napoca, Kogalniceanu 1
400084 Cluj - Napoca
Romania
E-mail: cgrosan@cs.ubbcluj.ro

Dr. Vitorino Ramos
CVRM-IST, IST
Technical University of Lisbon
Av. Rovisco Pais
1049-001, Lisboa
Portugal
E-mail: vitorino.ramos@alfa.ist.utl.pt

Library of Congress Control Number: 2006927548

ISSN print edition: 1860-949X
ISSN electronic edition: 1860-9503
ISBN-10 3-540-34689-9 Springer Berlin Heidelberg New York
ISBN-13 978-3-540-34689-0 Springer Berlin Heidelberg New York

Springer is a part of Springer Science+Business Media
springer.com

Printed in The Netherlands

Cover design: deblik, Berlin
Typesetting by the authors and SPi
Printed on acid-free paper SPIN: 11613657 89/SPi 5 4 3 2 1 0

Preface

Biologists studied the behavior of social insects for a long time. It is interesting how these tiny insects can find the shortest path for instance between two locations without any knowledge about distance, linearity, etc. After millions of years of evolution all these species have developed incredible solutions for a wide range of problems. Some social systems in nature can present an intelligent collective behavior although they are composed by simple individuals with limited capabilities. The intelligent solutions to problems naturally emerge from the self-organization and indirect communication of these individuals.

The word stigmergy was named about fifty years ago by *Pierre-Paul Grasse*, a biologist studying ants and termites. Self-Organization in social insects often requires direct and indirect interactions among insects. Indirect interactions occur between two individuals when one of them modifies the environment and the other responds to the new environment at a later time. Such an interaction is an example of stigmergy. A famous example of stigmergy is the pheromonal communication among ants, whereby ants engaging in certain activities leave a chemical trail which is then followed by their colleagues.

This book deals with the application of stigmergy for a variety of optimization problems. Addressing the various issues of stigmergic optimization using different intelligent approaches is the novelty of this edited volume. This Volume comprises 12 Chapters including an introductory chapter giving the fundamental definitions, inspirations and some research challenges. Chapters were selected on the basis of fundamental ideas/concepts rather than the thoroughness of techniques deployed. The twelve Chapters are organized as follows.

Grosan and Abraham in the introductory chapter summarize some of the well known stigmergic computational techniques inspired by nature. These techniques are mainly used for solving optimization related problems developed by mimicking social insects' behavior. Some facts about social insects namely ants, bees and termites are presented with an emphasis on how they could interact and self-organize for solving real world problems.

In Chapter two, *Cazangi et al.* propose two strategies for multi-robot communication based on stigmergy. In the deterministic strategy, the robots deposit artificial pheromones in the environment according to innate rules. In the evolutionary strategy, the robots have to learn how and where to lay artificial pheromones. Each robot is controlled by an autonomous navigation system based on a learning classifier system, which evolves during navigation from no a priori knowledge, and should learn to avoid obstacles and capture targets disposed in the environment.

Swaminathan and Minai in Chapter three introduce how the circle formation algorithm can be used as a means for solving formation and organization problems in multi-robot systems. The real challenge is in developing simple algorithms which the robots can execute autonomously, based on data from their vicinity, to achieve global behavior. Circle formation can be seen as a method of organizing the robots in a regular formation which can then be exploited. This involves identifying specific robots to achieve different geometric patterns like lines, semicircles, triangles and squares, and dividing the robots into subgroups, which can then perform specific group-wise tasks. The algorithms that achieve these tasks are entirely distributed and do not need any other intervention.

In Chapter four, *Wurr and Anderson* present the design and implementation of cooperative methods for reactive navigation of robotic agents. The proposed approach allows a team of agents to assist one another in their explorations through stigmergic communication. These methods range from simple solutions for avoiding specific problems such as individual local maxima, to the construction of sophisticated branching trails. The authors evaluated these methods individually and in combination using behavior-based robots in a complex environment.

Gerasimov et al. in the fifth Chapter explore how to bring swarm optimization methodology into the real-world self-assembly applications. Authors describe a software system to model and visualize 3D or 2D self-assembly of groups of autonomous agents. The system makes a physically accurate estimate of the interaction of agents represented as rigid cubic or tetrahedral structures with variable electrostatic charges on the faces and vertices. Local events cause the agents' charges to change according to user-defined rules or rules generated by genetic algorithms.

In Chapter six, *Ventrella* proposes a framework to evolve Cellular Automata (CA) transition functions that generate glider-rich dynamics. The technique is inspired by particle swarm optimization where the particles guide evolution within a single, heterogeneous 2D CA lattice having unique, evolvable transitions rules at each site. The particles reward local areas which give them a good ride, by performing genetic operators on the CA's transition functions while the CA is evolving.

Roth and Wicker in the seventh Chapter present a biologically inspired algorithm named *Termite* which addresses the problem of routing in the presence of a dynamic network topology. The stochastic nature of Termite is explored to find a heuristic to maximize routing performance. The pheromone decay rate is adjusted such that it makes the best possible estimate of the utility of a link to deliver a packet to a destination, taking into account the volatility or correlation time, of the network.

In Chapter eight, *Meyer et al.* present *stochastic diffusion search*, a novel swarm intelligence metaheuristic that has many similarities with ant and evolutionary algorithms. Authors explain the concept of partial evaluation of fitness functions, together with mechanisms manipulating the resource allocation of population based search methods. Empirical results illustrate that the stochastic process ensuing from these algorithmic concepts has properties that allow the algorithm to optimize noisy fitness functions, to track moving optima, and to redistribute the population after quantitative changes in the fitness function.

Mostaghim et al. in Chapter nine present a Linear Multi-Objective Particle Swarm Optimization (LMOPSO) algorithm. In the presence of the linear (equality and inequality) constraints, the feasible region can be specified by a polyhedron in the search space. LMOPSO is designed to explore the feasible region by using a linear formulation of particle swarm optimization algorithm. This method guarantees the feasibility of solutions by using feasibility preserving methods. LMOPSO is different from the existing constraint handling methods in multiobjective optimization in the way that it never produces infeasible solutions.

In the tenth Chapter, *El Abd and Kamel* survey the different cooperative models that have been implemented using particle swarm optimization algorithm for different applications and a taxonomy for classifying these models is proposed. Authors focus on the different parameters that can influence the behavior of such models and illustrate how the performance of a simple cooperative model can change under the influence of these parameters.

Chu et al. in the eleventh Chapter present a series of particle warm optimization algorithms is introduced namely the original particle swarm optimization algorithm, Particle Swarm Optimization algorithm with weighted factor (PSO) and Parallel Particle Swarm Optimization algorithm (PPSO). Further authors introduce a hybrid combination of simulated annealing with PPSO named Adaptive Simulated Annealing - Parallel Particle Swarm Optimization (ASA-PPSO). Experiment results obtained using the bench mark function sets indicate the usefulness of the proposed ASA-PPSO method.

In the last Chapter, *Iourinski et al.* theoretically prove that the swarm intelligence techniques, and the corresponding biologically inspired formulas are indeed statistically optimal (in some reasonable sense).

We are very much grateful to the authors of this volume and to the reviewers for their tremendous service by critically reviewing the chapters. The editors would like to thank Dr. Thomas Ditzinger (Springer Engineering Inhouse Editor, Studies in Computational Intelligence Series), Professor Janusz Kacprzyk (Editor-in-Chief, Springer Studies in Computational Intelligence Series) and Ms. Heather King (Editorial Assistant, Springer Verlag, Heidelberg) for the editorial assistance and excellent cooperative collaboration to produce this important scientific work. We hope that the reader will share our excitement to present this volume on '*Stigmergic Optimization*' and will find it useful.

April, 2006

Ajith Abraham, Chung-Ang University, Seoul, Korea
Crina Grosan, Cluj-Napoca, Babeş-Bolyai University, Romania
Vitorino Ramos, Technical University of Lisbon, Portugal

Contents

1

Stigmergic Optimization: Inspiration, Technologies and Perspectives

Crina Grosan[1] and Ajith Abraham[2]

[1]Department of Computer Science
Babeş-Bolyai University, Cluj-Napoca, 3400, Romania
[2]IITA Professorship Program,
School of Computer Science and Engineering
Chung-Ang University, Seoul 156-756, Korea
cgrosan@cs.ubbcluj.ro, ajith.abraham@ieee.org

Summary. This Chapter summarizes some of the well known stigmergic computational techniques inspired by nature, mainly for optimization problems developed by mimicking social insects' behavior. Some facts about social insects namely ants, bees and termites are presented with an emphasis on how they could interact and self organize for solving real world problems. We focused on ant colony optimization algorithm, bees behavior inspired algorithms, particle swarm optimization algorithm and bacterial foraging algorithm.

1.1 Introduction

Nature has inspired researchers in many different ways. Airplanes have been designed based on the structures of birds' wings. Robots have been designed in order to imitate the movements of insects. Resistant materials have been synthesized based on spider webs. The fascinating role that insects play in our lives is obvious. It is interesting how these tiny insects can find the shortest path for instance between two locations without any knowledge about distance, linearity, etc.

Biologists studied the behavior of social insects for a long time. For decades, entomologists have known that insect colonies are capable of complex collective action, even though individuals adhere to straightforward routines. When foraging, for example, workers appear to march to a drumbeat that dictates when to turn and when to lay down pheromone to guide other workers. As simple as these rules are, they create an effective dragnet to haul in food as efficiently as possible. In this manner, ants have been solving problems very skillfully every day of their lives for the last 100 million years [120].

After millions of years of evolution all these species have developed incredible solutions for a wide range of problems. Biologically inspired systems have been

C. Grosan and A. Abraham: *Stigmergic Optimization: Inspiration, Technologies and Perspectives*, Studies in Computational Intelligence (SCI) **31**, 1–24 (2006)
www.springerlink.com

gaining importance and it is clear that many other ideas can be developed by taking advantage of the examples that nature offers.

Some social systems in nature can present an intelligent collective behavior although they are composed by simple individuals with limited capabilities. The intelligent solutions to problems naturally emerge from the self-organization and indirect communication of these individuals. These systems provide important techniques that can be used in the development of distributed artificial intelligent systems [7].

Rest of this Chapter is organized as follows. Section 1.2 introduces Entomology and Stigmergy followed by some factual contents about social insects in Section 1.3. Some nature inspired computational algorithms are depicted in Section 1.4 and conclusions are provided in Section 1.5.

1.2 Entomology and Stigmergy

Entomology is the scientific study of insects. Hogue [79] noted that Entomology has long been concerned with survival (economic or applied Entomology) and scientific study (academic Entomology), but the branch of investigation that addresses the influence of insects (and other terrestrial Arthropoda, including Arachnids, Myriapods, etc) in literature, language, music, the arts, interpretive history, religion, and recreation has only recently been recognized as a distinct field. This is referred to as *cultural entomology*.

Over the last fifty years biologists have unraveled many of the mysteries surrounding social insects, and the last decade has seen an explosion of research in the fields variously referred to as *collective intelligence*, *swarm intelligence* and *emergent behavior*. Even more recently the swarm paradigm has been applied to a broader range of studies, opening up new ways of thinking about theoretical Biology, Economics and Philosophy.

The South African Scientist - Eugène Marais (1872-1936) - is considered as one of the first scientists who paid attention to the behavior of social insects. His work on termites led him to a series of stunning discoveries. He developed a fresh and radically different view of how a termite colony works, and indeed of what a termite colony is. In 1923 he began writing a series of popular articles on termites for the Afrikaans press and in 1925 he published a major article summing up his work in the Afrikaans magazine *Die Huisgenoot*. In 1925 Marais published an original research article and some conclusions about the white ant. In 1927, Maurice Maeterlinck (1862-1949), a Nobel Prize winner, lifted half of Marais's work and published it without any acknowledgement, as the book "*The Life of the White Ant*" [94]. This plagiarization may well have been a major factor in Marais's final collapse. Plagued for many years by ill-health and an addiction to morphine, he took his own life in March 1936. Marais's book "*The Soul of the White ant*" was published posthumous in 1937 [110].

Konrad Lorenz (1903-1989) is widely credited as being the father of Ethology, the scientific study of animal behavior, with his early work on imprinting and instinctive behavior. Although Marais had created a detailed document on termites, he was unaware of the mechanics of termite communication. How is it that a group of tiny, short-sighted, simple individuals are able to create the grand termite mounds, sometimes as high as six meters, familiar to inhabitants of dry countries? The answer to this question was first documented by the French Biologist, Pierre-Paul Grassé in his 1959 study of termites [76]. Grassé noted that termites tended to follow very simple rules when constructing their nests.

- First, they simply move around at random, dropping pellets of chewed earth and saliva on any slightly elevated patches of ground they encounter. Soon small heaps of moist earth form.
- These heaps of salivated earth encourage the termites to concentrate their pellet-dropping activity and soon the biggest heaps develop into columns which will continue to be built until a certain height, dependent on the species, is reached.
- Finally, if a column has been built close enough to other columns, one other behavior kicks in: the termites will climb each column and start building diagonally towards the neighboring columns.

Obviously, this does not tell the whole story but a key concept in the collective intelligence of social insects is revealed: the termites' actions are not coordinated from start to finish by any kind of purposive plan, but rather rely on how the termite's world appears at any given moment. The termite does not need global knowledge or any more memory than is necessary to complete the sub-task in hand; it just needs to invoke a simple behavior dependent on the state of its immediate environment. Grassé termed this *stigmergy*, meaning 'incite to work', and the process has been observed not just in termites, but also in ants, bees, and wasps in a wide range of activities. Writing about termites, he offered a more general definition of stigmergy - "the stimulation of the workers by the very performances they have achieved" [80].

Grassé quoted "*Self-Organization in social insects often requires interactions among insects: such interactions can be direct or indirect. Direct interactions are the "obvious" interactions: antennation, trophallaxis (food or liquid exchange), mandibular contact, visual contact, chemical contact (the odor of nearby nestmates), etc. Indirect interactions are more subtle: two individuals interact indirectly when one of them modifies the environment and the other responds to the new environment at a later time. Such an interaction is an example of stigmergy*".

Studying nest reconstruction in termites, Grassé showed that it doesn't rely on direct communication between individuals. The nest structure itself coordinates the workers' tasks, essentially through local pheromone concentrations. The state of the nest structure triggers some behaviors, which then modify the nest structure and trigger new behaviors until the construction is over [70].

According to Gordon [70], the application of stigmergy to computation is surprisingly straightforward. Instead of applying complex algorithms to static datasets, through studying social insects we can see that simple algorithms can often do just as well when allowed to make systematic changes to the data in question [70]. A famous example of stigmergy is pheromonal communication, whereby ants engaging in certain activities leave a chemical trail which is then followed by their colleagues.

This ability of ants to collectively find the shortest path to the best food source was studied by Jean-Louis Deneubourg ([36]-[44], [71]-[75], [104]-[108]). He demonstrated how the Argentine ant was able to successfully choose the shortest between the two paths to a food source. Deneubourg was initially interested in self organization, a concept which until then had been the fare of chemists and physicists seeking to explain the natural order occurring in physical structures such as sand dunes and animal patterns ([1]-[4], [5], [6], [8], [14]-[19], [26], [62], [84], [111], [133]-[135], [138]). Deneubourg saw the potential for this concept, which by 1989 had turned into a sizeable research project amongst Physicists, to be applied to Biology. In his experiments, a group of ants are offered two branches leading to the same food source, one longer than the other. Initially, there is a 50% chance of an ant choosing either branch, but gradually more and more journeys are completed on the shorter branch than the longer one, causing a denser pheromone trail to be laid. This consequently tips the balance and the ants begin to concentrate on the shorter route, discarding the longer one. This is precisely the mechanism underpinning an ant colony's ability to efficiently exploit food sources in sequential order: strong trails will be established to the nearest source first, then when it is depleted and the ants lose interest, the trails leading to the next nearest source will build up [70].

1.3 Facts About Social Insects

Among all social insects, the ants, social bees, social wasps, and termites, dominate the environment in most terrestrial habitats.

1.3.1 Facts about Ants

Shortly, while talking about ants, we can use Charlotte Sleigh's words: "Ants are legion: at present there are 11,006 species of ants known; they live everywhere in the world except the polar icecaps; and the combined weight of the ant population has been estimated to make up half the mass of all insects alive today" [123].

Like all insects, ants have six legs. Each leg has three joints. The legs of the ant are very strong so they can run very quickly. If a man could run as fast for his size as an ant can, he could run as fast as a racehorse. Ants can lift 20 times their own body weight. An ant brain has about 250,000 brain cells. Mushroom shaped brain appendages have function similar to the gray-matter of human brains. A human brain has 10,000 million so a colony of 40,000 ants has collectively the same size brain as a human [137].

The average life expectancy of an ant is 45-60 days. Ants use their antennae not only for touch, but also for their sense of smell. The head of the ant has a pair of large, strong jaws. The jaws open and shut sideways like a pair of scissors. Adult ants cannot chew and swallow solid food. Instead they swallow the juice which they squeeze from pieces of food. They throw away the dry part that is left over. The ant has two eyes; each eye is made of many smaller eyes called compound eyes. The abdomen of the ant contains two stomachs. One stomach holds the food for itself and second stomach is for food to be shared with other ants. Like all insects, the outside of their body is covered with a hard amour this is called the exoskeleton. Ants have four distinct growing stages, the egg, larva, pupa and the adult. Biologists classify ants as a special group of wasps - (*Hymenoptera Formicidae*). There are over 10000 known species of ants. Each ant colony has at least one or more queens. The job of the queen is to lay eggs which the worker ants look after. Worker ants are sterile; they look for food, look after the young, and defend the nest from unwanted visitors. Ants are clean and tidy insects. Some worker ants are given the job of taking the rubbish from the nest and putting it outside in a special rubbish dump. Each colony of ants has its own smell. In this way, intruders can be recognized immediately. Many ants such as the common Red species have a sting which they use to defend their nest. The common Black Ants and Wood Ants have no sting, but they can squirt a spray of formic acid. Some birds put ants in their feathers because the ants squirt formic acid which gets rid of the parasites. The Slave-Maker Ant (*Polyergus Rufescens*) raids the nests of other ants and steals their pupae. When these new ants hatch they work as slaves within the colony [81].

When searching for food, ants initially explore the area surrounding their nest in a random manner. While moving, ants leave a chemical pheromone trail on the ground. Ants are guided by pheromone smell. Ants tend to choose the paths marked by the strongest pheromone concentration . When an ant finds a food source, it evaluates the quantity and the quality of the food and carries some of it back to the nest. During the return trip, the quantity of pheromone that an ant leaves on the ground may depend on the quantity and quality of the food. The pheromone trails will guide other ants to the food source.

The indirect communication between the ants via pheromone trails enables them to find shortest paths between their nest and food sources as illustrated in Figure 1.1.

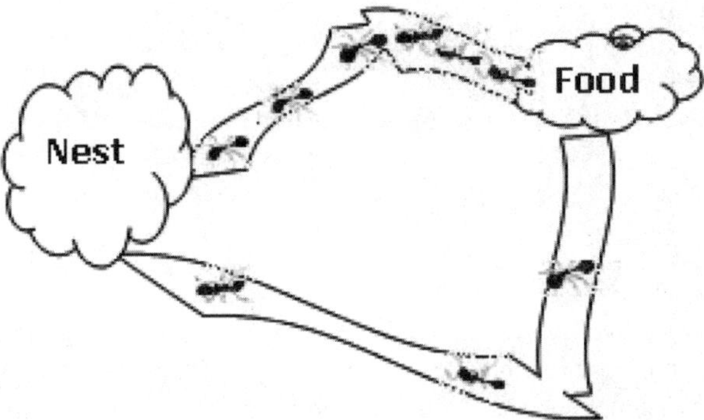

Fig. 1.1. The ants taking the shortest path can perform a greater number of trips between nest and food; implicitly the pheromone trail will be more than the one released by the ants following the longest path.

According to Truscio [137] both ants and humans share these endeavors:

- *livestock farming*: herd aphids and "milk" them for nectar-like food
- *cultivation*: growing underground gardens for food
- *childcare*: feeding young and providing intensive nursery care
- *education*: teaching younger ants the tricks of the trade
- *climate control*: maintaining a strict 77° F for developing ants
- *career specialization*: changing and learning new careers
- *civic duties*: responding with massive group projects
- *armed forces*: raising an army of specialized soldier ants
- *security*: warding off other ants, insects, and animals
- *earth movers*: move at least as much soil as earthworms
- *social planning*: maintain ratio of workers, soldiers, and reproductives
- *engineering*: tunnel from 2 directions and meet exactly midway
- *communications*: complex tactile, chemical communication system
- *flood control*: incorporate water traps to keep out rain
- *limited free will*: inter-relationships more symbiotic than coercive

1.3.2 Facts about Bees

There are two well known classes of bees: *European bees* and *Africanized bees*. A comparison between the two classes of bees is given in Table 1.1. The ancestors of the Africanized bee live throughout Africa, south of the Sahara Desert. African bees were accidentally introduced into the wild in South and North America during 1956. Brazilian scientists were attempting to create a new hybrid bee in the hopes of creating improved honey production. The Africanized bee escaped and began to attack the honey bees.

Table 1.1. African and European honey bees: a comparison

European Honey Bees	Africanized Bees
Pollinate flowers and crops	More aggressive
Calmed by smoke	Attack in larger groups
Swarm only when crowded	Make less honey
	Make less wax
	Hate high pitched sounds
	Swarm more often

Drones usually live five to ten weeks. Workers usually live about fifty days and all the workers are females. Queens live an average about three years and there is only one surviving queen bee in each colony. She mates over with many drones (male bees), and may lay 1500 eggs per day. The queen releases a pheromone that identifies her as the queen. When the beehive is overpopulated, Africanized Bees swarm to a local area to start a new hive. Too much warm or cold weather may cause swarming. Only one queen bee will rule. When the two queens reach the adult stage, they battle to the death for control of the hive. The cycle of swarming continues until the hive is worn out. An extremely aggressive Africanized bee colony may attack any 'threat' within 100 ft. and pursue for up to one-fourth a mile.

Africanized bees react to disturbance around the hive. They can stay angry for days after being disturbed. If one bee stings, it releases an alarm that smells like bananas. This pheromone causes the other bees to become agitated and sting. The Africanized bee, like the honey bee, dies when it stings. The tiny barbs on the stinger stick in the victim. When the bee tries to fly away, it rips its abdomen and eventually dies [82].

It is said that there is a relationship between bees and Fibonacci numbers. Fibonacci described the sequence "encoded in the ancestry of a male bee." This turns out to be the Fibonacci sequence. The following facts are considered:

- If an egg is laid by a single female, it hatches a male.
- If, however, the egg is fertilized by a male, it hatches a female.
- Thus, a male bee will always have one parent, and a female bee will have two.

If one traces the ancestry of this male bee (1 bee), he has 1 female parent (1 bee). This female had 2 parents, a male and a female (2 bees). The female had two parents, a male and a female, and the male had one female (3 bees). Those two females each had two parents, and the male had one (5 bees). If one continues this sequence, it gives a perfectly accurate depiction of the Fibonacci sequence. However, this statement is mostly theoretical. In reality, some ancestors of a particular bee will always be sisters or brothers, thus breaking the lineage of distinct parents [147].

1.3.3 Facts about Termites

Termites have been on Earth for over 50 million years. Some of their fossils date back to the Oligocene, Eocene, and Miocene periods. They have evolved into many different species. As of 1995 there were approximately 2,753 valid names of termite species in 285 genera around the world. The word 'termites' comes from the Latin word '*Tarmes*'. The Latin word was given to a small worm that makes holes in wood.

The *M. bellicosus* termites live in colonies but really they are more like families. Of some thirty or so insect orders, termites are the only one in which all species are categorized as highly social. They are very unique due to the fact that their colonies are based on monogamy. As far as entomologists know, they are the most sophisticated families ever to evolve in the universe. The termite colony has three separate stages: juvenile, adult, and senile. The survival of their species depends on their caste system. The smallest in size, yet most numerous of the castes are the workers. They are all completely blind, wingless, and sexually immature. Their job is to feed and groom all of the dependent castes. They also dig tunnels, locate food and water, maintain colony atmospheric homeostasis, and build and repair the nest. The soldiers' job is to basically defend the colony from any unwanted animals. Soldiers have larger heads that are longer and wider than that of the workers because it contains more muscles. The soldiers cannot feed themselves and must rely on the workers for this.

During swarming season the young males and females that are leaving the nest to make new colonies are exposed to birds, bats, reptiles and amphibians. One of the most dangerous predators of the *M. bellicosus* are driver ants. If their path crosses a mound of the *M. bellicosus* they will invade it by entering at the top where the building material is soft. Once inside the ants find little resistance since they have better eye sight and greater agility. It is rare that the colony is completely destroyed because some of the worker termites hide in the royal chamber. These termites continue the colony and before long life has returned to usual [82].

1.4 Nature Inspired Algorithms

Inspired by Nature became now a well known syntagma. Nature is offering models and humans are exploiting any interesting idea from this.

1.4.1 Ant colonies inspired algorithms

Ant colony optimization (ACO) was introduced around 1991-1992 by M. Dorigo and colleagues as a novel nature-inspired metaheuristic for the solution of hard combinatorial optimization problems [45], [50]-[54], [56]. Dorigo [56] was intrigued to learn how these virtually brainless creatures could create highly sophisticated messaging systems and build extremely complex architectural structures. Although an individual ant is quite small (measuring only 2.2 to 2.6 mm in length) and wanders quite aimlessly in isolation, a group of many ants exhibits extraordinarily intelligent behavior, recognizable to humans as meaningful pathways to food sources. This emergent intelligence can be summarized in the pseudocode below [83]:

1. At the outset of the foraging process, the ants move more or less randomly – this “random” movement is actually executed such that a considerable amount of surface area is covered, emanating outward from the nest.
2. If it is not carrying food, the ant “deposits” a nest pheromone and will prefer to walk in the direction of sensed food pheromone.
3. If it is carrying food, the ant deposits a food pheromone and will prefer to walk in the direction of sensed nest pheromone.
4. The ant will transport food from the source to the nest.

As a pheromone “trail” becomes stronger, the more ants follow it, leaving more pheromone along the way, which makes more ants follow it, and so on.

ACO is implemented as a team of intelligent agents which simulate the ants behavior, walking around the graph representing the problem to solve using mechanisms of cooperation and adaptation. ACO algorithm requires to define the following [57]:

- The problem needs to be represented appropriately, which would allow the ants to incrementally update the solutions through the use of a probabilistic transition rules, based on the amount of pheromone in the trail and other problem specific knowledge. It is also important to enforce a strategy to construct only valid solutions corresponding to the problem definition.

- A problem-dependent heuristic function η that measures the quality of components that can be added to the current partial solution.
- A rule set for pheromone updating, which specifies how to modify the pheromone value τ.
- A probabilistic transition rule based on the value of the heuristic function η and the pheromone value τ that is used to iteratively construct a solution.

According to Dorigo et al. [49], the main steps of the ACO algorithm are given below:

1. *pheromone trail initialization*
2. *solution construction using pheromone trail*
   Each ant constructs a complete solution to the problem according to a probabilistic
3. *state transition rule*
   The state transition rule depends mainly on the state of the pheromone [136]
4. *pheromone trail update.*

A global pheromone updating rule is applied in two phases. First, an evaporation phase where a fraction of the pheromone evaporates, and then a reinforcement phase where each ant deposits an amount of pheromone which is proportional to the fitness of its solution [136]. This process is iterated until a termination condition is reached. ACO was first introduced using the Traveling Salesman Problem (TSP) [?], [29]-[32], [46]-[49], [66], [67], [126]-[128], [130], [21]. Starting from its start node, an ant iteratively moves from one node to another. When being at a node, an ant chooses to go to a unvisited node at time t with a probability given by

$$p_{i,j}^{k}(t) = \frac{[\tau_{i,j}(t)]^{\alpha}[\eta_{i,j}(t)]^{\beta}}{\sum_{l \in N_i^k}[\tau_{i,j}(t)]^{\alpha}[\eta_{i,j}(t)]^{\beta}} \qquad j \in N_i^k \tag{1.1}$$

where N_i^k is the feasible neighborhood of the ant_k, that is, the set of cities which ant_k has not yet visited; $\tau_{i,j}(t)$ is the pheromone value on the edge (i,j) at the time t, α is the weight of pheromone; $\eta_{i,j}(t)$ is a priori available heuristic information on the edge (i,j) at the time t, β is the weight of heuristic information. Two parameters α and β determine the relative influence of pheromone trail and heuristic information. $\tau_{i,j}(t)$ is determined by

$$\tau_{i,j}(t) = \rho\tau_{i,j}(t-1) + \sum_{k=1}^{n}\Delta\tau_{i,j}^{k}(t) \qquad \forall(i,j) \tag{1.2}$$

where ρ is the pheromone trail evaporation rate $(0 < \rho < 1)$, n is the number of ants, Q is a constant for pheromone updating. A generalized version of the pseudo-code for the ACO algorithm with reference to the TSP is illustrated in Algorithm 1.1.

Algorithm 1.1 Ant Colony Optimization Algorithm

Initialize the number of ants n, and other parameters.
While (the end criterion is not met) do
    $t = t+1$;
    For k= 1 to n
        ant_k is positioned on a starting node;
        For m= 2 to *problem_size*
            Choose the state to move into
                according to the probabilistic transition rules;
            Append the chosen move into $tabu_k(t)$ for the ant_k;
        Next m
        Compute the length $L_k(t)$ of the tour $T_k(t)$ chosen by the ant_k;
        Compute $\Delta\tau_{i,j}(t)$ for every edge (i, j) in $T_k(t)$ according to Eq.(**??**);
    Next k
    Update the trail pheromone intensity for every edge (i, j) according to Eq.(1.2);
    Compare and update the best solution;
End While.

Other applications of the ACO algorithm include: sequential ordering problem [68], quadratic assignment problem [95]-[99], [69], [129], [132], vehicle routing problem [22]-[24], scheduling problems [33], [63], [65], [100], [10], [11], graph coloring [34], partitioning problems [92], [93], timetabling [124], shortest subsequence problem [101], constraint satisfaction problems [125], maximum clique problem [20], edge-disjoint paths problem [9].

Perrotto and Lopez use ant colonies optimization for reconstruction of phylogenetic trees, which are developed in order to help unveil the evolutionary relationships among species, taking into account the Darwinian principle of the natural evolution of species. That is, by analyzing a set of amino acid sequences (or proteins) of different species, it can be determined how these species probably have been derived during their evolution. A phylogenetic tree can be considered as a binary tree, whose leaf nodes represent the species to be analyzed and inner nodes the ancestral species from which the current species have evolved. Also, phylogenetic trees may or may not have a root that indicates the oldest ancestor. A tree is constructed using a fully connected graph and the problem is approached similarly to the traveling salesman problem [109].

ACO was successfully applied for routing and road balancing problems. Schoonderwoerd et al. [113]-[115] designed an ant based control system (ABC) was designed to solve the load-balancing problem in circuit-switched networks [117], [119].

One of the ramifications of the ABC system is the adaptation of Guérin's *smart ants* to solve the problems of routing and load-balancing in circuit-switched networks by Bonabeau et al. [13], [77]. While an ant in ABC updates only the entry corresponding to the source node in the pheromone table of each node it passes, Bonabeau smart ants update the pheromone table at each node, all entries corresponding to *every* node they pass. Two other ramifications of the ABC system are the work of Subramanian et al. [131] and Heusse et al. [78], [112].

Caro and Dorigo [27] proposed the AntNet algorithm, designed for routing in packet-switched networks. Unlike traditional routing algorithms (such as OSPF and RIP) which focused on minimal or shortest path routing, routing in AntNet was carried out with the aim of optimizing the performance of the entire network. In AntNet, routing was achieved by launching forward ants at regular intervals from a source node to a destination node to discover a feasible low-cost path and by backward ants that travels from to destination node to source node update pheromone tables at each intermediate node [117].

ACO algorithms are also applied in the bioinformatics field for the problems such as: protein folding [121], [122], to multiple sequence alignment [102], and to the prediction of major histocompatibility complex (MHC) class II binders [85].

Sim and Sun [118]proposed a *Multiple Ant Colony Optimization (MACO)* technique, in which more than one colony of ants are used to search optimal paths, and each colony of ants deposits a different type of pheromone represented by a different color.

1.4.2 Bees' behavior inspired algorithms

Farooq et al. [61], [139]-[145] developed a bee inspired algorithm for routing in telecommunication network. The work is inspired by the way these insects communicate. He is also using "dance" quality of the bees, as illustrated in the book The Dance Language and Orientation of Bees [64].

The worker bees in a honey bee colony are grouped as food-storer, scout and forager. The food collection is organized by the colony by recruiting bees for different jobs. The recruitment is managed by the forager bees which can perform dances to communicate with their fellow bees inside the hive and recruit them. At the entrance of the hive is an area called the dance-floor, where dancing takes place [116]. Different types of dances have been identified:

- *Waggle dance* - is an advertisement for the food source of the dancer. Another forager can leave her food source and watch out for a well advertised food source [116]. A forager randomly follows dances of multiple recruiting foragers and seems to respond randomly as well. Especially she does not compare several dances. A dance does not seem to contain any information that helps to choose a food source [143].
- *Tremble dance* - foragers are more likely to perform the tremble dance if they have to wait long for a food-storer bee to unload their nectar after their arrival at hive. Foragers perform the tremble dance on the dance-floor and in the brood nest as well, whereas the waggle dance is limited to the dance-floor. So maybe bees in the hive are addressed, too [143]. According to Seeley [116] worker bees in the hive are ordered by the tremble dancers to give up their jobs and to unload nectar.

Upon their return from a foraging trip, bees communicate the distance, direction, and quality of a flower site to their fellow foragers by making waggle dances on a dance floor inside the hive. By dancing zealously for a good foraging site they recruit foragers for the site. In this way a good flower site is exploited, and the numbers of foragers at this site are reinforced. A honey bee colony has many features that are desirable in networks:

- efficient allocation of foraging force to multiple food sources;
- different types of foragers for each commodity;
- foragers evaluate the quality of food sources visited and then recruit optimum number of foragers for their food source by dancing on a dance floor inside the hive;
- no central control;
- foragers try to optimize the energetic efficiency of nectar collection and foragers take decisions without any global knowledge of the environment.

For solving the routing problem [61] the following hypothesis are considered: if a honey bee colony is able to adapt to countless changes inside the hive or outside in the environment through simple individuals without any central control, then an agent system based on similar principles should be able to adapt itself to an ever changing network environment in a decentralized fashion with the help of simple agents who rely only on local information. Problem is modeled as a honey bee colony and as a population based multi-agent system, in which simple agents coordinate their activities to solve the complex problem of the allocation of labor to multiple forage sites in dynamic environments. The agents achieve this objective in a decentralized fashion with the help of local information that they acquire while foraging. The proposed algorithm for routing problem is called *BeeHive* and uses the following principles of a honey bee colony [61]:

1. Each node in the network is considered as being a hive that consists of bee agents. Each node periodically launches its bee agents to explore the network and collect the routing information that provides the nodes visited with the partial information on the state of the network. These bee agents can be considered as scouts that explore and evaluate the quality of multiple paths between their launching node and the nodes that they visit.
2. Bee agents provide to the nodes which they visit, with the information on the propagation delay and queuing delay of the paths they explored. These lead to their launching node from the visited nodes. One could consider the propagation delay as distance information, and the queuing delay as a direction information (please remember bee scouts also provide these parameters in their dances): this reasoning is justified because a data packet is only diverted from the shortest path to other alternate paths when large queuing delays exist on the shortest path.
3. A bee agent decides to provide its path information only if the quality of the path traversed is above a threshold. The threshold is dependent on the number of hops that a bee agent is allowed to take. Moreover, the agents model the quality of a

path as a function of the propagation delay and the queuing delay of the path; lower values of the parameters result in higher values for the quality parameter.
4. The majority of the bee agents in the BeeHive algorithm explore the network in the vicinity of their launching node and very few explore distant part of the network. The idea is borrowed from honey bee colony resulting in not only reducing the overhead of collecting the routing information but also helping in maintaining smaller/local routing tables.
5. We consider a routing table as a dance floor where the bee agents provide the information about the quality of the paths they traversed. The routing table is used for information exchange among bee agents, launched from the same node but arriving at an intermediate node via different neighbors. This information exchange helps in evaluating the overall quality of a node as it has multiple pathways to a destination) for reaching a certain destination.
6. A nectar forager exploits the flower sites according to their quality while the distance and direction to the sites is communicated to it through waggle dances performed by fellow foragers on the dance floor. In our algorithm, we map the quality of paths onto the quality of nodes for utilizing the bee principle. Consequently, we formulate the quality of a node, for reaching a destination, as a function of proportional quality of only those neighbors that possibly lie in the path toward the destination.

Data packets are interpreted as foragers. Once they arrive at a node, they access the information in the routing tables, stored by bee agents, about the quality of different neighbors of the node for reaching their destinations. They select the next neighbor toward the destination in a stochastic manner depending upon its goodness. As a result, not all packets follow the best paths. This will help in maximizing the system performance although a data packet may not follow the best path.

Craig [35] borrowed the following idea from bees colonies behavior and used it for Internet Server Optimization: each colony must collect extra nectar during the warm season to make and store enough honey – usually 20 to 50 kg – in order to survive the winter. Efficient nectar collection is thus crucial for the colony survival. It is inefficient, in general, for all of the colony's foragers to collect from the same flower patch. A large number of bees at one patch can "swamp out" the flowers' capacity to generate nectar. On the other hand, some flower patches are richer or more productive than others. To maximize nectar intake, the honey bee colony must 'decide' in some decentralized but intelligent fashion how many bees will forage at each flower patch.

1.4.3 Particle swarm optimization algorithm

The Particle Swarm Optimization (PSO) model [58]-[60], [86]-[91] consists of a swarm of particles, which are initialized with a population of random candidate solutions. They move iteratively through the d-dimension problem space to search the new solutions, where the fitness, f, can be calculated as the certain qualities measure. Each particle has a position represented by a position-vector $\mathbf{x}_i$ (i is the index of the particle), and a velocity represented by a velocity-vector $\mathbf{v}_i$. Each particle remembers its own best position so far in a vector $\mathbf{x}_i^{\#}$, and its j-th dimensional value is $x_{ij}^{\#}$. The best position-vector among the swarm so far is then stored in a vector $\mathbf{x}^*$, and its j-th dimensional value is x_j^*. During the iteration time t, the update of the velocity from the previous velocity to the new velocity is determined by Eq.(1.3). The new position is then determined by the sum of the previous position and the new velocity by Eq.(1.4).

$$v_{ij}(t+1) = wv_{ij}(t) + c_1r_1(x_{ij}^{\#}(t) - x_{ij}(t)) + c_2r_2(x_j^*(t) - x_{ij}(t)). \tag{1.3}$$

$$x_{ij}(t+1) = x_{ij}(t) + v_{ij}(t+1). \tag{1.4}$$

where w is called as the inertia factor, r_1 and r_2 are the random numbers, which are used to maintain the diversity of the population, and are uniformly distributed in the interval [0,1] for the j-th dimension of the i-th particle. c_1 is a positive constant, called as coefficient of the self-recognition component, c_2 is a positive constant, called as coefficient of the social component. From Eq.(1.3), a particle decides where to move next, considering its own experience, which is the memory of its best past position, and the experience of its most successful particle in the swarm. In the particle swarm model, the particle searches the solutions in the problem space with a range $[-s, s]$ (If the range is not symmetrical, it can be translated to the corresponding symmetrical range.) In order to guide the particles effectively in the search space, the maximum moving distance during one iteration must be clamped in between the maximum velocity $[-v_{max}, v_{max}]$ given in Eq.(1.5):

$$v_{ij} = sign(v_{ij})min(|v_{ij}|, v_{max}). \tag{1.5}$$

The value of v_{max} is $p \times s$, with $0.1 \leq p \leq 1.0$ and is usually chosen to be s, i.e. $p = 1$. The pseudo-code for particle swarm optimization algorithm is illustrated in Algorithm 1.2.

The end criteria are usually one of the following:

- Maximum number of iterations: the optimization process is terminated after a fixed number of iterations, for example, 1000 iterations.
- Number of iterations without improvement: the optimization process is terminated after some fixed number of iterations without any improvement.
- Minimum objective function error: the error between the obtained objective function value and the best fitness value is less than a pre-fixed anticipated threshold.

1.4.4 Bacteria foraging algorithm

Since selection behavior of bacteria tends to eliminate animals with poor foraging strategies and favor the propagation of genes of those animals that have successful foraging strategies, they can be applied to have an optimal solution through methods for locating, handling, and ingesting food. After many generations, a foraging animal takes actions to maximize the energy obtained per unit time spent foraging. That is,

Algorithm 1.2 Particle Swarm Optimization Algorithm

Initialize the size of the particle swarm n, and other parameters.
Initialize the positions and the velocities for all the particles randomly.
While (the end criterion is not met) do
  $t = t + 1$;
  Calculate the fitness value of each particle;
  $\mathbf{x}^* = argmin_{i=1}^{n}(f(\mathbf{x}^*(t-1)), f(\mathbf{x}_1(t)), f(\mathbf{x}_2(t)), \cdots, f(\mathbf{x}_i(t)), \cdots, f(\mathbf{x}_n(t)))$;
  For i= 1 to n
    $\mathbf{x}_i^\#(t) = argmin_{i=1}^{n}(f(\mathbf{x}_i^\#(t-1)), f(\mathbf{x}_i(t))$;
    For j = 1 to *Dimension*
      Update the j-th dimension value of $\mathbf{x}_i$ and $\mathbf{v}_i$
        according to Eqs.(1.3), (1.4), (1.5);
    Next j
  Next i
End While.

poor foraging strategies are either eliminated or shaped into good ones. To perform social foraging an animal needs communication capabilities and it gains advantages that can exploit essentially the sensing capabilities of the group, so that the group can gang-up on larger prey, individuals can obtain protection from predators while in a group, and in a certain sense the group can forage a type of collective intelligence [103].

Escherichia Coli (E. Coli) normally lives inside the intestines, where it helps to body break down and digest the food. Its behavior and movement comes from a set of six rigid spinning (100-200 r.p.s) flagella, each driven as a biological motor. An E. coli bacterium alternates through running and tumbling. When the flagella rotate clockwise (counterclockwise), they operate as propellers and hence an E. Coli may run or tumble. Passino et al. [103] has modeled the chemotactic actions of the bacteria as follows:

- If in neutral medium, alternate tumbles and runs it is considered as search.
- If swimming up a nutrient gradient (or out of noxious substances), swim longer (climb up nutrient gradient or down noxious gradient)then it is considered as seeking increasingly favorable environments.
- If swimming down a nutrient gradient (or up noxious substance gradient), then search is considered as avoiding unfavorable environments.

In this way, it can climb up nutrient hills and at the same time avoid noxious substances. E. coli occasionally engages in a conjugation that affects the characteristics of a population of bacteria. There are many types of taxes that are used by bacteria. For instance, some bacteria are attracted to oxygen (aerotaxis), light (phototaxis), temperature (thermotaxis), or magnetic lines of flux (magnetotaxis). Some bacteria can change their shape and number of flagella based on the medium to reconfigure so as to ensure efficient foraging in a variety of media. Bacteria can form intricate stable spatio-temporal patterns in certain semisolid nutrient substances and they can eat radially their way through a medium if placed together initially at its center. Moreover, under certain conditions, they will secrete cell-to-cell attractant signals so that they will group and protect each other.

1.5 Conclusions and Some Potential Areas for Exploration

This chapter reviewed some of the well known nature inspired stigmergic computational models which has evolved during the last few decades. We presented some facts about social insects namely ants, bees and termites and how they could interact and self organize for solving real world problems. We focused on ant colony optimization algorithm, bees behavior inspired algorithm, particle swarm optimization algorithm and bacterial foraging algorithm.

The subject of copying, imitating, and learning from biology was coined *Biomimetics* by Otto H. Schmitt in 1969 [28]. This field is increasingly involved with emerging subjects of science and engineering and it represents the studies and imitation of nature's methods, designs and processes. Nature, through billions of years of trial and error, has produced effective solutions to innumerable complex real-world problems. Even though there are several computational nature inspired models, there is still a lot of room more research, at least in the form of finding some collaborations and interactions between the existing systems as well as developing new systems by borrowing ideas from nature. Butler [25] suggests some potential research areas:

1. Spiders spin silk that is stronger than synthetic substances developed by man but require only insects as inputs.
2. Diatoms, microscopic phytoplankton responsible for a quarter of all the photosynthesis on Earth, make glass using silicon dissolved in seawater.
3. Abalone, a type of shellfish, produces a crack-resistant shell twice as tough as ceramic from calcium found in seawater using a process known as biomineralization.
4. Trees "turn sunlight, water, and air into cellulose, a sugar stiffer and stronger than nylon, and bind it into wood, a natural composite with a higher bending strength and stiffness than concrete or steel," as noted by Paul Hawken, Amory and L. Hunter Lovins in *Natural Capitalism*.
5. Countless plants generate compounds that fight off infection from fungi, insects, and other pests.

References

1. Aron S, Pasteels JM, Deneubourg JL, Boevé JL (1986). Foraging recruitment in Leptothorax unifasciatus: The influence of foraging area familiarity and the age of nest-site. Insectes Sociaux, 33, 338-351.

2. Aron S, Pasteels JM, Deneubourg JL (1989). Trail-laying behaviour during exploratory recruitment in the argentine ant, Iridomyrmex humilis (Mayr). Biology of Behaviour, 14, 207-217.
3. Aron S, Deneubourg JL, Goss S, Pasteels JM (1990). How Argentine ants establish a minimal-spanning tree to link different nests. In Social Insects and the Environment, Eds. G.K. Veeresh, B. Mallik and C.A. Viraktamath, Oxford & IBH, New Delhi, 533-534.
4. Aron S, Deneubourg JL, Goss S, Pasteels JM (1990). Functional self-organisation illustrated by inter-nest traffic in the argentine ant Iridomyrmex humilis. In Biological Motion, Eds W. Alt and G. Hoffman. Lecture Notes in Biomathematics, Springer-Verlag, Berlin, 533-547.
5. Beckers R, Goss S, Deneubourg JL, Pasteels JM (1989). Colony size, communication and ant foraging strategy, Psyche, 96, 239-256.
6. Beckers R, Deneubourg JL, Goss S (1990). Changing foraging strategies inherent to trail recruitment. In Social Insects and the Environment, Eds. G.K. Veeresh, B. Mallik and C.A. Viraktamath, Oxford & IBH, New Delhi, 549.
7. Benzatti D, Collective Intelligence (Ants), Available at: http://ai-depot.com/CollectiveIntelligence/Ant.html
8. de Biseau JC, Deneubourg JL, Pasteels JM (1992). Mechanisms of food recruitment in the ant Myrmica sabuleti : an experimental and theoretical approach. In Biology and Evolution of Social Insects, Ed. J. Billen. Leuven University Press, Leuven (B), 359-367.
9. Blesa M, Blum C (2004) Ant colony optimization for the maximum edge-disjoint paths problem. In: Raidl GR, Cagnoni S, Branke J, Corne DW, Drechsler R, Jin Y, Johnson CG, Machado P, Marchiori E, Rothlauf R, Smith GD, Squillero G, editors. Applications of evolutionary computing, proceedings of EvoWorkshops 2004. Lecture Notes in Computer Science, vol. 3005. Berlin: Springer, 160-9.
10. Blum C. Beam (2005) ACO-Hybridizing ant colony optimization with beam search: An application to open shop scheduling. Computers & Operations Research, 32(6), 1565-1591.
11. Blum C, Sampels M (2004) An ant colony optimization algorithm for shop scheduling problems. Journal of Mathematical Modelling and Algorithms, 3(3), 285-308.
12. Blum C (2005) Ant colony optimization: Introduction and recent trends. Physics of Life Reviews, 2, 353-373.
13. Bonabeau, E, Hénaux F, Guérin S, Snyer D, Kuntz P, Théraulaz G (1998) Routing in telecommunications networks with ant-like agents. In Proceedings of Intelligent Agents Telecommunications Applications, Berlin, Germany.
14. Bonabeau E, Dorigo M, Theraulaz G (2000) Inspiration for optimization from social insect behavior, Nature, 406, 39-42.
15. Bonabeau E, Theraulaz G, Deneubourg JL (1995). Phase diagram of a model of self-organizing hierarchies. Physica A, 217, 373-392.
16. Bonabeau E, Theraulaz G, Deneubourg JL (1996). Mathematical model of self-organizing hierarachies in animal societies. Bulletin of Mathematical Biology, 58, 661-717.
17. Bonabeau E, Theraulaz G, Deneubourg JL (1996). Quantitative study of the fixed threshold model for the regulation of division of labour in insect societies. Proceedings of the Royal Society of London Series B, 263, 1565-1569.
18. Bonabeau E, Theraulaz G, Fourcassié V, Deneubourg JL (1998). Phase-ordering kinetics of cemetery organization in ants. Physical Review E, 57, 4568-4571.
19. Bonabeau E, Theraulaz G, Deneubourg JL, Lioni A, Libert F, Sauwens C, Passera L (1998). Dripping faucet with ants. . Physical Review E, 57, 5904-5907.

20. Bui TN, Rizzo JR (2004) Finding maximum cliques with distributed ants. In: Deb K, et al., editors. Proceedings of the genetic and evolutionary computation conference (GECCO 2004). Lecture Notes in Comput Sci, vol. 3102. Berlin: Springer, 24-35.
21. Bullnheimer B, Hartl RF, Strauss C (1999). A New Rank Based Version of the Ant System: A Computational Study, Central European Journal for Operations Research and Economics, 7(1):25-38.
22. Bullnheimer B, Hartl RF, Strauss C (1997) An Improved Ant system Algorithm for the Vehicle Routing Problem. Sixth Viennese workshop on Optimal Control, Dynamic Games, Nonlinear Dynamics and Adaptive Systems, Vienna (Austria).
23. Bullnheimer B., R.F. Hartl and C. Strauss (1999) Applying the Ant System to the Vehicle Routing Problem. In Voss S., Martello S., Osman I.H., Roucairol C. (eds.), Meta-Heuristics: Advances and Trends in Local Search Paradigms for Optimization, Kluwer:Boston.
24. Bullnheimer B. (1999). Ant Colony Optimization in Vehicle Routing. Doctoral thesis, University of Vienna, January 1999.
25. Butler R, Biomimetics, technology that mimics nature, available online at: mongabay.com
26. Calenbuhr V, Deneubourg JL (1990). A model for trail following in ants: Individual and collective behaviour. In Biological Motion, Eds. W. Alt and G. Hoffman. Lecture Notes in Biomathematics, Springer-Verlag, Berlin, 453-469.
27. Caro GD, Dorigo M (1998) AntNet: Distributed stigmergetic control for communications networks, Journal of Artificial Intelligence Research, 9, 317-365.
28. Cohen YB (2005), Biomimetics: Biologically Inspired Technologies, CRC Press.
29. Colorni A, Dorigo M, Maffioli F, Maniezzo V, Righini G, Trubian M (1996). Heuristics from Nature for Hard Combinatorial Problems. International Transactions in Operational Research, 3(1):1-21.
30. Colorni A, Dorigo M, Maniezzo V (1992). Distributed Optimization by Ant Colonies. Proceedings of the First European Conference on Artificial Life, Paris, France, F.Varela and P.Bourgine (Eds.), Elsevier Publishing, 134-142.
31. Colorni A, Dorigo M, Maniezzo V (1992). An Investigation of Some Properties of an Ant Algorithm. In Proceedings of the Parallel Problem Solving from Nature Conference (PPSN 92), Brussels, Belgium, R.Männer and B.Manderick (Eds.), Elsevier Publishing, 509-520.
32. Colorni A, Dorigo M, Maniezzo V (1995). New Results of an Ant System Approach Applied to the Asymmetric TSP. In Proceedings of the Metaheuristics International Conference, Hilton Breckenridge, Colorado, I.H.Osman and J.P. Kelly (Eds.), Kluwer Academic Publishers, 356-360.
33. Colorni A, Dorigo M, Maniezzo V, Trubian M (1994). Ant system for Job-shop Scheduling. - Belgian Journal of Operations Research, Statistics and Computer Science, 34(1):39-53.
34. Costa D, Hertz A (1997). Ants Can Colour Graphs.Journal of the Operational Research Society, 48, 295-305.
35. Tovey CA (2004) HONEY BEE Algorithm: A Biologically Inspired Approach to Internet Server Optimization, Engineering Enterprise, Spring, 13-15
36. Deneubourg JL, Aron S, Goss S, Pasteels JM (1990) The self-organizing exploratory pattern of the argentine ant, Journal of Insect Behavior, 3, 159-168.
37. Deneubourg JL, de Palma A (1980) Self-organization and architecture in human and animal societies. In International Conference on Cybernetics and Society, IEEE, 1126-1128.

38. Deneubourg JL, Pasteels JM, Verhaeghe JC (1983). Probabilistic behaviour in ants: a strategy of errors? Journal of Theoretical Biology, 105, 259-271.
39. Deneubourg JL, Aron S, Goss S, Pasteels JM, Duerinck G (1986). Random behaviour, amplification processes and number of participants: how they contribute to the foraging properties of ants. Physica D, 22, 176-186.
40. Deneubourg JL, Aron S, Goss S, Pasteels JM (1987). Error, communication and learning in ant societies. European Journal of Operational Research, 30, 168-172.
41. Deneubourg JL, Goss S, Pasteels JM, Fresneau D, Lachaud JP (1987). Self-organization mechanisms in ant societies (II): learning in foraging and division of labor. In From individual to collective behavior in social insects, Eds J.M. Pasteels and J.L. Deneubourg. Experientia Supplementum, 54, Birkhaüser, Bâle, 177-196.
42. Deneubourg JL, Aron S, Goss S, Pasteels JM (1990). The self-organizing exploratory pattern of the argentine ant. Journal of Insect Behavior, 3, 159-168.
43. Deneubourg JL, Theraulaz G, Beckers R (1991). Swarm made architectures. In Proceedings of the 1st European Conference on Artificial Life, Paris 1991, Eds P. Bourgine and E. Varela. MIT Press, Cambridge (Mass.), 123-133.
44. Deneubourg JL, Camazine S, Detrain C (1999) Self-organization or individual complexity: a false dilemna or a true complementarity In Information processing in social insects, Detrain C., Deneubourg J.L. and Pasteels J.M. (eds), Birkhauser Verlag, 401-408.
45. Dorigo M, Blum C (2005) Ant colony optimization theory: A survey, Theoretical Computer Science, 344, 243-278.
46. Dorigo M, Maniezzo V, Colorni A (1991) Positive feedback as a search strategy,Tech. Report 91-016, Dipartimento di Elettronica, Politecnico di Milano, Italy.
47. Dorigo M, Maniezzo V, Colorni A (1996) Ant system: optimization by a colony of cooperating agents, IEEE Transaction on Systems, Man and Cybernetics-Part B, 26(1), 29-41.
48. Dorigo M, Gambardella LM (1997) Ant Colony System: A Cooperative Learning Approach to the Traveling Salesman Problem. IEEE Transactions on Evolutionary Computation, 1(1):53-66.
49. Dorigo M, Gambardella LM (1997) Ant Colonies for the Traveling Salesman Problem. BioSystems, 43, 73-81.
50. Dorigo M, Caro GD, Gambardella LM (1999) Ant algorithms for discrete optimization,Artificial Life, 5(2), 137-172.
51. Dorigo M, Bonabeau E, Theraulaz G (2000) Ant algorithms and stigmergy, Future Generarion Computer Systems, 16(8) 851-871.
52. Dorigo M, Caro GD (1999) The ant colony optimization metaheuristic, New Ideas in Optimization, D. Corne, M. Dorigo, and F. Glover, Eds. New York: McGraw-Hill.
53. Dorigo M, Stutzle T (2002) The ant colony optimization metaheuristic: Algorithms, applications, and advances, Handbook of Metaheuristics, F. Glover and G. Kochenberger, Eds. Norwell, MA: Kluwer.
54. Dorigo M, Maniezzo V, Colorni A (1991) Positive Feedback as a Search Strategy, Dipartimento Elettronica, Politecnico Milano, Italy, Tech. Rep. 91-016.
55. Dorigo M, Gambardella LM (1997) Ant colony system: A cooperative learning approach to the travelling salesman problem, IEEE Transaction on Evolutionary Computation, 1, 53-66.
56. Dorigo M, Caro DG (1999) Ant colony optimization: a new meta-heuristic. Proceeding of the 1999 Congress on Evolutionary Computation (CEC), 1470-1477.
57. Dorigo M, Bonaneau E, Theraulaz G (2000) Ant algorithms and stigmergy, Future Generation Computer Systems, 16, 851-871.

58. Eberhart RC, Kennedy J (1995) A new optimizer using particle swarm theory. In Proceedings of the Sixth International Symposium on Micromachine and Human Science, Nagoya, Japan, 39-43.
59. Eberhart RC, Shi Y (2001) Particle swarm optimization: developments, applications and resources. In Proceedings of the IEEE Congress on Evolutionary Computation (CEC), Seoul, Korea
60. Eberhart RC, Simpson PK, Dobbins RW (1996) Computational Intelligence PC Tools, Boston, MA: Academic Press Professional
61. Farooq M (2006) From the Wisdom of the Hive to Intelligent Routing in Telecommunication Networks: A Step towards Intelligent Network Management through Natural Engineering, PhD Thesis, University of Dortmund, Germany.
62. Focardi S, Deneubourg JL, Chelazzi G (1989). Clustering in intertidal gastropods and chitons: models and field observations, Mémoire de la Société Vaudoise des Sciences Naturelles, 18(3) 1-15.
63. Forsyth P, Wren A (1997) An Ant System for Bus Driver Scheduling. Presented at the 7th International Workshop on Computer-Aided Scheduling of Public Transport, Boston.
64. von Frisch K (1967) The Dance Language and Orientation of Bees. Harvard University Press, Cambridge.
65. Gagné C, Price WL, Gravel M (2002) Comparing an ACO algorithm with other heuristics for the single machine scheduling problem with sequencedependent setup times. Journal of Operation Research Society, 53, 895-906.
66. Gambardella LM, Dorigo M (1995) Ant-Q: A Reinforcement Learning Approach to the Traveling Salesman Problem. In Proceedings of ML-95, Twelfth International Conference on Machine Learning, Tahoe City, CA, A. Prieditis and S. Russell (Eds.), Morgan Kaufmann, 252-260.
67. Gambardella LM, Dorigo M (1996) Solving Symmetric and Asymmetric TSPs by Ant Colonies. In Proceedings of IEEE Intenational Conference on Evolutionary Computation, IEEE-EC 96, Nagoya, Japan, IEEE Press, 622-627.
68. Gambardella LM, Dorigo M (1997) HAS-SOP: An Hybrid Ant System for the Sequential Ordering Problem. Technical Report No. IDSIA 97-11, IDSIA, Lugano, Switzerland.
69. Gambardella LM, Taillard E, Dorigo M (1999) Ant Colonies for the Quadratic Assignment Problem. Journal of the Operational Research Society, 50, 167-176.
70. Gordon D, Collective Intelligence in Social Insects, AI-depot essay, available online at: http://ai-depot.com/Essay/SocialInsects.html
71. Goss S, Deneubourg JL, Pasteels JM (1985) Modelling ant foraging systems; the influence of memory and the ants' size. In The Living State-II. Ed. R.K. Mishra. World Scientific Publishing Co., Singapore, 24-46.
72. Goss S, Aron S, Deneubourg JL, Pasteels JM (1989) Self-organized shortcuts in the argentine ant. Naturwissenschaften, 76, 579-581.
73. Goss S, Deneubourg JL, Pasteels JM, Josens G (1989) A model of noncooperative foraging in social insects. The American Naturalist, 134, 273-287.
74. Goss S, Fresneau D, Deneubourg JL, Lachaud JP Valenzuela-Gonzalez J (1989) Individual foraging in the ant Pachycondyla apicalis, Oecologia, 80, 65-69.
75. Goss S, Beckers R, Deneubourg JL, Aron S, Pasteels JM (1990) How trail laying and trail following can solve foraging problems for ant colonies. In Behavioural Mechanisms of Food Selection, Ed. R.N. Hughes. NATO ASI Series, G 20, Springer-Verlag, Berlin, 661-678.

76. Grasse, PP (1959) La reconstruction du nid et les coordinations interindividuelles chez bellicositermes natalensis et cubitermes sp. La theorie de la stigmergie: essai d'interpretation du comportament des termites constructeurs. Insects Sociaux, 6, 41-81.
77. Guérin S (1997) Optimization Multi-Agents en Environment Dynamique: Application au Routage Dans les Réseaux de Telecommunications, DEA, Univ. Rennes I, Ecole Nat. Supér. Télécommun. Bretagne, , Bretagne, France.
78. Heusse M, Snyers D, Guérin S, Kuntz P (1998) Adaptive Agent-Driven Routing and Load Balancing in Communication Networks, ENST Bretagne, Brest, France, Tech. Rep. RR-98 001-IASC.
79. Hogue C (1987) Cultural entomology, Annual Review of Entomology, 32
80. http://institute.advancedarchitecture.org/Research/Ants/Optimization
81. http://www.lingolex.com/ants.htm
82. http://www.insecta-inspecta.com
83. http://institute.advancedarchitecture.org/Research/Ants/Optimization
84. Jaffé K, Deneubourg JL (1992) On foraging, recruitment systems and optimun number of scouts in eusocial colonies. Insectes Sociaux, 39, 201-213.
85. Karpenko O, Shi J, Dai Y (2005) Prediction of MHC class II binders using the ant colony search strategy. Artificial Intelligence in Medicine, 35(1-2), 147-156.
86. Kennedy J, Eberhart RC (1995) Particle Swarm Optimization. In Proceedings of IEEE International Conference on Neural Networks, Perth, Australia, IEEE Service Center, Piscataway, NJ, Vol.IV, 1942-1948
87. Kennedy J (1997) Minds and cultures: Particle swarm implications. Socially Intelligent Agents. Papers from the 1997 AAAI Fall Symposium. Technical Report FS-97-02, Menlo Park, CA: AAAI Press, 67-72
88. Kennedy J (1998) The Behavior of Particles, In Proceedings of 7th Annual Conference on Evolutionary Programming, San Diego, USA.
89. Kennedy J (1997) The Particle Swarm: Social Adaptation of Knowledge. In Proceedings of IEEE International Conference on Evolutionary Computation, Indianapolis, Indiana, IEEE Service Center, Piscataway, NJ, 303-308
90. Kennedy J (1992) Thinking is social: Experiments with the adaptive culture model. Journal of Conflict Resolution, 42, 56-76
91. Kennedy J, Eberhart R (2001) Swarm Intelligence, Morgan Kaufmann Academic Press
92. Kuntz P, Snyers D (1994) Emergent Colonization and Graph Partitioning. IN Proceedings of the Third International Conference on Simulation of Adaptive Behavior: From Animals to Animats 3, MIT Press, Cambridge, MA.
93. Kuntz P, Layzell P, Snyers D (1997) A Colony of Ant-like Agents for Partitioning in VLSI Technology. In Proceedings of the Fourth European Conference on Artificial Life, P. Husbands and I. Harvey, (Eds.), MIT Press, 417-424.
94. Maeterlinck M (1930) The life of the white ant, Dodd, Mead & Co.
95. Maniezzo V, Colorni A, Dorigo M (1994). The Ant System Applied to the Quadratic Assignment Problem. Technical Report IRIDIA/94-28, Université Libre de Bruxelles, Belgium.
96. Maniezzo V, Muzio L, Colorni A, Dorigo M (1994). Il sistema formiche applicato al problema dell'assegnamento quadratico. Technical Report No. 94-058, Politecnico di Milano, Italy.
97. Maniezzo V (1998). Exact and approximate nondeterministic tree-search procedures for the quadratic assignment problem. Research Report CSR 98-1, Scienze dell'Informazione, UniversitÂ di Bologna, Sede di Cesena, Italy.
98. Maniezzo V, Colorni A (1999). The Ant System Applied to the Quadratic Assignment Problem. IEEE Transactions on Knowledge and Data Engineering.

99. Maniezzo V, Carbonaro A (2001), Ant Colony Optimization: an overview, in C.Ribeiro (eds.) Essays and Surveys in Metaheuristics, Kluwer, 21-44.
100. Merkle D, Middendorf M, Schmeck H (2002) Ant colony optimization for resource-constrained project scheduling. IEEE Transaction on Evolutionary Computation, 6(4), 333-46.
101. Michel R, Middendorf M (1998) An island model based ant system with lookahead for the shortest supersequence problem. In: Eiben AE, Bäck T, SchoenauerM, Schwefel H-P, editors. Proceedings of PPSN-V, fifth International Conference on Parallel Problem Solving from Nature, Lecture Notes in Comput Science, Berlin, Springer, 692-701.
102. Moss JD, Johnson CG (2003) An ant colony algorithm for multiple sequence alignment in bioinformatics. In: Pearson DW, Steele NC, Albrecht RF, editors. Artificial neural networks and genetic algorithms, Berlin Springer, 182-186.
103. Passino KM (2005), Biomimicry for Optimization, Control, and Automation,Springer-Verlag, London, UK.
104. Pasteels JM, Verhaeghe JC, Deneubourg JL (1982). The adaptive value of probabilistic behavior during food recruitment in ants : experimental and theoretical approaches. In Biology of Social Insects. Eds M.D. Breed, C.D. Michener AND H.E. Evans, Westview Press, Boulder, Co, 297-301.
105. Pasteels JM, Deneubourg JL, Verhaeghe JC, Boevé JL, Quinet Y (1986). Orientation along terrestrial trails by ants. In Mechanisms in insect olfaction. Eds T.L. Payne, M.C. Birch ANDC.E.J. Kennedy, Oxford University Press, Oxford, 131-138.
106. Pasteels JM, Deneubourg JL, Goss S (1987). Transmission and amplification of information in a changing environment: The case of insect societies. In Law of Nature and Human conduct, Eds I. Prigogine and M. Sanglier, Gordes, Bruxelles, 129-156.
107. Pasteels, JM Deneubourg, JL, Goss S (1987) Self-organization mechanisms in ant societies (I) : trail recruitment to newly discovered food sources. In From individual to collective behavior in social insects, Eds J.M. Pasteels and J.L. Deneubourg. Experientia Supplementum, 54, Birkhaüser, Bâle, 155-175.
108. Pasteels JM, Roisin Y, Deneubourg JL, Goss S (1987) How efficient collective behaviour emerges in societies of unspecialized foragers: the example of Tetramorium caespitum. In Chemistry and Biology of Social Insects, Eds J. Eder and H. Rembold. Verlag Peperny, Munich, 513-514.
109. Perretto M, Lopes HS (2005) Reconstruction of phylogenetic trees using the ant colony optimization paradigm. Genetics and Molecular Research, 4(3), 581-589.
110. Preface to the book "the Soul of a White Ant" published in 1937, available online at http://journeytoforever.org/farm_library/Marais1/whiteantToC.html
111. Saffre F, Deneubourg JL (2002) Swarming strategies for cooperative species. Journal of Theoretical Biology, 214, 441-451.
112. Schatz B, Chameron S, Beugnon G, Collett TS (1999) The use of path integration to guide route learning in ants, Nature, 399(6738), 769-777.
113. Schoonderwoerd R, Holland O, Bruten J, Rothkrantz L (1996) Antbased load balancing in telecommunications networks,Adaptive Behavior, 5(2), 169-207.
114. Schoonderwoerd R, Holland O, Bruten J Ant-like agents for load balancing in telecommunications networks, In Proceedings of Agents, Marina del Rey, CA, 209-216.
115. Schoonderwoerd R, Holland O (1999) Minimal agents for communications networks routing: The social insect paradigm. In Software Agents for Future Communication Systems, A. L. G. Hayzeldean and J. Bingham, Eds. New York: Springer-Verlag.
116. Seeley TD, Towne WF (1992) Tactics of dance choice in honey bees: Do foragers compare dances? Behav. Ecol. Sociobiol. 30, 59-69.

117. Sim KM, Sun WH (2003) Ant Colony Optimization for Routing & Load-Balancing: Survey and New Directions. IEEE Transaction on Systems, Man and Cybernetics, Part A, 33(5), 560-572.
118. Sim KM, Sun WH (2002) Multiple ant-colony optimization for network routing, In Proceedings of 1st International Symposium on Cyberworld, Tokyo, Japan, November, 277-281.
119. Sim KM, Sun WH (2001) A comparative study of ant-based optimization for dynamic routing, In Proceedings of Conference on Active Media Technology, Lecture Notes Computer Science, Hong Kong, 153-164.
120. Shouse B (2002) Getting the Behavior of Social Insects to Compute. Science, 295, 2357.
121. Shmygelska A, Aguirre-Hernández R, Hoos HH (2002) An ant colony optimization algorithm for the 2D HP protein folding problem. In: Dorigo M, Di Caro G, Sampels M, editors, Ant algorithms-Proceedings of ANTS2002-Third international workshop. Lecture Notes in Computer Science, 40-52.
122. Shmygelska A, Hoos HH (2005) An ant colony optimisation algorithm for the 2D and 3D hydrophobic polar protein folding problem. BMC Bioinformatics, 6(30), 1-22.
123. Sleigh C (2003) Ant, Reaktion Books Ltd.
124. Socha K, Sampels M, Manfrin M (2003) Ant algorithms for the university course timetabling problem with regard to the state-of-the-art. In: Cagnoni S, Romero Cardalda JJ, Corne DW, Gottlieb J, Guillot A, Hart E, Johnson CG, Marchiori E, Meyer A, Middendorf M, Raidl GR, editors. Applications of evolutionary computing, proceedings of EvoWorkshops 2003. Lecture Notes in Computer Science, Berlin Springer, 334-345.
125. Solnon C (2002) Ant can solve constraint satisfaction problems. IEEE Transaction on Evolutionary Computation 6(4), 347-57.
126. Stützle T, Hoos H (1997) The MAX-MIN Ant System and Local Search for the Traveling Salesman Problem. Proceedings of ICEC'97 - 1997 IEEE 4th International Conference on Evolutionary Computation, IEEE Press, 308-313.
127. Stützle T, Hoos H (1997) The MAX-MIN Ant System and local Search for Combinatorial Optimization Problems: Towards Adaptive Tools for Global Optimization. 2nd Metaheuristics International Conference (MIC-97), Sophia-Antipolis, France, 21-24.
128. Stützle T, Dorigo M (1999) ACO Algorithms for the Traveling Salesman Problem. In K. Miettinen, M. Makela, P. Neittaanmaki, J. Periaux, editors, Evolutionary Algorithms in Engineering and Computer Science, Wiley.
129. Stützle T, Dorigo M (1999) ACO Algorithms for the Quadratic Assignment Problem. In D. Corne, M. Dorigo and F. Glover, editors, New Ideas in Optimization, McGraw-Hill.
130. Stutzle T, Hoos HH (2000) MAX-MIN ant system, Future Generation Computer Systems Journal, 16(8), 889-914.
131. Subramanian D, Druschel P, Chen J (1997) Ants and reinforcement learning: A case study in routing in dynamic networks, In Proceedings of International Joint Conference on Artificial Intelligence IJCAI-97, Palo Alto, CA, 832-838.
132. Taillard E, Gambardella LM (1997) An Ant Approach for Structured Quadratic Assignment Problems. 2nd Metaheuristics International Conference (MIC-97), Sophia-Antipolis, France.
133. Theraulaz G, Goss S, Gervet J, Deneubourg JL (1990). Self-organisation of behavioural profiles and task assignment based on local individual rules in the eusocial wasp Polistes dominulus Christ. In : Social Insects and the Environment, Eds. G.K. Veeresh, B. Mallik & C.A. Viraktamath. Oxford & IBH, New Delhi, 535-537.

134. Theraulaz G, Deneubourg JL (1994) Swarm Intelligence in social insects and the emergence of cultural swarm patterns. In The Ethological roots of Culture, Eds R.A. Gardner, A.B. Chiarelli, B.T. Gardner & F.X. Ploojd. Kluwer Academic Publishers, Dordrecht, 1- 19.
135. Theraulaz G, Bonabeau E, Deneubourg JL (1995) Self-organization of hierarchies in animal societies: The case of primitively eusocial wasp Polistes dominulus Christ. Journal of Theoretical Biology, 174, 313-323.
136. Toksari, MD (2005) Ant colony optimization for finding the global minimum. Applied Mathematics and Computation.
137. Truscio R, Stranger than fiction, available on-line at: http://home.att.net/~B-P.TRUSCIO/STRANGER.htm
138. Verhaeghe JC, Deneubourg JL (1983) Experimental study and modelling of food recruitment in the ant Tetramorium impurum (Hym. Form.) Insectes Sociaux, 30, 347-360.
139. Wedde HF, Farooq M, Zhang Y (2004) BeeHive: An Efficient Fault Tolerant Routing Algorithm under High Loads Inspired by Honey Bee Behavior. In Proceedings of the Fourth International Workshop on Ant Colony and Swarm Intelligence (ANTS 2004), Brussels, Belgium, Lecture Notes in Computer Science, 83-94.
140. Wedde HF, Farooq M (2005) A Performance Evaluation Framework for Nature Inspired Routing Algorithms. In Proceedings of EvoWorkshops 2005, Lausanne, Switzerland, Lecture Notes in Computer Science, 136-146.
141. Wedde HF, Farooq M (2005) BeeHive: Routing Algorithms Inspired by Honey Bee Behavior. K"unstliche Intelligenz. Schwerpunkt: Swarm Intelligence, 18-24.
142. Wedde HF, Farooq M (2005) BeeHive: New Ideas for Developing Routing Algorithms Inspired by Honey Bee Behavior In Handbook of Bioinspired Algorithms and Applications, Albert Zomaya and Stephan Olariu, Ed. Chapman & Hall/CRC Computer and Information Science, 321-339.
143. Wedde HF, Farooq M, BeeHive: An Efficient, Scalable, Adaptive, Faulttolerant and Dynamic Routing Algorithm Inspired from the Wisdom of the Hive. Technical Report 801, Department of Computer Science, University of Dortmund.
144. Wedde HF, Farooq M (2005) The Wisdom of the Hive Applied to Mobile Ad-Hoc Networks. In Proceedings IEEE Swarm Intelligence Symposium (SIS 2005), Pasadena, California, USA , 341-348.
145. Wedde HF, Timm C, Farooq M (2006) BeeHiveGuard: A Step Towards Secure Nature Inspired Routing Algorithms. In Rothlauf et al., editors, Proocedings of the EvoWorkshops 2006, Lecture Notes in Computer Science, Budapest, 2006
146. Wagner IA, Linderbaum M, Bruckstein AM (2000) ANTS: Agents, networks, trees, and subgraphs, Future Generation ComputerSystems Journal, M. Dorigo, G. D. Di Caro, and T. Stutzle, Eds. Amsterdam, The Netherlands: North Holland, 16, 915-926.
147. http://en.wikipedia.org

2

Stigmergic Autonomous Navigation in Collective Robotics

Renato R. Cazangi[1], Fernando J. Von Zuben[1], and Mauricio F. Figueiredo[2]

[1] LBiC - DCA - FEEC - University of Campinas (Unicamp)
PO Box 6101 CEP 13083-970
Campinas, SP, Brazil
{renato, vonzuben}@dca.fee.unicamp.br

[2] DIN - State University of Maringá (UEM)
Av. Colombo, 5790 CEP 87020-900
Maringá, PR, Brazil
mauricio@din.uem.br

Summary: Unknown environments, noisy information and no human assistance provide several challenges and difficulties in autonomous robot navigation, especially when multiple robots operate together. Dealing with autonomous navigation in collective robotics requires sophisticated computational techniques, being the biologically-inspired approaches the most frequently adopted. One of them, named Swarm Intelligence, explores features of social insects like those of ants. This Chapter proposes two strategies for multi-robot communication based on stigmergy, that is, the ants communication by means of pheromones. In the deterministic strategy, the robots deposit artificial pheromones in the environment according to innate rules. On the other hand, in the strategy called evolutionary, the robots have to learn how and where to lay artificial pheromones. Each robot is controlled by an autonomous navigation system (ANS) based on Learning Classifier System, which evolves during navigation from no *a priori* knowledge, and should learn to avoid obstacles and capture targets disposed on the environment. Aiming to validate the ANS in collective scenarios and also to investigate the stigmergic communication strategies, several experiments are simulated. The results show that the robot's communication by artificial pheromones implies a higher effectiveness in accomplishing the navigation tasks and the minimization of distances traveled between targets.

2.1 Introduction

The problems related to autonomous robot navigation have shown to be very attractive for researchers of diverse areas because of the challenging technical and conceptual issues involved. Without human intervention, autonomous robots must

Renato R. Cazangi et al.: *Stigmergic Autonomous Navigation in Collective Robotics*, Studies in Computational Intelligence (SCI) **31**, 25–63 (2006)
www.springerlink.com

carry out tasks in unknown and unpredictable environments by means of noisy sensors and actuators.

The complexity is still augmented when more than a single autonomous robot operates in the environment. When many robots navigate together, each one is a moving obstacle to the others. The dynamical states and the possibilities of interaction are countless. That is why most research efforts have been concentrated on single robot systems. Just recently multi-robot systems and collective robotics are receiving a great deal of attention [1]. The growth of interest is also related to limitations of single robots and advantages associated with multi-robots applications. Several tasks, like transport of objects, may require the use of multiple robots. In other tasks, a single robot can suffice but not as well as a collection of robots would do, especially if they cooperate.

Some of the approaches for collective robotics, possibly the most frequent, are the biologically-inspired ones. Such approaches are based on social characteristics of insects and vertebrate animals, mainly ants, bees, and birds. Viewing social organisms of nature as metaphors for developing artificial multi-robot systems is one of the main foundations of the swarm intelligence field. Roughly, the success of swarm intelligence methodologies when applied to collective robotics can be attributed to the notion of emergent complexity derived from the interactions of elementary individuals [2].

Being the interaction so relevant, the exchange of information among the individuals becomes prominent. Hence, communication issues are inherent attributes of collective artificial system. In the literature, there is an extensive discussion about the way communication should be established, with a general agreement: direct communication should be avoided in favor of indirect communication [3]. Such a conclusion is mainly supported by the amazing property of stigmergy, as Grasse [4] named the indirect communication performed by social insects.

Much of the ant societies' sophistication is attributed to their recurrent use of stigmergy by means of pheromones. Though the stigmergic effects seen in nature are powerful, its working mode is as simple as its definition: individual behavior modifies the environment, which in turn modifies the behavior of other individuals [5]. In robotics, the stigmergic communication is appealing because it requires no encoding or decoding, no structural knowledge, no memory, and follows very simple rules. All that it requires is that a robot passes close enough to the location where the communication entity was placed to be affected by it [6].

Stigmergy also plays a decisive role in this work, motivating the proposition of two different strategies for indirect multi-robot communication. In fact, the authors have developed in previous works [7, 8] an autonomous navigation system (ANS), primarily devoted to single robot navigation. In this work, the ANS is extended to collective robotics applications, with communication strategies based on stigmergy. The system is based on an evolutionary approach, more specifically learning [9], which are able to synthesize adaptive inference mechanisms capable of operating in time-varying conditions.

Though not being the case here, in general the robot controllers in collective robotics are developed as a behavior-based approach. This approach is centered on the idea of providing the robot with a finite set of simple basic behaviors (usually hand designed) and letting the environment determine which basic behavior should be chosen at any given time [10]. Unfortunately, such a necessity of *a priori* knowledge is very restrictive when dealing with autonomous robots, given that it relies heavily on the intuition of the programmer to be properly specified and designed [11].

The top-down approach to selecting basic behaviors greatly constraints the possibilities of interactions, which could, in principle, arise from stigmergy or self-organization. The behavior being built by a designer is produced from the viewpoint of an observer, and not from the perspective of the robot. Just breaking down a global behavior, as it is seen by an external observer into a set of arbitrary basic behaviors, may bring inconsistency and incompleteness to the robot's controller.

An evolutionary procedure is possibly capable of relaxing many of those constraints by relying on a bottom-up approach that deals directly with the global behavior of the system. The policy of leaving the system development on its own, releases the designer from the burden of deciding how to break the desired global behavior down into simple basic behaviors [12].

When a learning classifier system is applied to synthesize an ANS, all knowledge necessary to compose the robot behavior is evolved according to environmental feedback during navigation. No *a priori* basic or complex behavior is inserted into the system. Without external assistance, the robots should learn to accomplish their task, that can be composed of multiple and possibly conflicting objectives such as avoiding obstacles and capturing targets. Notice that, in such problems, the robots usually present a poor initial performance, since the learning takes place on-line. By the way, the ANS is an open-ended solution and so the learning phase never stops.

To enable indirect communication among robots, they will be able to mark regions of the environment with artificial pheromones, according to past experiences, assisting one another in a cooperative and indirect way. With the presence of pheromone trails, improvements are expected in accomplishing the navigation task. In addition to a higher effectiveness in avoiding collisions and capturing targets, the minimization of distances traveled between targets is also desired as a consequence of the accumulated deposition of pheromone. Again, the robots have no *a priori* ability to properly explore the pheromone trails.

There are two strategies for stigmergic communication. In the first, named deterministic, the pheromones are deposited according to innate rules. On the other hand, in the strategy called evolutionary, the robots also have to learn how and where to lay pheromones. The former can be associated with the behavior-based paradigm, and the latter is clearly akin to evolutionary approaches. Aiming to investigate both communication strategies and also to validate the ANS in collective scenarios, several experiments and simulations are performed.

The remainder of this chapter is organized as follows. Section 2.2 reviews some pertinent biological topics and Section 2.3 describes several works related to collective robotics. In Section 2.4 the stigmergic strategies for communication are proposed and a detailed description of the autonomous navigation system developed is presented in Section 2.5. Section 2.6 outlines and discusses a series of experimental results obtained, and concluding remarks are delineated in Section 2.7.

2.2 Biological Background

Collective phenomena in nature and self-organizing systems are the most frequent inspirations when dealing with collective robotics. Such approach generates a set of techniques based on social characteristics of insects and vertebrate animals, and are investigated under the perspective of the field called swarm intelligence. In nature, collective phenomena are very numerous and diversified, and reproducing self-organizing systems in computational environments can be interpreted as the engineering side of self-organization research [13].

According to Bonabeau *et al.* [14], self-organization does not rely on individual complexity to account for complex spatiotemporal features that emerge at the colony level. Essentially, the main hypothesis is that interactions among simple individuals can produce highly structured collective behaviours.

Several examples of natural self-organizing processes may be depicted. In terms of inanimate objects, an example is sand grains forming rippled dunes. Regarding living beings, even on microscopic organisms these phenomena can be observed: via countless interactions of its components (cells), the human immune system is able to self-organize in response to a wide range of internal and external stimuli [15]. Scaling up, it is also possible to detect self-organization in several insect species and in larger animals such as fish and birds, both presenting abilities to establish and maintain complex formations while moving as a social unit [13].

Regardless of the few cited examples and a lot of others not mentioned, the most studied processes of self-organization are related to insect societies, more specifically ant colonies. Owing to the remarkable fact that from such simple individuals, complex collective abilities emerge, ants have attracted huge interest in swarm intelligence. In addition to the complexity of ant colonies, there are three main features that motivate the study of their behavior [16]:

- Ants are very common in nature, thus their observation and experimentation is easily done;
- They are readily visible and respond well to testing in laboratory situations;
- Ants behave robustly, implying that serious changes do not occur when few significant modifications are performed in their environment for experimentation.

Despite the great number of amazing issues related to ants, there is only one that motivates and is employed as the foundation of this work: ant communication. The major types of communication used by ants are chemical, tactile, visual, and acoustical. For any purpose, chemical communication, by means of pheromones, is by far the most common form used by ants. Holldobler and Wilson [17] argued that pheromones play the central role in the organization of ant societies. It is supposed that, on average, an ant colony has from 10 to 20 kinds of signals, most chemical. By the way, some socially advanced species of ants may make use of up to 20 pheromone types, each one with a different meaning [18].

The chemical communication performed by means of pheromonal deposition and sensing is classified as indirect or implicit. Pheromone trails are laid intending to express some information toward other ants, but the communication is indirect in the sense that a trail is neither individual-specific nor time-specific. It can be detected at anytime (until it evaporates) and by any individual ant. It is also considered implicit because the information is transferred through the environment, using the ground as communication means. This very important kind of communication is named stigmergy [4].

Aiming to emphasize the role stigmergy and, therefore, pheromone trails play in the self-organization of ant societies, such a process is described next. In general, every self-organizing process has interactions among individuals as the most important mechanism. Leaving aside for a while the randomness involved, there are two possible forces able to influence these interactions: positive and negative feedback [13].

Positive feedback can be observed in the following situation: when single ants find a food source, they deposit a pheromone trail while going to the nest. Such a trail attracts more ants which, again, collect food and reinforce the trail and so on. As a result of positive feedback, the pheromone concentration of the trail increases rapidly, as does the number of ants leaving the nest [17].

The role of negative feedback is to balance and stabilize the effects of positive feedback, among other regulatory activities. In ant societies, negative feedback takes place as a consequence of some circumstances such as: a limited number of individuals, food source exhaustion, competition between food sources, and/or pheromone evaporation [2].

Despite the central role of these forces to trigger and control the intensity of ant behavior, alone they are not enough to make possible the admirable and surprisingly intricate features observed in self-organized ant systems. A ubiquitous natural phenomenon, namely, random fluctuations, allies to them, and so their acting together promotes conditions for emerging orderly behaviors. Dynamics involving randomness are common and essential in nature because they allow the discovery of new solutions. It is well-known that ants follow trails imperfectly, especially trails with low pheromone concentrations [19]. When an individual loses the trail, it can find a potentially better alternative source of food (e.g., closer, richer, larger) than that currently exploited by the colony [20].

2.3 Multi-robots Basis and Related Work

Even though collective robotics and multi-robot systems are the focus of many works nowadays, these fields were just minor topics in mobile robots research up to a few decades ago [1]. The previous efforts were dedicated almost exclusively to single robot systems. Recently, this tendency has shifted to collectivity. This growth of interest in multi-robotic systems has some important motivations [21]:

- Many tasks are inherently too complex for a single robot to accomplish;
- Multiple simple robots may be a cheaper and easier solution than only one powerful robot;
- Multi-robot systems are generally more flexible and robust than single robots working alone.

Regarding robots communication – the main issue in this work and a fundamental aspect of multi-robots architectures – there are two possibilities: direct and indirect. Analogously to the biological counterpart, the direct communication consists of sending and receiving messages that are intended to convey information. The indirect version, also known as stigmergy, occurs through the environment and is neither time nor individual specific.

Balch and Arking [3] discussed both methods and concluded that the use of direct or intentional communication may be avoided when similar information can be transmitted by modifications imposed to the environment. Such a conclusion is attested considering that in most domains there is a cost associated with explicit communication. At least, communicating directly requires a common representation and specific communication protocol to transmit and receive messages [22]. Stigmergy is appealing because it implies storing information in the environment rather than in an internal world model, so that agent memory requirements may be minimized. Moreover, maintaining consistency between the world and a world model is not necessary given that the world serves as its own best model [10].

In face of theadvantages related to indirect communication, or stigmergy, many works in collective robotics have been developed based on such a biological inspiration. In what follows, a non-exhaustive list of relevant works is briefly presented.

Among other works concerning collective robotics, Wagner and Bruckstein [23] developed an algorithm for cleaning a dirty area where multiple robots cooperate by leaving trails on the ground. Ding *et al.* [24] employed artificial pheromone as quantifiers of task difficulty, that is, the robots should try to carry out tasks and, after that, deposit an amount of pheromone proportional to the task difficulty. Harder tasks, associated with higher pheromone concentration, attract more robots to cooperate with each other.

When dealing with stigmergy, most works intended to artificially reproduce some biological tasks, being the ants foraging the most usual one. The fundamental definition of foraging is the location and collection of objects (or targets) [25]. In biological systems, the targets are generally considered to be food or prey. For robotic implementation, the objects to be collected are application-dependent.

Edelen [13] made an investigation of intelligent foraging behavior via indirect communication among simple individual agents. Based on an extensive analysis of theoretical models, trail-laying and trail-following techniques are employed to produce the required stigmergic cooperation. Experiments by means of simulation and real robots are performed, guiding to results in which the stigmergy provides increased collective efficiency and ability to adapt to dynamic environments.

Other works addressed foraging from different perspectives. Based on differential equations, Sugawara and Watanabe [26] developed a model for single source robotic systems, while Nicolis and Deneubourg [27] tackled the problem of inter-source competition between more than three sources. In [2] and [28], probabilistic foraging models were proposed and computational simulations were performed to experiment with them. Although cellular automata are not able to model a complete foraging system, they have been applied to simulate network trails in several works [5, 29].

A different trail-laying approach is employed in [30] for navigating between places of interest through unknown environments with a team of robots (localization and path planning). In the approach proposed, there are trails of virtual landmarks, which are not deposited in the real environment, but they are recorded in a shared space and based on each robot's local coordinate system. The experiments indicated that the approach is robust to failure and is able to converge to the best route found by one of the robots.

Path planning is also the problem considered in [31]. The authors developed a system for controlling robots using synthetic pheromones which can be sensed, deposited and followed according to equations composed of several parameters. Thus, the main focus of the work is tuning such parameters by means of evolutionary algorithms.

Object clustering is a common task in ant societies too, inspiring many multi-robot works. For instance, Deneubourg *et al.* [32] studied the performance of a distributed sorting algorithm, inspired by how ant colonies sort their brood. The authors make use of simulated robot teams that move randomly, without direct communication. The same purpose was extended for real robots experiments in [33]. Holland and Melhuish [6] did a similar work but involving different types of sorting tasks.

Stigmergic communication by means of pheromone trails is not the only approach employed in multi-robot systems. There are other works that make use of different environmental cues, instead of trails, for indirect communication among robots. For instance, the cues may be signs, landmarks or suchlike signals. Moreover, in some works the cues are not disposed by the robots, but they are common elements contained in the environment.

The milestone workof Kube and Bonabeau [34] tackled the problem of cooperative transport of boxes. The authors implemented a robotic system that mimics such behavior of ants by means of locally sensed cues. They adopted three types or perceptual cues: for obstacle detection (threshold distances), for box detection (a light on top of the box) and for goal orientation (a spotlight near the goal). Each cue, when detected, triggers specific and innate behaviors. Moreover, the authors investigated the role of several parameters in the system (e.g., number of robots) along the experiments, concluding that a coordinated group effort is possible without use of direct communication or robot differentiation.

Rybski *et al.* [35] also employed cues from the environment aiming at solving a foraging problem. The system proposed is based on random walk and reactive behaviors to avoid obstacles. The authors were interested in verifying the improvements in task performance provided by the addition of simple communication capabilities. That is, the robots are equipped with light beacons that, when activated, are used as cues for the other robots. The results obtained show that such a type of indirect communication, depending on the distance from other robots and the moment the signal is activated, helps to reduce the time spent searching for targets.

Now, consider two situations: a robot can see its goal, but must first move away from the goal to ultimately reach it; and, the robot is unable to see any goal and thus it is reduced to wandering. In such problems, pure reactive agents do not work properly. In [36], a reactive system was proposed (schema-based behaviors) with a number of different stigmergic techniques that involve the deployment and exploitation of physical markers in order to assist in team navigation. The techniques include marking bottlenecks, local maxima and local minima. Marking bottlenecks involves creating marker trails through constricted areas (i.e. hallways and doorways), which draw agents to new areas. Local maxima markers assist agents in negotiating areas devoid of sensory stimuli by first attracting agents (drawing them from walls of rooms to areas of greater visibility) and then repelling them outward toward potential exits. Local minima markers are unique in that they prescribe an action to undertake in response to a specific situation at a particular place in the environment, rather than a more basic attractive/repulsive force. The simulations performed show that the developed techniques increase the frequency and facilitate the discovery and navigation toward hidden goals.

Note that most works cited, perhaps all of them, are behavior-based. In such approaches, the basic behaviors are innate and hand designed. The challenge for the system is to properly manage or switch among a finite set of basic behaviors in order to select the one that will be adopted in the current situation. Differently, in the present work the main contributions are focused on acquiring from the environment the knowledge necessary to promote the emergence of the navigation and communication behaviors (not designed *a priori*). All details are depicted in the next section.

2.4 Evolutionary Approach for Collective Robotics

The approach proposed here is an extension of the autonomous navigation system (ANS) presented in [7, 8, 37], which was directed to robots operating in isolation. In order to adapt the system to multi-robot navigation, an indirect communication mechanism is proposed and simple modifications are performed in the ANS, specifically related to the scheme of decision making. Both the original and the new aspects involved will be described below. At first, the adopted robot model is described, then the communication mechanism is presented, and finally the core of the ANS is detailed.

2.4.1 Robot Model

The authors' research group owns one unity of the scientific mini-robot Khepera II, shown in the right side of Figure 2.1, and has adopted its characteristics to implement the model of the virtual robot employed in the simulations of this work. As the left side of Figure 2.1 shows, there are eight infrared sensors disposed around the robot, being six at the front and two in the hinder part. They are responsible for measuring the distance from obstacles and target direction and intensity. As the robot will not be allowed to move backward, the navigation system will ignore, in the applications to be presented, the measurements of obstacle distance provided by the two rear sensors. One new sensor is introduced: it reads the type and concentration of artificial pheromone present in the current position of the robot.

The robot belongs to the classic and popular differential-drive category. It consists of two motors and drive wheels mounted co-linearly with the center of the base. It can turn about its center and move forward and backward. Thus, it is a 2 degree-of-freedom base. In this work, the adjustment of direction defined by the navigation system can range from 15 degrees (clockwise) to 15 degrees (counter-clockwise). In addition to the actuators in the wheels, there is also a new actuator that is responsible for depositing artificial pheromone in the environment.

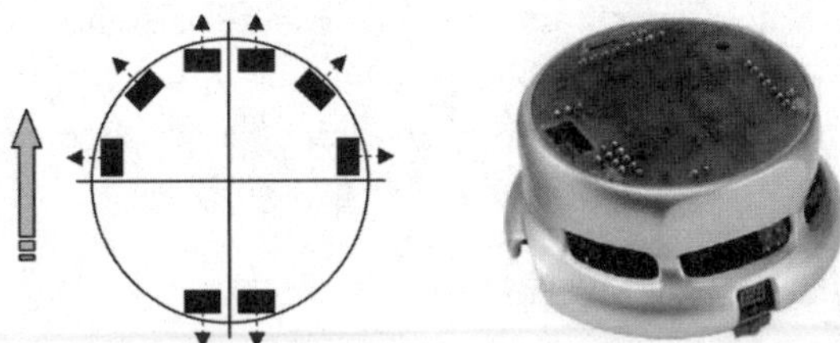

Fig. 2.1. Mini-robot Khepera II and its sensors organization. These sensors are responsible for measuring the distance to obstacles (proximity sensors) and targets (luminosity sensors). The virtual pheromone sensor is not presented.

Although the standard Khepera II is unable to leave marks on the ground and sense them, other robots with such functionalities have already been conceived. Some works have developed and applied physical mechanisms that make marks on the floor which, though not exactly based on pheromones, represent properly the role trails have in the simulations performed here. One simpler alternative is tracing lines on the ground with a pen or chalk. A more sophisticated way involves laying down solvent substances (thinner) that discolor the floor surface (i.e., a black paper) [38].

2.4.2 Multi-robot Communication based on Stigmergy

In contrast with single robot systems in which one single robot is responsible for performing tasks and the only possibility of interaction is with the environment, in multiple robot systems there exist several potential agents for interaction and such relations may be of diverse types.

Ant societies present great complexity in terms of interactions, hierarchy, distribution of tasks and so on. By the way, not all elements involved are actually understood nowadays. One exception is their communication mechanism by means of pheromones trails laid on the ground, known as stigmergy.

Table 2.1. Analogies between ants and robots in this work.

Features	Ants	Robots
Decision Making	Autonomous	Autonomous
Sensorial Capability	Reduced[1]	Reduced (8 sensors)[2]
Communication	Direct and indirect[3]	Indirect
Hierarchy	Decentralized	Decentralized
Tasks	Search for sources, terrain exploration, etc.	Search for targets and obstacle avoidance

[1]Some species are very restricted and others are more sophisticated.
[2]Infra-red sensors with short-range for obstacles (10 cm).
[3]Direct by means of contact and indirect by means of pheromones.

Inspired by such a natural process and regarding some analogies between multiple robots and ant societies, presented in Table 2.1, a mechanism for indirect communication among autonomous robots has been designed in this work. The proposal is outlined in what follows.

2.4.3 Artificial Pheromone Mechanism

In general terms, the idea of the pheromone trail mechanism is to benefit from previous circumstances experienced by the robots for assisting them in taking more suitable decisions when navigating. In this sense, the robots are able to lay artificial pheromone in any position of the environment in which they are situated. Such pheromones are used for indirect communication among robots and for cooperating to improve their performance.

Although some ant species have a large number of pheromones that may represent different information and promote different reactions according to additional factors, the robots in this work deal with the artificial pheromones in a much simpler mode. There is only one type of pheromone and its existence or absence in regions of the environment has well defined meanings, causing specific effects in the robot decision making system.

On the one hand, the presence of pheromone intends to indicate attractive regions of the environment, which are interesting for the robots to explore. As occurs with ants, the most relevant regions are expected to be more highlighted and, hence, more visited by the robots (and vice-versa), resulting in performance improvements. In practice, relevant regions are those located close to targets and the paths connecting them. Robots being attracted and visiting such areas will probably capture a higher number of targets.

On the other hand, regions not marked with pheromones are probably less promising and may even represent dangerous areas. For instance, regions close to obstacles, characterized by a higher risk of collision, will have, over the iterations, fewer visits of robots and consequently will not be marked significantly (and vice-versa). Also areas of the environment that are scarce in resources (targets) will not be signalized.

The artificial pheromone trails can have different degrees of concentration, as an effect of repeated deposition of pheromone in coincident positions, and evaporation. The concentration is incremented by a constant value each time the respective position receives a pheromone deposit. There is a maximum concentration of pheromone allowed that is fixed and denoted σ_{max}.

In nature, the pheromone evaporation is the process whereby minuscule particles of a substance in liquid state gain sufficient energy to enter the gaseous state. In the simulation, such a phenomenon is modeled as the reduction of the artificial pheromone concentration by a constant rate. The lower limit of concentration is zero.

An additional aspect related to the computational implementation of the mechanism is worthy of mention. Though the environment and the robots' movements are continuous, the computational pheromone representation is discrete. Each pheromone position of the environment is modeled as a cell in a grid fashion arrangement. For robot laying or detecting pheromone, each cell is considered as a unity, that is, it responds for any point inside the cell or grid square.

2.4.4 Pheromone and Robot Decision Making

After the description of the main aspects of the pheromone mechanism, it is time to explain an essential issue: the role pheromonal stimuli play within the robot's control system. Despite the complex dynamic of the chemical processing involving biological agents, artificial pheromone trails affect just one function of the robots' navigation system: the way the control actions (robot direction and speed) that will be carried out at each sensory-motor cycle are chosen. Such a function is detailed in sub-section 2.5, which is dedicated to the presentation of the autonomous navigation system. It is possible to say beforehand that the pheromonal information is employed in a straightforward manner, influencing the decision making procedure in favor of more suitable and efficient control actions.

2.4.5 Strategies for Pheromone Deposition

Along the work, two distinct strategies were developed for the robots laying pheromone on the environment: the deterministic and the evolutionary. Both schemes, described next, are experimented and compared in the section 2.6 devoted to the results. Note that such strategies just affect how the pheromone is deposited; all the other system features already discussed remain valid.

Deterministic

In this strategy, the pheromone deposition is not decided by the robot, but it is done based on two *a priori* rules (suchlike behavior-based approaches) and depends on the occurrence of specific events. The first case is when the robot captures targets: immediately after any capture, the robot repeats backwardly the last 10 movements, marking each point of the trajectory (performed just before the capture) with artificial pheromone.

The second case is more sophisticated and involves an important concept that will be recalled many times along the text: the *intact trajectories*. An *intact trajectory* is a trajectory free from collisions and monotony events performed by one robot going from a captured target to another one.

A trajectory is called intact exactly because no other events occur between two consecutive target captures. Such routes are considered ideal given that they meet perfectly the primary navigation objectives: capture targets and do not collide in obstacles. Thus, the number of intact trajectories is a relevant criterion for measuring the robots' performance. Moreover, computing the length of every intact trajectory automatically furnishes the mean distance traveled by robots while capturing targets, which is an optimality measure. The mean distance has a general unit denoted d.u. (distance unit).

There is an important relation between intact trajectories and pheromone deposition. In the deterministic strategy, after an intact trajectory is detected, it is automatically transformed in a pheromone trail. That is, every point of the referred trajectory is marked with pheromone as if the robot had deposited it.

Evolutionary

Conversely to the previous strategy, the strategy with evolution does not make use of predefined rules. The robots themselves are responsible for deciding how and where to lay pheromone. In each iteration, besides determining the direction and speed adjustments, the robot navigation system also decides whether it is going to deposit pheromone in its current position or not.

As emphasized before, the navigation control system learns how to capture targets and avoid obstacles by means of an evolutionary approach. In this strategy, the system also learns to deposit pheromones evolutionarily. So, initially, the robot performs deposition randomly and the system is expected to learn and become able to use the pheromone properly along the iterations. The intact trails are also computed here but just for performance reasons, that is, intact trails do not influence over the pheromone deployment process. The whole evolutionary process and pertinent details are further presented (sub-section 2.5.3).

2.5 Autonomous Navigation System

The structure and dynamics of the autonomous navigation system (ANS) is presented in this section. It is important to make clear that the entire developed system, except the parts associated with the pheromones, has been extensively investigated and experimented for single robot navigation in [8, 7, 37]. Therefore, it is not the intention of this work to explore the system from the viewpoint of robots in isolation, but it is aimed at studying the system aspects related to collective robotics.

Each robot is controlled by an autonomous navigation system (ANS) that works based only on instantaneous stimuli captured from the environment and contains no *a priori* knowledge. So, the ANS is reactive because it takes decisions based only on the current situation of the environment. On the other hand, it is autonomous since it is able to learn navigation strategies in an independent manner. To accomplish the primary navigation objectives, the robot has to present different behaviors in time, sometimes searching for targets, sometimes avoiding obstacles. The nature of the behavior emerges as a consequence of the interaction between the robot and its environment. The proper behavior to be adopted during navigation cannot always be established in a straightforward manner in face of the frequent occurrence of conflicting situations. Therefore, coordination of behaviors is also necessary to be acquired by the system.

The ANS is based on the learning classifier system (LCS) paradigm proposed by Holland [9], an evolutionary approach to synthesize adaptive inference mechanisms capable of operating in time-varying conditions. The LCS interacts with the environment by means of detectors and actuators. Detectors receive and encode incoming messages from the environment. Actuators provide means to act on the environment, decoding and performing actions defined by the system. After acting, the system also receives an environmental feedback.

A set of classifiers composes the LCS. They are $\langle condition \rangle - \langle action \rangle$ rules with an if-then inference mechanism. The antecedent and consequent parts of each classifier are usually binary strings. Each classifier has an associated strength that is related to its capability to act toward the achievement of predefined objectives.

There are three sub-systems in a LCS: rule and message sub-system, apportionment of credit sub-system and rule discovery sub-system. They interact, in brief words, as follows. When a detector captures messages from the environment, they are sent to the rule and message sub-system. Then all the classifiers try to match its antecedent part with the environment message. Those classifiers with better matching take part in a competition process. Among them, the classifier that offers the highest bid (depending mainly on the classifier strength) wins and acts on the environment. The action causes an environmental feedback that is used by the apportionment of credit sub-system to readjust the strength of the winner classifier. Thereafter, the environment emits a message with its new current state that is received again by the rule and message sub-system. The process continues until an epoch of iterations is concluded. After that, the rule discovery sub-system runs aiming at producing new and improved classifiers. Usually, a genetic algorithm is responsible for the process of rule discovery, taking the individual strength as the fitness of each classifier. Detailed information about the original LCS can be found in [9].

Owing to functional purposes, the classifier system implemented in the ANS differs in some relevant aspects from Holland's LCS. At first, each classifier has two distinct antecedent parts and three distinct consequent parts, instead of just one. Moreover, not all parts are composed of binary values. Integers are also used in the codification. When the classifiers compete to act on the environment, the winner is the one that presents the best matching with the received message. There is no bid and the classifier strength does not influence the competition process. The strength is only used for the rule discovery sub-system, when computing the fitness value. This sub-system is triggered every time one of three possible events is detected during navigation: collision, target capture, or monotony (virtual event detected when the robot presents monotonous behavior). Furthermore, for each event there are different evolutionary procedures with specific fitness functions that produce a new generation of rules. The complete details of the whole system are described in the following sections.

Returning to the description of the autonomous navigation system, it can be said that it interacts with the environment by means of the robot sensors and actuators, and it is arranged in four main components: population of classifiers, evaluation module, reproduction module and competition module (see Figure 2.2). Notice that the actuators determine the control actions of the robot at each navigation step (iteration). The population of classifiers represents the knowledge base and evolves during the robot navigation. The competition module receives stimuli captured by target and obstacle sensors, performs a matching with $\langle condition \rangle - \langle action \rangle$ rules, and defines which one is going to act on the environment. This process is repeated every time a control action is required, forming a loop that is only interrupted to give rise to an evolutionary update of the population of classifiers, denoted here as the evolution phase.

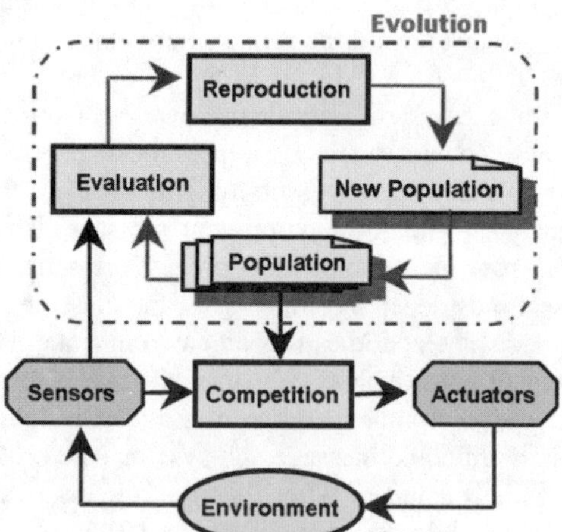

Fig. 2.2. The main structure of the autonomous navigation system (ANS).

It is important to highlight that the evolution of the set of classifiers (knowledge base) depends on the interactions with the environment, and that these interactions will be determined by the sequence of classifiers selected to provide the control actions at each instant of time.

2.5.1 Population of Classifiers

As in the original LCS, each individual of the population is represented by a $\langle condition \rangle - \langle action \rangle$ rule, with a *modus-ponens* inference mechanism. Each classifier (rule) can be described by a list of attributes, called a chromosome. Each chromosome, shown in Figure 2.3, is composed of five vectors: obstacle distance (RO: vector of integers), target light intensity (RA: vector of integers), direction adjustment (RD: binary vector), speed adjustment (RV: binary vector) and pheromone deposition (RP: binary vector). Note that the RP vector, related to pheromone deposition, is enabled just for the strategy with evolution. In the deterministic one, such a vector does not make sense because the pheromone laying is performed by predefined laws (sub-section 2.4.5).

Therefore, RO and RA comprise the antecedent part and contain, respectively, six and eight components. The former corresponds to the number of proximity sensors, and the latter to the number of luminosity sensors of the robot. The antecedent part of each classifier is supposed to represent a possible situation of the environment that, if actually happens, will imply carrying out the consequent part of the respective classifier.

In the consequent part of the classifier, the RD vector has nine components that are decoded to determine the adjustment of the robot direction. The four less significant bits of the vector are converted in an absolute value (integer) and the five more significant bits are converted to the signal of the direction. The signal conversion is done based on a majority procedure: if most of the bits are zero, the signal is negative, otherwise the signal is positive. Also in the case of the RV and RP vectors, the same conversion procedure is employed, determining the speed and pheromone decisions as shown in Table 2.2.

Table 2.2. Conversion of binary vectors to robot actions based on majority procedure.

	Most bits are ZERO	Most bits are ONE
Direction signal (RD[1])	Negative (-)	Positive (+)
Speed (RV)	Decrease	Increase
Pheromone (RP)	Do not deposit	Deposit

[1] The subset composed of the five more significant bits.

Several bits, together with a majority voting procedure, are adopted simply to provide a smooth transition between states in antagonism. The classifiers are initially constructed with totally random values in both antecedent and consequent parts.

The ANS was referred to as having no *a priori* knowledge, because the classifiers, representing the navigation system knowledge, are created with random content at

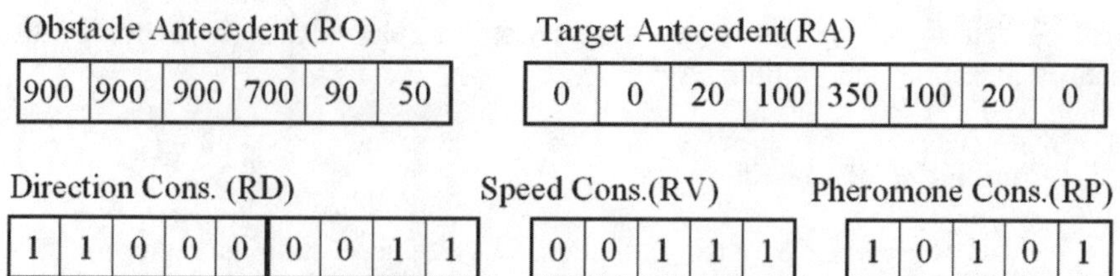

Fig. 2.3. A chromosome representation where the consequent parts, RA, RV, and RP, determine direction adjustment equal to 3**?**, speed increment and pheromone deployment (ignored in deterministic strategy), respectively.

the beginning of the navigation. As a consequence, the robots do not know how to avoid obstacles, how to capture targets and, in the evolutionary communication strategy, how to deposit pheromone. Such behaviors must be learnt by the robot system during navigation (autonomously).

2.5.2 Competition Module

This module is responsible for determining the classifier which, among all individuals of the population, owns the antecedent part that best matches with the current environmental situation faced by the robot. Such a classifier will control the robot in the current iteration.

At each robot movement, the obstacle, target and pheromone sensors capture stimuli from the environment in the position the robot is located. Such stimuli represent, respectively, the distance from the robot to the obstacles, the luminosity of the target and the concentration of pheromone.

The obstacle and target stimuli are coded in two vectors, EO and EA, respectively. Both vectors and the pheromone stimulus (CA) are sent to the competition module. At this stage, all classifiers compete in a winner-takes-all process. The winner will be the classifier whose antecedent vectors (RO and RA) present the best matching with the captured stimuli (EO and EA), taking into account the pheromone influence (CA). The matching of each classifier $r, S(r)$, is given by:

$$S(r) = \frac{\|RO(r) - EO\|}{MaxO} + \frac{\|RA(r) - EA\|}{MaxA} CA \tag{2.1}$$

where $MaxO$ and $MaxA$ mean suitable values for normalization of the first and second terms of $S(r)$, respectively, and $\|.\|$ is the Euclidean norm.

While deciding the winner classifier, the role of CA is to favor candidates that are more suitable to the sensorial information associated with targets. As the CA value is proportional to the pheromone concentration, the most frequently visited regions of the environment are intended to stimulate the robot to search targets. Note that the CA represents an innovative aspect of the proposal when compared with the original system for single robot navigation [7, 8].

At the end of the competition, the consequent of the winner classifier is then used to command the actuators. The RD consequent determines the adjustment in direction. The RV consequent establishes the variation in speed for the next movement. The speed is always modified by a constant value. This way, the winner classifier just indicates if the speed increases or decreases. If the robots are set with the strategy with evolution, the RP consequent defines whether the robot will deposit pheromone or not on the ground. Otherwise, if the deterministic strategy is set, the RP consequent is ignored.

2.5.3 Evolution

The evolutionary process, responsible for evolving the population of classifiers, is composed of evaluation and reproduction modules. Evolution takes place every time one of the following events is detected: collision in obstacle, capture of target or monotony. A monotony event is triggered if the robot does not capture a target for a long time, or if the sum of direction adjustments of previous iterations exceeds a predefined threshold (to indicate situations when the robot is moving in circles, for example).

It is important to make additional considerations involving the collision event. When such an event is detected, the robot did not in fact hit an obstacle. In order to avoid eventual damage in the robot structure, there is a predefined minimum distance that, if measured by any sensor, immediately triggers a collision event and stops the robot. Even though the occurrence of collisions may represent failures in the navigation system, they are essential for its evolution. If the robot does not collide, it will never learn how to avoid obstacles, given that such a behavior is not innate.

Still referring to collisions, in collective robotics several robots navigate in the environment and each one is an obstacle to the others. Consequently, because of their size and limited sensorial capability, collisions between robots are frequent and may be caused by a large variety of circumstances. This way, two eventual consecutive evolution processes tend to be separated by small intervals of time, with no chance for a proper evaluation of the current population of classifiers. Because of that, events of collision between robots tend to degenerate the population and therefore will not trigger evolution processes.

Evidently, collisions and also monotony events are caused by deficiencies in the navigation system, being derived from an immature state of the system or sets of misadjusted classifiers. Both cases demand evolutionary processes aiming to repair or improve the classifiers population, setting up a negative reinforcement scenario. On the other hand, the occurrence of target captures is an indication of qualified classifiers controlling the robot. In this case, a positive reinforcement scenario arises because the evolutionary process intends to spread the characteristics of such classifiers along the population.

Once again it is necessary to reinforce that all evolutionary procedures described next, involving the pheromone consequent part (RP) of the classifiers, are performed only in the strategy with evolution. Because the deterministic strategy has fixed rules for robots depositing pheromones, no evolutionary operation is done concerning the vector RP.

Considering that the events have different nature, each event-specific evolutionary process is guided by distinct objectives associated with dedicated fitness functions and genetic operators. Thus, the evolutionary procedures are depicted next, organized by event and divided in two parts: evaluation and reproduction.

2.5.4 Evaluation Module

Basically, the rating and sorting of the whole population of classifiers is performed in the evaluation module based on a particular fitness function, specific for each event. As in the evolutionary theory, the individuals with higher fitness (rating) have higher probability of being selected and recombined in the reproduction module.

Collision

The aim of evolution just after each collision event is to improve the skill for obstacle avoidance. Independent evaluations are performed, one for the antecedent of the classifiers, and another for the consequent part. The first one rates the individuals according to their similarity to the instantaneous collision situation (last stimuli set read before collision). Equation 2.1 is used here, where EO and EA correspond to the stimuli captured at the collision instant and CA=1 (the pheromone information does not influence here). In the consequent evaluation, the criterion for rating the individuals is being in accordance with the instinctive reflex for the robot after each collision, given by $T(r)$ in Equation 2.2. The idea is that the output proposed by the selected classifiers be altered, by means of crossover, by a fixed amount of 15, forcing the robot to point to a direction that tends to move the robot away from the obstacle. For example, if the robot collides with its left side, the instinctive reflex will be a value representing a turn right action.

$$T(r) = \begin{cases} |[RD(r)]_d - 15|, \text{if left collision;} \\ |[RD(r)]_d + 15|, \text{otherwise.} \end{cases} \tag{2.2}$$

where $[RD(r)]_d$ is a real value that represents the adjustment in direction defined by the consequent part of classifier r.

It is clear that in risky situations (imminent collision) the robot need to reduce its speed. Because of that, the consequent part of each classifier, related to speed, is evaluated based on a Hamming distance between the $RD(r)$ vector and a pattern vector that represents speed reduction (all elements with value one). The same procedure is done when evaluating the RP vector, related to pheromone deposition: the better classifiers are those that do not indicate pheromone deposition (most bits are zero).

Capture

The evaluation procedure here is an analogue of the collision one, aiming at improving the capture efficiency. Nevertheless, the evaluation of the antecedent part takes into account the instantaneous capture situation (last stimuli set read before capture). The evaluation of $D(r)$ (consequent RD) is dependent on the sensor that detected the capture, as shown in Equation 2.3. Such a criterion rates higher the individuals whose control action causes the robot to be better aligned with the target.

$$D(r) = |[RD(r)]_d - \alpha| \tag{2.3}$$

where $[RD(r)]_d$ is the same as in Equation 2.2, and α is the angle of the sensor that detected the event (sensor indicating higher luminosity).

When capturing a target, it is important to decrease the speed (see the previous section), since it is generally related to a place where a task must be carried out, such as object collecting. So, the evaluation is identical to the one associated with the collision case. On the other hand, the evaluation for pheromone deposition vector (RP) is opposed to the case of collision. That is, in events of capture the best classifiers will be those that do indicate pheromone deposition (it is expected that the system learns to lay pheromone in regions attractive for capturing targets). Thus, classifiers owning actions that imply in pheromone deposition will have higher rating.

Monotony

The monotony events are characterized by the robot presenting navigation behaviors that do not produce collision or target capture events. Such unproductive behaviors are common in initial phases of the robot navigation and also as a consequence of repetitive sequences of navigation steps.

Monotony events are generally measured as a function of a predefined number of iterations. Thus, every time that monotony is detected, all classifiers that have been selected to provide the control action during such interval (for example, moving the robot consistently away from the target) will be evaluated. The rating is based on the classifier action: if the effect was to reduce the robot alignment with the target, it is marked to participate in the reproduction stage; the remaining ones are kept unaltered.

2.5.5 Reproduction Module

This module is responsible for selecting and applying genetic operators to the classifiers (previously rated and sorted by the evaluation module) in order to produce offspring. Although it is not described in an independent way, the recombination and mutation operations are performed separately for antecedent and consequent parts of the individuals.

Collision and Capture

In response to collision or capture, the reproductive process is very similar for both. The only differences are the evaluation procedures performed before reproduction (details in evaluation module). Considering that the candidates for reproduction were rated, the next step is parent selection in pairs (roulette wheel) and application of one-point crossover operation in each classifier vector. As a result, two offspring are generated. Finally, they are mutated according to a small probability rate. These procedures are done independently for antecedent and consequent sub-populations.

Concluding the offspring generation, new antecedent parts are randomly combined with new consequent parts, originating the final individuals that replace the corresponding parents. The remaining individuals are kept and the new generation is finally obtained.

It is important to remark that the amount of offspring produced per generation is variable and assumes, at most, 10 % (procreation rate) of the population's total size. This policy reduces the probability of occurring harmful interference among different kinds of sequential evolution processes.

The procreation rate is an important mechanism that plays the role of keeping the population balance among individuals associated with obstacle avoidance and target capture behaviors. A balance mechanism is necessary because there is a natural trend toward the predominance of classifiers associated with target capturing, simply due to the consecutive occurrence of target captures.

Table 2.3. Relation between the number of consecutive captures and the procreation rate.

Consecutive Captures	1	2	3	4	5	6	7	8	>8
Procreation Rate (%)	10	8	7	6	5	4	3	2	1

Classifiers associated with obstacle avoidance must not be discarded as a consequence of the reduction in the number of collision events. In order to implement the aforementioned balance, the procreation rate is decreased gradually, while consecutive captures take place. The progeny size is adjusted as described in Table 2.3. If a collision or monotony event is detected, the procreation rate is set immediately to 10%. Although the number of descendants generated in evolution processes can be reduced decreasing the procreation rate, the system never stops evolving. The smallest possible rate is 1%. Therefore, at least this amount of offspring classifiers will be always produced. Such a fact characterizes the ANS as an open-ended solution.

Monotony

Assuming that the evaluation module has found all classifiers responsible for the monotonous behavior, just those with worst evaluation are modified. The modification consists in removing those classifiers and inserting new random individuals in replacement. This way, the classifiers responsible for monotonous behavior tend to be eliminated, thus suppressing the anomalous navigation behavior.

2.6 Experiments and Results

The following experiments were performed with the autonomous navigation system (ANS) using the pheromone trail strategies proposed as means of indirect communication among several robots. The results obtained are derived from comparisons among three communication strategies: evolutionary, deterministic and absence of communication.

The experiments to be presented intend to explore, compare and validate the proposals of this work. For this purpose, several environments and diverse situations are employed to verify the implications of the stigmergic communication strategies in the multi-robot system, as well as to analyze the achievement of the objectives (obstacle avoidance, target capturing and trajectory optimization).

All experiments were performed using a simulator implemented by the authors, whose robot model is based on the Khepera II robot. Of course, simulated research does not eliminate the necessity for real-world experimentation. In addition to the absence of the required number of robots in the authors' laboratory, there are also many other reasons in favor of simulations in a computational environment. Ziemke [39] argues that simulations are powerful tools for the realization and analysis of experiments devoted to the study of agent-environment interaction, because in many cases they make possible more extensive, systematic experimentation, which simply takes less time in simulation, as well as experiments that can only be carried out in very limited form on real robots (e.g., involving pheromone laying and sensing). Hence, instead of focusing on few experiments with a real robot, many questions are more suitably addressed through a large number of experiments allowing more variations of agent morphologies, control architectures and/or environments.

Every simulation in this work was repeated three times and the average of the resulting values was considered for analysis. To measure the experiment performance, six criteria were adopted: total number of collisions in obstacles, collisions between robots, capture of targets, monotony events, amount of intact trails, and mean distance between targets. Valuable results are those that present as many events of capture as possible and as few events of collision and monotony as possible. Of course it is desirable that the frequency of intact trails occurrence is high and that the mean distance is the shortest possible.

Although several robots navigate simultaneously through the environment, each one is controlled by their own autonomous navigation system. Furthermore, the robots do not communicate directly. Every environment set up is closed and has obstacles (black rectangles), targets (circles) and robots (triangles) disposed arbitrarily. Robots must capture the targets in a fixed and sequential manner. When robots complete the sequence of captures, it is restarted. For clarity sake, the environments shown in the Figures along this section have their targets labeled with numbers. Thus, it is important to make clear that the robots do not compete for targets; they just have to capture them in a pre-established sequence. No target is removed or inserted during simulation. Moreover, the maximum concentration of pheromone (σ_{max}) is set to 30.

It is important to highlight that the robots do not have initial knowledge, in the sense that *a priori* behaviors or navigation strategies have not been incorporated. Consequently, they are expected to suffer numerous events of collision with obstacles and monotony until they become able to navigate appropriately. Such an evolution of basic behaviors is best presented and explored in [8].

2.6.1 Sensitivity to Parameters Variation

Some parameters involved in the system have important roles in the simulations and results, so that an investigation about their influence seems to be of great relevance. To date, the results are analyzed considering the system sensitivity while varying the following parameters: rate of pheromone evaporation, population size (number of classifiers), initial environmental conditions and number of robots.

To clarify the interpretation of the results, note that abbreviations are used in the following tables. The strategy with evolution is represented by "Ev." and the deterministic strategy is indicated by "Det.". If both stigmergic mechanisms are disabled, then the results are denoted by "No".

Rate of Pheromone Evaporation

In nature, the pheromone evaporation is dependent on some factors such as the characteristics of the weather, the substrate and the ant colony itself. It is known that the pheromone trails may persist from several hours up to several months [2]. Hence, there is no specific rate for evaporation, assuring the plausibility when varying such a rate in artificial systems.

High evaporation rates may imply in more random trajectories, on the other hand, low rates or no evaporation may cause convergence to sub-optimal routes. These circumstances evoked tests to determine the most suitable evaporation rate in the proposed system.

The decay rate is modeled as a simple subtraction from the trail concentration per iteration. Three possibilities are considered: $1\text{x}10^{-5}$, $5\text{x}10^{-5}$ and $1\text{x}10^{-4}$. In the environment presented in Figure 2.4, the simulations were performed three times for each case and for each one of the communication strategies: evolutionary, deterministic and disabled. There were four robots and the simulations lasted 150 thousand iterations. The final results are shown in Table 2.4.

Analyzing Table 2.4, it is evident that increasing the evaporation rate causes a reduction of performance in terms of the frequency of target capture and mean distance of trajectories between targets. The fact is that higher rates hinder the occurrence of trails with enough pheromonal concentration, not allowing the convergence toward shorter trajectories and thus degrading the capture behavior. Regarding collisions and monotony events, the influence of the variation in the rate is not significant.

In addition to identifying that the evaporation rate of $1x10^{-5}$ per iteration provided the best results (and it is adopted in all further experiments), this first set of simulations demonstrates clear advantages when using stigmergic communication among the robots. In all runs, both communication strategies produced better results than when there was no communication. Although the deterministic strategy seems to be less sensitive to the evaporation rate (in terms of mean distance), the strategy with evolution proved to be, in general, the most efficient one.

The shortest mean distance obtained by the strategy with evolution was 461.78, and the final pheromone trails of its simulation is presented in Figure 2.4. In the environment, there are two ways from a target to the other; the robots clearly visited more often the shortest one, which got higher pheromone concentration. In the same figure, the graphic at right shows the curve of mean distance variation along the iterations (each circle represents an intact trail occurrence). Initially, the captures occurred rarely, and, as their frequency was increased (the pheromone concentration got higher), the mean distance converged to reduced values.

Table 2.4. Results of the experiments performed varying the pheromone evaporation rate.

Evaporation Rate	$1x10^{-5}$		$5x10^{-5}$		$1x10^{-4}$		
Version	Ev.	Det.	Ev.	Det.	Ev.	Det.	No
Monotony	110.33	152.67	146.33	128.67	166	213.67	176.67
Robot Collision	77.33	173.67	65.33	77.33	128.67	121	54.67
Obstacle Collision	167.67	161.33	154	135.33	167	180.33	167
Capture	861	695.33	713	704.67	567.33	493.67	454.67
Intact Trails	740.67	562.33	614.33	601.67	435.33	359.33	333
Mean Distance	491.23	563.72	501.80	560.62	604.67	574.06	650.36

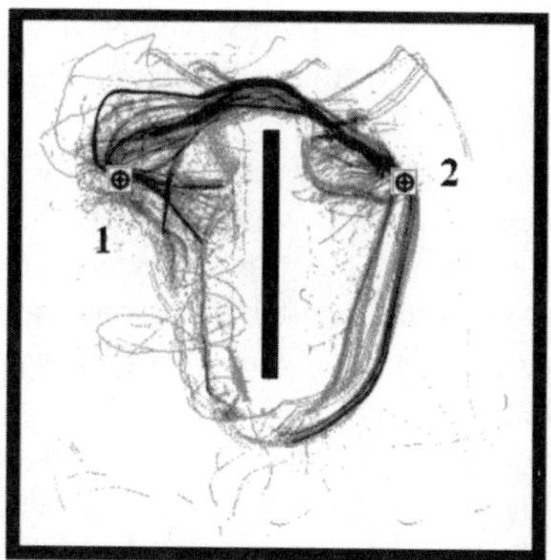

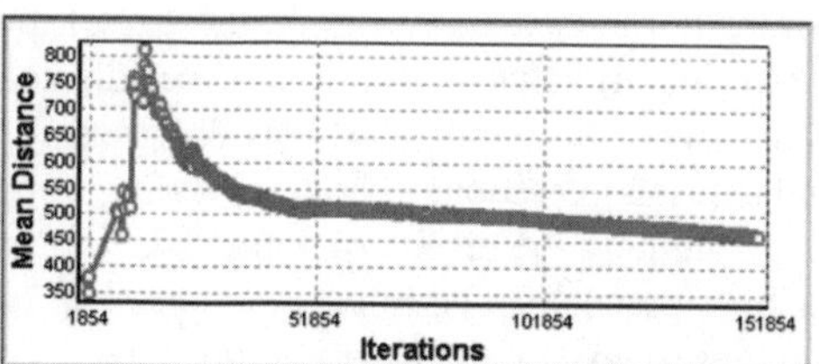

Fig. 2.4. Environment with the pheromone trail concentration (left) and graphic of mean distance variation (right) for the best simulation using the strategy with evolution.

Figure 2.5 depicts the best result achieved by the deterministic strategy (mean distance = 530.16). In this simulation, the shortest path did not prevail significantly. Comparing Figures 2.4 and 2.5, it is possible to note the different effects promoted by both communication strategies. While the evolutionary one tends to converge to a single, more concentrated trail, the deterministic one produces several trails with high concentration. That is, the indirect interaction among the robots is weaker in the latter case, but it is still better than no interaction (no communication).

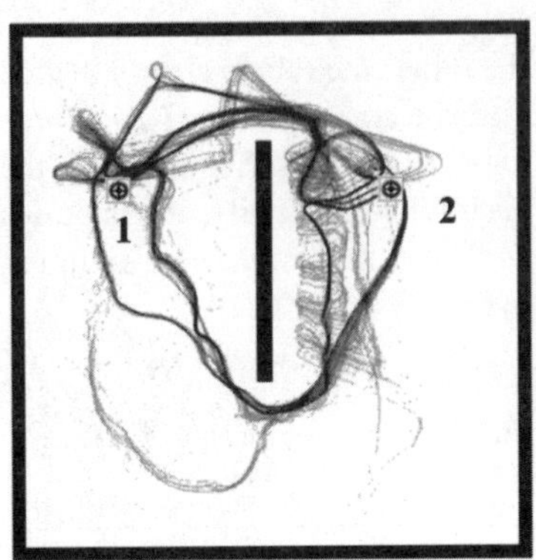

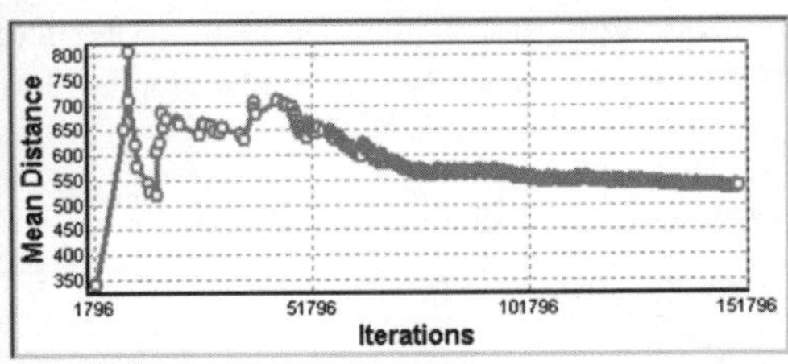

Fig. 2.5. Environment with the pheromone trail concentration (left) and graphic of mean distance variation (right) for the best simulation using the deterministic strategy.

System Complexity

Most of the admiration social insects raise is associated with their capability of doing complex things, even being such simple organisms. In this sense, one question arises: can the collective performance be maintained by means of stigmergy even reducing the level of individual complexity? Such a question is considered for the artificial system proposed in this section.

One parameter of the ANS that is directly linked to the system complexity is the number of classifiers. To examine the performance of collective navigation when "simpler" robots are navigating, the size of the classifiers population is ranged from 100 (original) down to 40 in four experiments. An environment with four targets and four robots (see Figure 2.4) was designed for the simulations whose duration was 150 thousand iterations. The results are grouped in Table 2.5.

Examining the results of Table 2.5, some events are clearly not much dependent on the amount of classifiers. For instance, collisions, monotony events and also captures seem to be hardly influenced by reducing the number of classifiers. Conversely, the intact trails and mean distance presented an evident general degradation, except for the case of no communication.

When the robots do not communicate, even when decreasing the number of classifiers, the performance was maintained. In both communication strategies, the

Table 2.5. Results obtained for navigation systems containing from 40 to 100 classifiers.

Classifiers	100			80		
Version	Ev.	Det.	No	Ev.	Det.	No.
Monotony	201.67	199	201	153.80	190.33	191.33
Robot Collision	77.33	63.33	83.67	39.60	45.67	164.33
Obstacle Collision	208.67	178.67	175.33	161.40	157	171.33
Capture	733	679.33	589	869.40	669.67	667.33
Intact Trails	585.67	546	456	745.40	544.67	531.33
Mean Distance	381.84	414.66	426.72	388.10	415.36	440.97
Classifiers	60			40		
Version	Ev.	Det.	No	Ev.	Det.	No.
Monotony	185.67	172.60	176.50	172	216.67	198.60
Robot Collision	60	69.33	48.75	27	85.33	55.60
Obstacle Collision	178	160.67	153.25	159	180	167.80
Capture	658.33	616.67	582.50	674.67	470	553.80
Intact Trails	523.33	481.33	468.75	536.33	343.33	428.80
Mean Distance	420.15	449.32	441.79	437.15	459.33	431.28

performance obtained with 100 classifiers is sustained with 80 classifiers. With 60 and 40 classifiers, however, both strategies suffered serious performance degradation, being the deterministic one more sensitive (mainly in intact trails).

Again, an important conclusion achieved from those experiments is the superiority of the evolutionary over the deterministic strategy. Such superiority is significantly higher when comparing with the strategy without communication, but just with 100 and 80 classifiers. The best mean distance obtained was 346.68 with the strategy with evolution, whose final environment state is depicted in Figure 2.6. Concerning the simpler systems (60 and 40 classifiers), the attempt to produce stigmergy fails in face of the better results presented by no communication.

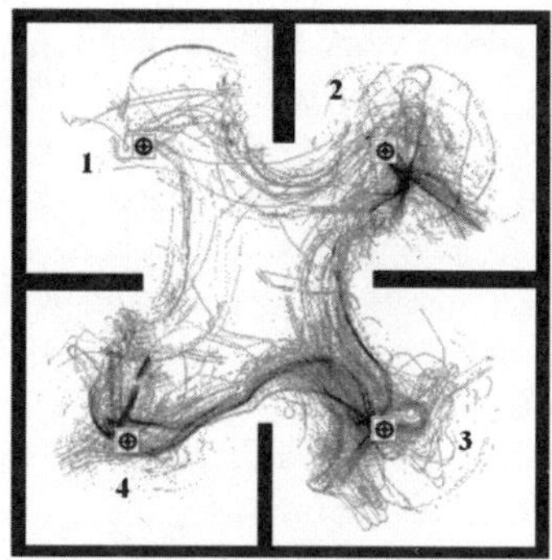

Fig. 2.6. Final pheromone trails for the best result obtained by means of evolutionary version.

Initial Position

In this section, the initial conditions of the experiments are discussed. Using the same environment of the previous section (four targets, four robots and 150 thousand iterations), four slightly different experiments were performed always employing the strategy with evolution. In each one, the robots were initially positioned in a distinct quadrant, in which the sequence of targets also started. The resulting pheromone trails are shown in Figure 2.7.

Analyzing Figure 2.7, it is possible to observe that the regions corresponding to the quadrants the robots were positioned initially received less pheromone deposition than the other quadrants. There is a simple explanation for that. The region where the robots start navigating is, in general, the local they experience the first collisions, and that collisions are necessary to develop the ability to avoid obstacles. As mentioned in the system description, learning by means of collisions produces classifiers that tend not to lay pheromones (taking into account that the region is not attractive due to the collisions). Thus, when the robots navigate in such a region, those classifiers are activated often and then the robots deposit pheromone seldom. On the other hand, regions associated with the other quadrants have a more extensive exploration and consequently a more intensive pheromone deposition.

Although the initial condition has proved to influence at least in how the pheromone trails are built, the effects in the navigation performance are insignificant. The four experiments obtained very similar results. Just to cite the number of captures ranged from 1011 up to 1254, and the mean distance varied from 332.25 (obtained in the 1st quadrant experiment) up to 346, with less than 5% of deviation. That is, the system is robust to different initial conditions.

Number of Robots

Collectivity is everything for social insects. There are ant species whose number of individuals living in the same nest may reach up to 300 million [18]. Of course in robotics it is inconceivable to work with so many robots. Anyway, performing experiments with different number of robots is a widespread practice in literature and is also the subject of this section.

The environment set up for these experiments is the same used in the section of evaporation rate. There are two targets that must be captured alternately and the simulations had 150 thousand iterations. Experiments with 4, 6, 8, 12 and 20 robots were carried out and their results are presented in Table 2.6 (except for the four robots case which is in the two first columns of Table 2.4). It is pertinent to highlight that the simulation with 20 robots took place in a larger but analogous environment (see Figure 2.9) and lasted 300 thousand iterations.

Firstly, the higher the number of robots, the more collisions between robots occurred, as can be seen in Table 2.6. As expected, having more robots in an environment with fixed size implies in having less free space as well as higher frequency of collisions between them. In average, the number of collisions between robots for the strategy with evolution increased 41% when additional robots

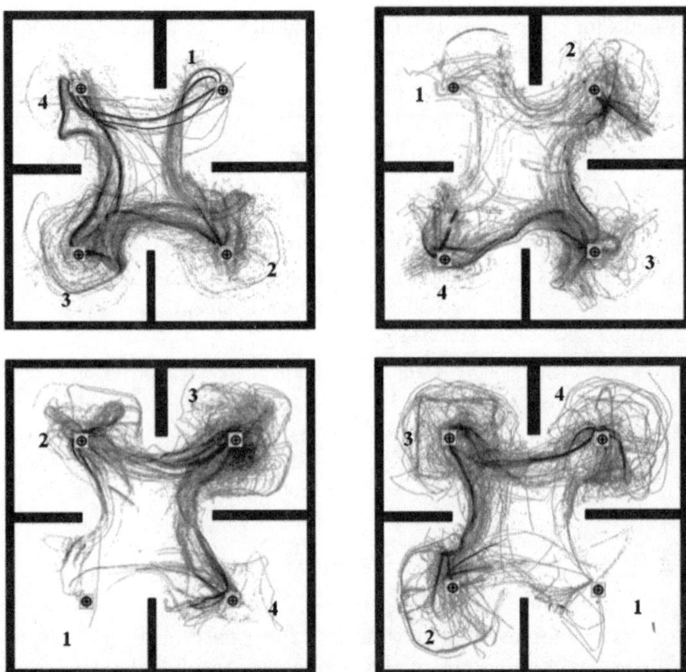

Fig. 2.7. Environments where the robots were initially positioned in the first quadrant (left top), second quadrant (right top), third quadrant (left bottom) and fourth quadrant (right bottom).

were inserted in the environment; 31% for the deterministic, and 80% for no communication. Thus, it is clear that the absence of communication is responsible for an increment in collisions between robots.

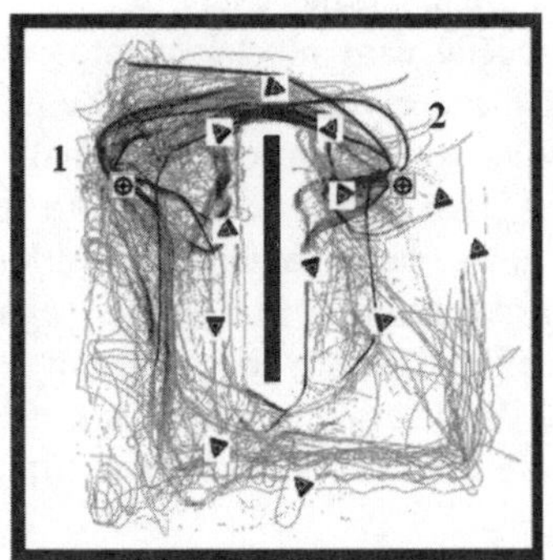

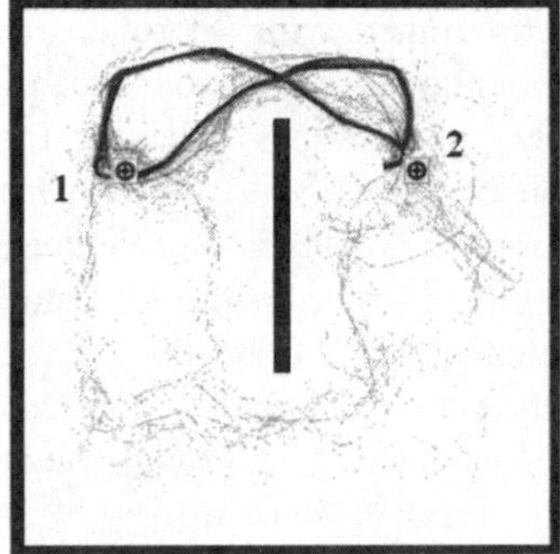

Fig. 2.8. Evolutionary (left) and deterministic (right) experiments. Best results obtained with 12 robots.

Table 2.6. Results for experiments with 6, 8, 12, and 20 robots. A larger navigation scenario was adopted in the case of 20 robots.

Robots		6			8	
Version	Ev.	Det.	No	Ev.	Det.	No.
Monotony	217.33	201	207	266.67	290.33	338
Robot Collision	144.33	90.67	145	175	174	322.67
Obstacle Collision	290.33	268	258	362.33	352.67	350.33
Capture	946	826.67	849.67	1367.33	1218.33	955.67
Intact Trails	737.67	629.67	644.33	1089.33	959	698.67
Mean Distance	541.23	620.61	650.36	508.31	555.94	615.30
Robots		12			20	
Version	Ev.	Det.	No	Ev.	Det.	No.
Monotony	496.33	485.33	416.33	1964	2517	2355
Robot Collision	331	313.67	483	213.33	232.67	303
Obstacle Collision	772.67	614	498.33	503.67	588.67	540
Capture	1640.33	1585	1506.67	3548.33	2144.33	2541.33
Intact Trails	1154.33	1101	1154.67	3042.33	1624.67	2029.33
Mean Distance	568.36	596.25	628.72	705.94	732.83	756.18

More robots in the environment and less free space also imply an increase in obstacle collisions and monotony events, being such results relatively similar for all three strategies. Considering all cases in terms of capture behavior, once again the strategy with evolution was significantly more successful than the others, always reaching the shortest distances between targets. It is interesting to note that when a higher number of robots navigate through the environment, there is a general tendency toward shorter trajectories than when smaller groups are considered. Such a tendency is stronger in the communication strategies because more robots laying pheromones on the ground enhance the probability to be established promising trails. The best results obtained by 12 robots for evolutionary and deterministic strategies correspond to the left and right images of Figure 2.8, respectively.

The experiment with 20 robots is a special case not taken into account in the aforementioned discussion because of the different size of the environment and number of iterations. Obeying the respective proportions, the analysis done above can be extended to this experiment. That is, the strategy with evolution outperformed the others by far. Figure 2.9 shows the pheromonal deposition for the best evolutionary run, on the left, and best deterministic run, on the right (mean distances equal 667.34 and 691.02, respectively). Notice that both have stronger trails in the shortest side of the obstacle. In the deterministic case, however, many robots also navigate often through the opposite side of the environment, contributing to make the mean distance worse

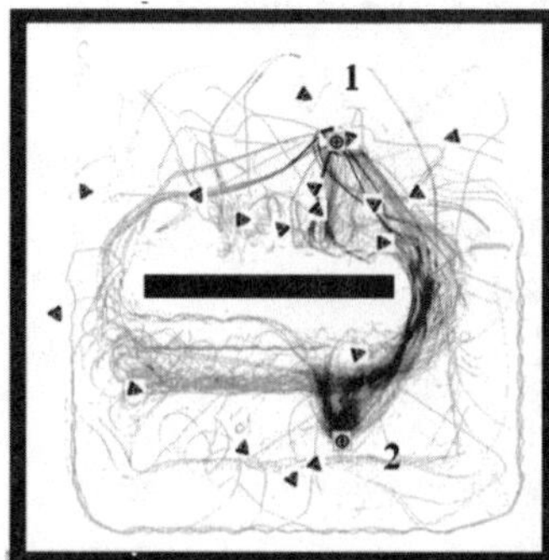

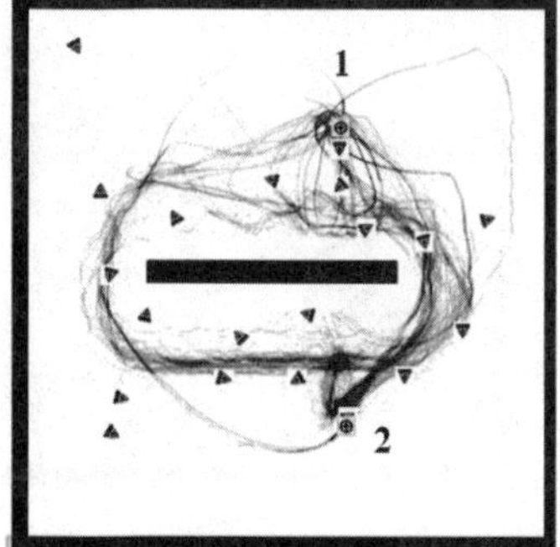

Fig. 2.9. Evolutionary (left) and deterministic (right) experiments best results obtained with 20 robots.

2.6.2 Additional Experiments

In this section three other experiments are considered. In all of them, the evaporation rate is 10^{-5} and the navigation system of each robot is composed of 100 classifiers. The other parameters are customized.

Experiment 1

The environment of Figure 2.10 was configured with 16 targets, six robots, and the simulations had 300 thousand iterations. The robots should capture an outside target followed by an inside target, and so on. Analyzing the obtained results, presented in Table 2.7, it can be seen that the events of obstacle collision and monotony occurred in an equivalent manner for the three strategies. Nevertheless, the strategy with absence of communication almost duplicates the number of collisions between robots. One possible explanation is that in the stigmergic strategies the robots form and pursue distinct trails when facing conflicting situations (e.g., two robots navigating in contrary directions), which helps to avoid robot-robot collisions. Related to mean distance, it is redundant to say that the strategy with evolution outperformed the others.

Table 2.7. Results of the experiment performed in the environment presented in Figure 2.10.

Version	Ev.	Det.	No
Monotony	403	415.33	467
Robot Collision	70	77.67	123
Obstacle Collision	866.33	835.33	803.67
Capture	2588.33	2626.33	2072
Intact Trails	1936	2008.33	1498
Mean Distance	369.04	380.91	418.9

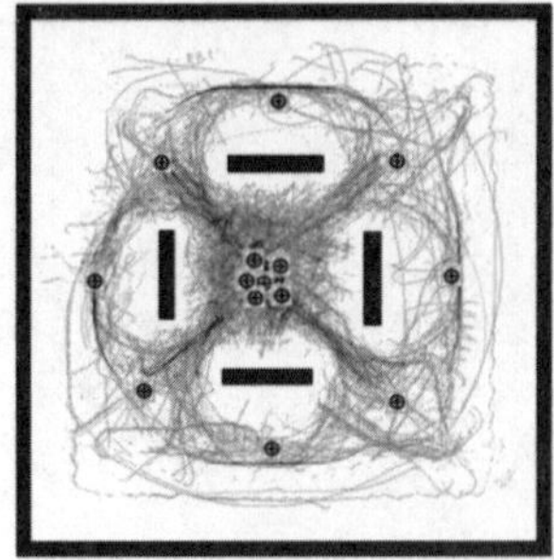
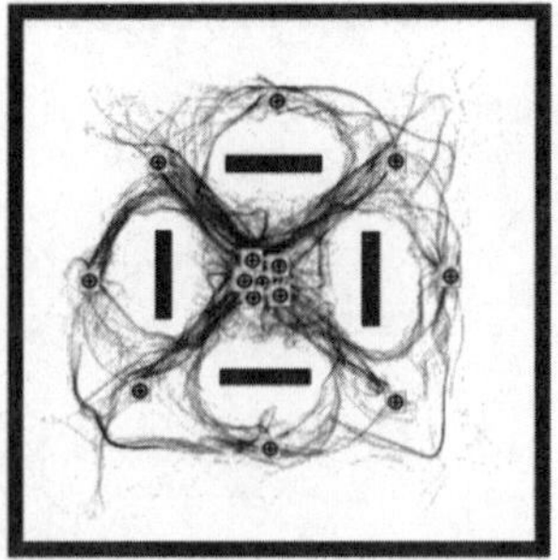

Fig. 2.10. Best simulations for the evolutionary (left) and deterministic (right) experiments.

Experiment 2

In this experiment the environment of Figure 2.11 is employed, containing four targets and ten robots. Clearly the best path to capture the targets is located in the diagonal and the aim is to verify if the robots really converge to such a route. The simulations lasted 500 thousand iterations and the results are summarized in Table 2.8, with favorable scores when the strategy with evolution is adopted.

Table 2.8. Results of the experiment performed in the environment presented in Figure 2.11.

Version	Ev.	Det.	No
Monotony	768	1018.33	888
Robot Collision	83.33	186.67	101.67
Obstacle Collision	723.67	884	773.67
Capture	6272	4807	4651.33
Intact Trails	5663.67	4086.33	3991.33
Mean Distance	512.82	549.38	606.82

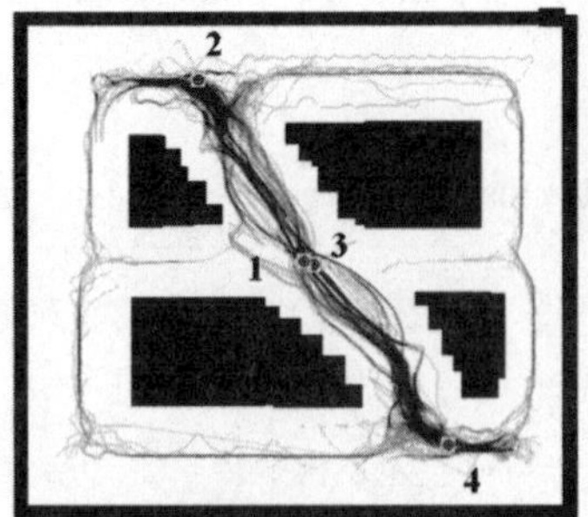

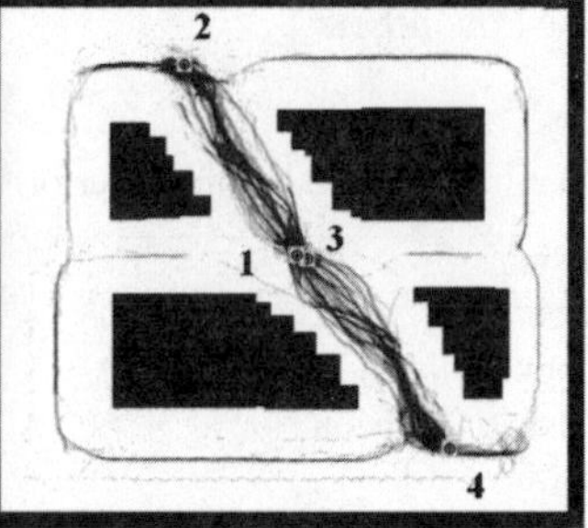

Fig. 2.11. Best simulations for the evolutionary (left) and deterministic (right) experiments.

In Figure 2.11, it is attested that in fact the robots visited very often the diagonal of the environment, laying pheromones in large scale. Of course, there exist trails around the obstacles as a result of random fluctuations, discussed in the next experiment, and also due to the fact that all ten robots do not fit simultaneously in the diagonal corridor. Therefore, some of them are forced to capture the targets using the external corridors.

Experiment 3

This is a long term experiment (1.2 million iterations) carried out in the environment exhibited in Figure 2.12. There are six targets and four robots. The simulation results presented in Table 2.9 indicate the higher efficiency of the strategy with evolution in comparison with the others. Differently from what has happened in the previous experiments, in this experiment the stigmergic versions obtained more obstacle collisions than the version without communication. Owing to the long simulation time, there were many pheromone trails accumulated close to the obstacles, resulting in more imminent collisions. Perhaps, in long term simulations the use of a higher evaporation rate would be more appropriate to avoid such a tendency.

Table 2.9. Results of the experiment performed in the environment presented in Figure 2.12.

Version	Ev.	Det.	No
Monotony	1447	1431	1523.67
Robot Collision	267.67	211.67	282.33
Obstacle Collision	1641.67	1462.67	1377.33
Capture	5209.33	3876.33	4259
Intact Trails	3924	2757	3110.67
Mean Distance	551.53	604.02	617.7

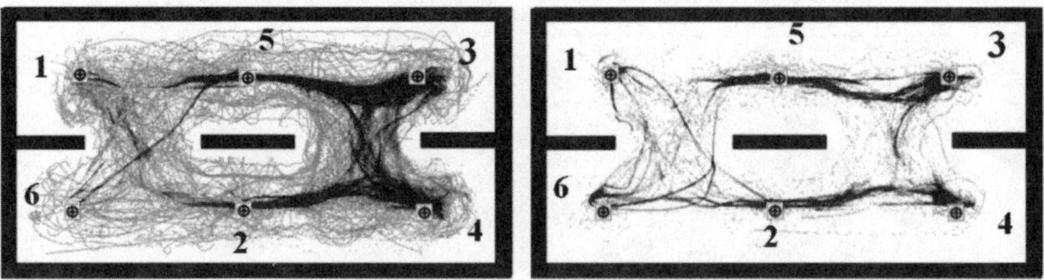

Fig. 2.12. Best simulations for the evolutionary (left) and deterministic (right) experiments.

Taking a closer look at Figure 2.12, it is noticeable that the more concentrated trails got larger than usually observed. Moreover, numerous trails with low pheromonal concentration, apparently disordered, can be seen, mainly in the evolutionary version. This phenomenon can be ascribed to system intrinsic randomness and may be associated with the random fluctuations usually viewed in social insects [19]. Such an implication could also be derived from experiments previously performed, and the strategy with evolution always presents a wider number of disordered trails. This may be one of the reasons why such a strategy has shown to be more efficient in getting the shortest routes.

2.6.3 Adaptability

The purpose of this section is to study the navigation system capability to adapt to changes in the environment. The experiment shown in Figure 2.13 consisted of 100 thousand iterations, being the initial (left) environment modified exactly in the meantime of the simulation by inserting an obstacle between the two targets. Such an event has forced the four robots to readapt in order to remain capturing targets by means of intact trails. In this experiment, the strategy with evolution is employed.

Looking at the left side of Figure 2.13, it is evident the fast convergence achieved by the robots, finishing the first 50 thousand iterations with 990 intact trails and mean distance equal to 186.30. When the central obstacle was inserted, the graphic at the right side indicates a period of readaptation (about 30 thousand iterations) before the robots restart reducing the mean distance, until 401.04. The trails around the new obstacle confirm that the robots were able to find new alternative trajectories between the targets.

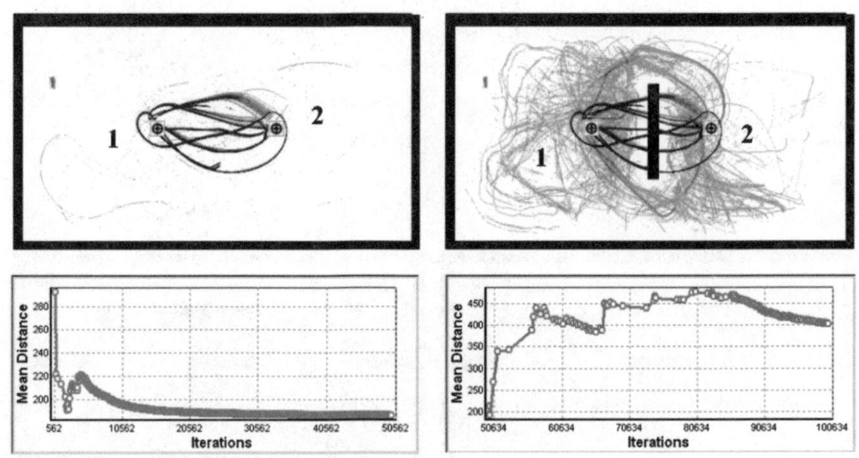

Fig. 2.13. Simulation for the evolutionary experiment before (left) and after (right) the new obstacle has been inserted between the two targets.

An identical experiment was performed when the four robots had no pheromonal communication. Figure 2.14 presents the curves of mean distance along the iterations. The left graphic corresponds to the first half of the simulation and, again, a fast convergence was obtained. Nevertheless, a noticeable distinction can be noticed in the graphic associated with the second half of the simulation (when the obstacle was introduced between the two targets). After a few 5 thousand iterations, the robots readapted to the environmental change and reached a good performance again. That is, the readaptation was quicker without communication among the robots.

Table 2.10 presents the results for both experiments, divided into before and after the environmental modification. Such results corroborate and give details on how the robots carried out the readaptation. For the strategy with evolution, 61 monotony events, 2 robot collisions and 75 obstacle collisions were faced by the robots until they familiarize with the new environment. On the other hand, in the no communication experiment, the robots re-adapted easily: there were 52 monotony events and only 33 obstacle collisions after inserting the new obstacle. As a result, they were able to complete the experiment with a final performance, in terms of captures and mean distance, better than the evolutionary counterpart.

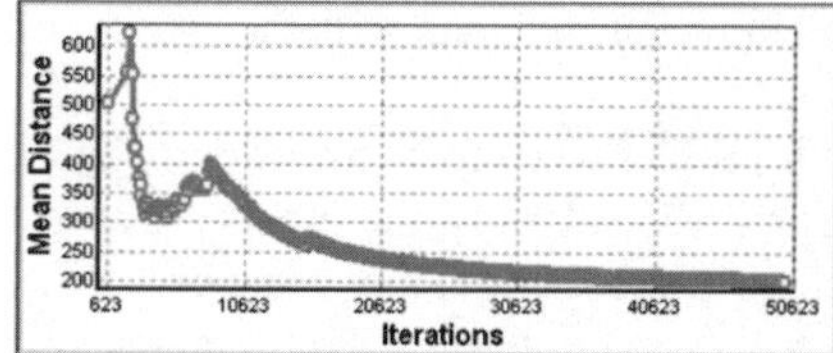

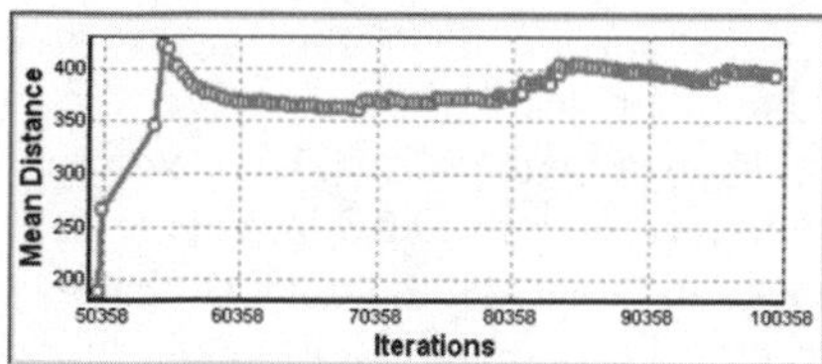

Fig. 2.14. Mean distance curves for the no communication experiment before (left) and after (right) the new obstacle be inserted between the two targets.

The best adaptability of the robots without communication may be attributed to the fact that their navigation system are more reactive because they work completely independent of any pheromonal information. As the stigmergic system is evolved based on the pheromone trails, environmental changes allied with a low rate of evaporation of the trails imply a slower adaptation.

The more relevant conclusion obtained in this section is that the autonomous navigation system proposed with stigmergic communication is effective in learning and adapting to unknown and dynamic environments. When the dynamic is given by a step transition, the existence of a fading memory associated with previously high-quality trails seems to postpone the achievement of proper alternative trails. So, when the transition in the environmental conditions happens in a rate higher than the rate of pheromonal evaporation, the absence of stigmergic communication guides to a more effective and prompt reaction.

2.6.4 Tour Optimization

The previously presented set of experiments provided convincing results regarding the ability of the ANS with stigmergic communication for minimizing trajectories between targets. To explore deeply such optimization proficiency, multiple robots

Table 2.10. Results obtained for the experiments with environmental changes.

Version	Evolutionary		No Communication	
Obstacle Insertion	Before	After	Before	After
Monotony	4	61	4	52
Robot Collision	73	2	15	0
Obstacle Collision	18	75	22	33
Capture	999	220	932	222
Intact Trails	990	170	917	195
Mean Distance	186.30	401.06	200.60	393.30

have been adopted to solve the traveling salesperson problem (TSP), which is one of the most studied problems in the combinatorial optimization field.

Given a collection of cities, the standard version of TSP consists of finding the shortest way of visiting all of the cities once and returning to the starting one. The navigation system is applicable to the TSP, considering that the targets represent the cities and the robots must capture them making a closed tour, with the total length to be minimized.

Though there are analogies between both problems, some modifications had to be performed in the ANS in order to deal with the TSP. Firstly, the robots are able to detect all targets all the time, so that they can capture them in any sequence. When a target is captured, it is disabled (only for that robot) until the robot completes the entire tour. In this way, each robot has its own list of active targets (they do not compete for targets, but every target is candidate to be captured). Once the robot has captured all the targets, only the one that was caught first is enabled, forcing the robot to complete the tour. After that, all targets are enabled again, except the first one. Furthermore, the total length of the tour is computed in spite of the mean distance between targets. The sequence of targets that composes the tour is also considered, given that it is an important indicator of the optimal tour.

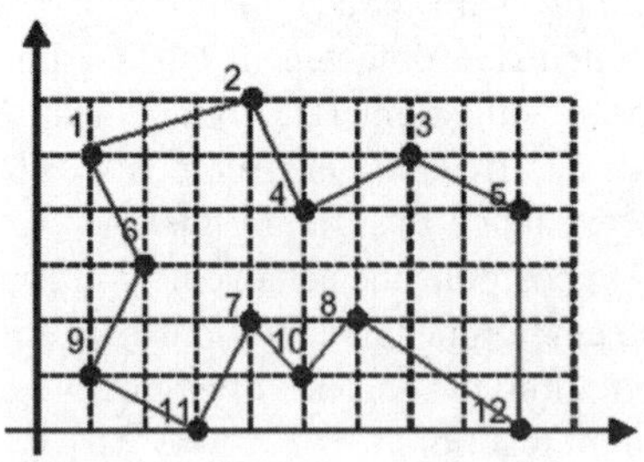

Fig. 2.15. Artificial instance of the TSP with 12 cities.

An artificial instance of the TSP was employed in this experiment. It contains 12 cities (targets) arbitrarily disposed, as presented in Figure 2.15. The optimal tour is highlighted, connecting all the cities in the shortest way. Such an instance was adopted to synthesize a collective robot navigation scenario by converting the cities in targets, as shown in Figure 2.16. Four robots controlled by the ANS with stigmergic communication were used in the experiment. The strategy with evolution was chosen because it has presented a superior performance in the previous experiments with a static environment.

The best result obtained by the robots among three trials of 500 thousands iterations is shown in Figure 2.16. The tour that the robots have highlighted by means of pheromone deposition is exactly the optimal order of cities (1-2-4-3-5-12-8-10-7-11-9-6-1). In terms of the optimal tour length, the result suggests a good solution, but not best, given that the shortest length is 2340 and the length actually obtained was 2673. However, taking into account the nature of each solution (one obtained just by straightforwardly connecting the correct cities, and the other by robots navigating in a dynamic and unknown environment) the 14.23% of increase in length is not so significant. Moreover, in addition to avoiding collisions with obstacles and other robots, the coordinates of the targets are unknown; the robots sense all the targets simultaneously and have to decide which one will be captured next.

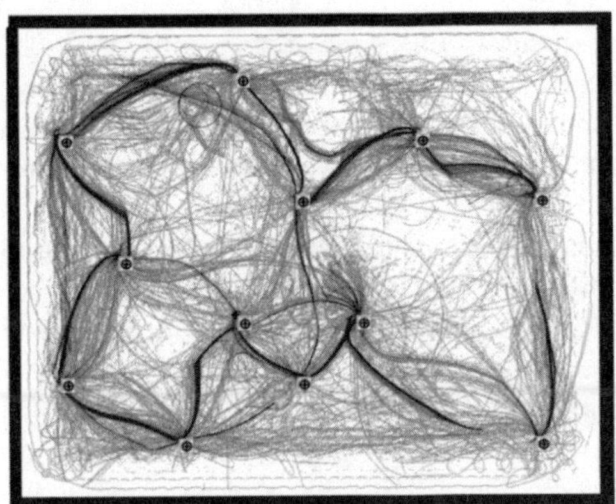

Fig. 2.16. Optimal tour for TSP with 12 cities obtained in a collective robot navigation scenario using the ANS with the stigmergic communication (strategy with evolution). There were 64 robot collisions, 32 collisions with obstacles, 182 monotony events, and 5434 captures.

Of course there exist several other methods for solving traveling salesperson problems. To date, some well succeeded techniques are based on ant stigmergy [40]. It is also obvious that a 12-cities instance does not represent a significant challenge for such approaches. It is interesting to note, however, that the navigation scenario is not based on a graph and every kind of trajectory is admissible. The ANS with stigmergic communication was able to accomplish not only the navigation objective, but it was also able to solve the TSP instance at optimality.

2.7 Conclusions

This work proposes the extension of an autonomous navigation system (ANS) for collective robotics, involving complex, dynamic and unknown environments. Based on learning classifier systems, the ANS comprehends open-ended controllers, one independent controller for each robot, without any kind of initial knowledge, and capable of acquiring knowledge during operation.

The ANS expansion to deal with multiple robots consists, basically, in the development of two communication strategies aiming at improving the collective performance. Such strategies are based on stigmergy, an indirect form of communication commonly performed by social insects, in special by ants. The robots are able to sense and deposit artificial pheromones on the environment in order to mark attractive regions. By means of evolutionary operators, the robots learn, during the simulation, how and where to lay pheromones on the environment. On the other hand, the second strategy for communication is deterministic in the sense that the pheromones are deposited according to hand designed rules.

Intending to investigate and compare the ANS with and without the communication strategies in collective scenarios, several experiments were carried out involving approximately 130 simulations, with 4 to 20 robots navigating simultaneously in the arena. The results obtained provided enough details to indicate the advantages and drawbacks of the proposals, which are discussed next.

One very important conclusion achieved along this work concerns the stigmergic communication among the robots. As already attested in the literature, the use of pheromone trails in the environment has proved to lead to an overall performance consistently better than that obtained when the robots do not communicate. Owing to the continuous learning capability, the strategy with evolution demonstrated to be more flexible and hence more efficient than the deterministic strategy in assisting the robots to find the shortest trajectories between targets.

It was detected, however, that high pheromone evaporation rates imply performance degradation. Thus, it is very relevant to properly define such a rate given that a mistaken choice makes the pheromonal communication useless or yet prejudicial to the system. Another factor that influences the evolutionary approach, affecting the manner the pheromones are used, is the robots initial positioning. But it is not so critic as the evaporation rate, with negligible effects on the navigation performance.

The number of classifiers of the ANS, associated with its complexity, was also examined in the experiments. The conclusion is that reducing the number of classifiers causes a general degradation in performance when there is communication. The robots without communication presented better results for the simpler navigation systems tested. Moreover, the robots with pheromonal communication disabled also showed to be able to readapt quicker to sudden environmental changes than the stigmergic version. When a higher number of robots are navigating together, however, the absence of communication is extremely harmful in terms of increasing the quantity of collisions between robots.

With more robots in the environment, the number of collisions may increase. However, the probability of achieving shorter trajectories between targets is higher, given that the chance to find a better trajectory increases with the number of robots in the environment. The discovery of shortest trajectories between targets was not the only optimization ability expressed by the navigation system. The ANS demonstrated that is also able to find high-quality sequences of targets that represent city tours in traveling salesperson problems. This result is relevant because in addition to dealing successfully with the problem of encountering the shortest length of a tour, the ANS is also able to achieve the navigation objectives and present efficient performance as well.

Summarizing the experimental results, it can be said that the autonomous system proposed without innate basic behaviors, but with a powerful learning engine and an evolutionary approach for stigmergic communication, showed to be very effective in avoiding obstacles, capturing targets and optimizing trajectories. Of course, there are some drawbacks that deserve detailed investigation, and the virtues raised along this work serve as a vigorous motivation for this undertaking.

Although it can be considered just an indirect consequence, the system capability to minimize distances is promising, especially because the ANS was not directly designed toward this end. Thus, some future perspectives are associated with preparing the ANS for application in clustering and real optimization problems. Furthermore, in further works the intention is also to deal with other collective tasks (e.g., box-pushing and co-evolutionary scenarios). Even though the design of the physical devices for artificial pheromone deposition is a technical challenge, it has been performed successfully in the literature by means of some feasible solutions [38]. This initiative motivates the most ambitious perspective of this work: the implementation of the stigmergic communication in teams of real robots.

References

1. T. Arai, E. Pagello, and L. Parker. Guest editorial, advances in multi-robot systems. *IEEE Transactions on Robotics and Automation*, 18(5):655–661, 2002.
2. E. Bonabeau, M. Dorigo, and G. Theraulaz. *Swarm Intelligence: From Natural to Artificial Systems*. 1999.
3. T. Balch and R. C. Arkin. Communication in reactive multiagent robotic systems. *Autonomous Robots*, 1(1):27–52, 1994.
4. P. Grasse. La reconstruction du nid et les coordinations inter-individuelle chez bellicoitermes natalenis et cubitermes sp la theorie de la stigmergie: Essai d'interpretation des termites constructeurs. *Insectes Sociaux*, 6, 1959.
5. S. Camazine, N. R. Franks, J. Sneyd, E. Bonabeau, J-L Deneubourg, and G. Theraula. *Self-Organization in Biological Systems*. Princeton University Press, 2001.
6. O. Holland and C. Melhuish. Stimergy, self-organization, and sorting in collective robotics. *Artificial Life*, 5(2):173–202, 1999.
7. R. R. Cazangi, F. J. Von Zuben, and M. F. Figueiredo. A classifier system in real applications for robot navigation. In *Proceedings of the 2003 Congress on Evolutionary Computation*, volume 1, pages 574–580, Canberra, Australia, 2003. IEEE Press.

8. R. R. Cazangi. Uma proposta evolutiva para controle inteligente em navegação autônoma de robôs. Master's thesis, Faculdade de Engenharia Elétrica e de Computação, Universidade Estadual de Campinas, 2004.
9. J. Holland. Escaping brittleness: The possibilities of general purpose learning algorithms applied to parallel rule-based systems. In R. Michalsky, J. Carbonell, and T. Mitchell, editors, *Machine Intelligence II*. Morgan Kaufmann, 1986.
10. R. A. Brooks. Intelligence without reason. In *Proceedings of the 1991 International Joint Conference on Artificial Intelligence*, pages 569–595, 1991.
11. C. R. Kube, C. Parker, T. Wang, and H. Zhang. Biologically inspired collective robotics. In L. N. de Castro and F. J. Von Zuben, editors, *Recent Developments in Biologically In-spired Computing*. Idea Group Inc., 2004.
12. S. Nolfi and D. Floriano. *Evolutionary Robotics*. The MIT Press, 2000.
13. M. Edelen. Swarm intelligence and stigmergy: Robotic implementation of foraging behavior. Master's thesis, University of Maryland, 2003.
14. E. Bonabeau, G. Theraulaz, J-L Deneubourg, S. Aron, and S. Camazine. Self-organization in social insects. *Trends in Ecology and Evolution*, 12:188–193, 1997.
15. L. N. de Castro and J. Timmis. *Artificial Immune Systems: A New Computational Intelligence Paradigm*. SpringerVerlag, 2002.
16. M. Resnick. *Turtles, termites, and traffic jams: Explorations in massively parallel microworlds*. Bradford Books/MIT Press.
17. B. Holldobler and E. Wilson. *The ants*. Belknap Press of Harvard University Press, 1990.
18. F. H. Caetano, J. Klaus, and F. J. Zara. *Formigas: Biologia e Anatomia*. Editora da UNESP, 2002.
19. J. L. Deneubourg, J. M. Pasteels, and J. C. Verhaeghe. Probabilistic behaviour in ants: A strategy of errors? *Journal of Theoretical Biology*, 105:259–271, 1983.
20. J. Pasteels, J. L. Deneubourg, and S. Goss. Self-organization mechanisms in ant societies (i): Trail recruitment to newly discovered food sources. *Experientia Supplementum*, 54:155–175, 1987.
21. Y. Cao, A. Fukunaga, and A. Kahng. Cooperative mobile robotics: Antecedents and directions. *Autonomous Robots*, 4(1):7–27, 1997.
22. M.Tambe, J. Adabi, Y. Al-Onaizan, A. Erden, G. Kaminka, S. C. Marsella, and I. Muslea. Building agent teams using an explicit teamwork model and learning. *Artificial Intelligence*, 110(2):215–239, 1999.
23. Israel A. Wagner and Alfred M. Bruckstein. Cooperative cleaners: a study in ant-robotics. *Communications, Computation, Control, and Signal Processing*, pages 298–308, 1997.
24. Y. Ding, Y. He, and J. Jiang. Multi-robot cooperation method based on the ant algorithm. In *Proceedings of the 2003 IEEE Swarm Intelligence Symposium*, pages 14–18, Indianapolis, USA, 2003. IEEE.
25. A. Drogoul and J. Ferber. From tom thumb to the dockers: Some experiments with foraging robots. In *Proceedings of the Second International Conference on Simulation of Adaptive Behavior*, pages 451–459, Honolulu, USA, 1992.
26. K. Sugawara and T. Watanabe. Swarming robots - foraging behavior of simple multirobot system. In *Proceedings of the IEEE/RSJ International Conference on Intelligent Robots and Systems*, pages 2702–2707, Lausanne, Switzerland, 2002.
27. S. C. Nicolis and J. L. Deneubourg. Emerging patterns and food recruitment in ants: an analytical study. *Journal of Theoretical Biology*, 198:575–592, 1999.
28. J. L. Deneubourg ans S. Aron, S. Goss, and J. M. Pasteels. The self-organizing exploratory pattern of the argentine ant. *Journal of Insect Behavior*, 3:159–168, 1990.
29. G. Ermentrout and L. Edelstein-Keshet. Cellular automata approaches to biological modeling. *Journal of Theoretical Biology*, 160:97–133, 1993.

30. R. Vaughan, K. Stoy, G. Sukhatme, and M. Mataric. Lost: Localization-space trails for robot teams. *IEEE Transactions on Robotics and Automation*, 18(5).
31. J. Sauter, R. Matthews, H. Parunak, and S. Brueckner. Evolving adaptive pheromone path planning mechanisms. In *Proceedings of the First International Joint Conference on Autonomous Agents and Multi-Agent Systems*, pages 434–440, 2002.
32. J. L. Deneubourg, S. Goss, N. Franks, A. Sendova-Franks, C. Detrain, and L. Christien. The dynamics of collective sorting robot-like ants and ant-like robots. In *Proceedings of the first international conference on simulation of adaptive behavior on From animals to animats*, pages 356–363, Paris, France, 1991. MIT Press.
33. R. Beckers, O. E. Holland, and J. L. Deneubourg. From local actions to global tasks: Stigmergy and collective robotics. In R. A. Brooks and P. Maes, editors, *Proceedings of the 4th International Workshop on the Synthesis and Simulation of Living Systems*, pages 181–189, Cambridge, USA, 1994. MIT Press.
34. C. R. Kube and E. Bonabeau. Cooperative transport by ants and robots. *Robotics and Autonomous Systems*, 1:85–101, 2000.
35. P. Rybski, A. Larson, H. Veeraraghavan, M. LaPoint, and M. Gini. Communication strategies in multi-robot search and retrieval. In *Proceedings of the 7th International Symposium on Distributed Autonomous Robotic Systems*, pages 301–310, Toulouse, France, 2004.
36. A.Wurr. Robotic team navigation in complex environments using stigmergic clues. Master's thesis, University of Manitoba, 2003.
37. R. R. Cazangi, F. J. Von Zuben, and M. F. Figueiredo. Autonomous navigation system applied to collective robotics with ant-inspired communication. In *Proceedings of the 2005 Conference on Genetic and Evolutionary Computation*, volume 1, pages 121–128, Washington DC, USA, 2005. ACM Press.
38. J. Svennebring and S. Koenig. Towards building terrain-covering ant robots. In *Ant Algorithms*, pages 202–215, 2002.
39. T. Ziemke. On the role of robot simulations in embodied cognitive science. *Artificial Intelligence and Simulation of Behaviour*, 1(4):1–11, 2003.
40. M. Dorigo and T. Stützle. *Ant Colony Optimization*. MIT Press/Bradford Books, 2004.

3

A General Approach to Swarm Coordination using Circle Formation

Karthikeyan Swaminathan[1] and Ali A. Minai[2]

[1] Department of Biomedical Engineering, University of Cincinnati, Cincinnati, Ohio 45221 swamink@email.uc.edu

[2] Department of Electrical & Computer Engineering and Computer Science, University of Cincinnati, Cincinnati, Ohio 45221 aminai@ececs.uc.edu

Summary The field of collective robotics exploits the use of technologically simple robots, deployed in large numbers, to collectively perform complex tasks. Here, the real challenge is in developing simple algorithms which the robots can execute autonomously, based on data from their vicinity, to achieve global behavior. One such global task that many researchers (including the authors) have developed algorithms for is the formation of a circle. In this chapter, we discuss how the circle formation algorithm can be used as a means for solving other formation and organization problems in multi-robot systems. The idea behind this approach is that circle formation can be seen as a method of organizing the robots in a regular formation which can then be exploited. This involves identifying specific robots to achieve different geometric patterns like lines, semicircles, triangles and squares, and dividing the robots into subgroups, which can then perform specific group-wise tasks. The algorithms that achieve these tasks are entirely distributed and do not need any manual intervention. The results from these studies are presented here.

3.1 Collective Robotics

Collective robotics is the study of groups of relatively simple robots that are capable of moving around and accomplishing tasks collaboratively. The number of robots used can vary from tens to tens of thousands, based on the application. The goal is to use robots that are as simple — and therefore, as cheap — as possible, deploy them in large numbers and coordinate them to achieve complex tasks.

Groups of robots create a very complex coordination and control problem because of the extremely large configuration space created by numerous interacting agents. Centralized control methods are not feasible in this situation, and the goal of much research in collective robotics has been to find efficient methods of distributed and decentralized control [1, 2, 3, 4, 5, 6, 7, 8, 9, 10, 11]. In particular, the idea of *self-organized control* arising from each robot/agent following its own set of behavioral rules has gained almost universal currency [1, 12, 13, 14, 15, 16]. In extremely large

K. Swaminathan and Ali A. Minai: *A General Approach to Swarm Coordination using Circle Formation*, Studies in Computational Intelligence (SCI) **31**, 65–84 (2006)
www.springerlink.com

collections of robots, called swarms, the focus is on finding very simple behaviors at the robot level that still lead to the emergence of sophisticated behavior at the group level [18, 19, 20, 21, 22], but even with relatively small groups of robots, the self-organized approach has proved to be the most effective.

The key issue in collective robotics is how to specify rules of behavior and interaction at the level of individual robots such that coordination can be achieved automatically at the global level. This is called the coordination problem.

3.2 Pattern Formation in Collective Robotics

A fundamental problem in collective robotics is to have the group organize into global formations or patterns. These include simple patterns like circles, lines, uniform distribution within a circle or square, etc. In the presence of a central controller, these tasks are trivial, but this is not the case in a distributed system. The goal here is to have an entirely decentralized system in which each robot performs tasks *autonomously* based on information gathered by itself, preferably from its neighborhood. An important feature in most of these systems is that the individual entities (robots) are not explicitly aware that they are involved in pattern formation, and certainly cannot direct it globally. This also means that only a global observer can truly assess the performance of these systems, and the robots themselves must rely on limited — possibly incorrect — estimates to make their decisions. This last aspect has important implications for convergence in robot collectives and other decentralized systems.

The formation of patterns in multi-robot systems has several applications, some of which are listed below. These highlight the importance of pattern formation in the real world.

- Surrounding an object or feature in the environment. [23].
- Forming a uniform distribution of robots in a given area for protecting the area or surveillance. [23, 24].
- Election of a leader or follow-the-leader situations. [25, 26].
- Gathering to share information or for some other task. [25].
- Removal of mines or bomb disposal [27].
- Exploration and mapping in space, underwater, or in other hazardous environments. [28, 29].
- Formation of sensing grids. [30, 31].
- Managing processes in a manufacturing unit. [32].
- Carrying, moving and assembling objects. [33].
- Aiding emergency response and decongestion of traffic on highways. [34].

3.3 Background

This section briefly discusses the work of other researchers in the area of pattern formation. The reader is cautioned that the amount of research done in this field is immense and this section covers only a few papers that have directly influenced the work presented in this chapter. In particular, the formation of a circle has been studied by many. The work of Suzuki et al. stands out in this area [28, 35, 36]. Their multi-robot system (henceforth termed the Suzuki model) consists of around 50 robots, each of which is autonomous and mobile. Each is equipped with sensors that can detect the positions of *all* other robots instantaneously in the unlimited visibility case, and only its neighbors in the limited visibility case. Each robot is anonymous, meaning that it does not have any label, and executes the same algorithm as all other robots. The robots do not possess any communication capabilities and each has its own local coordinate system. In the unlimited visibility case, each robot senses the positions of all other robots and based on the observed coordinates, makes a move. Each robot determines its nearest and farthest robots and assumes the center of these robots to be the center of the circle to be formed and moves towards or away from it. All other robots similarly execute these simple tasks which eventually lead them to form a circle. In some cases, the original algorithm can lead to the robots forming a triangular shape instead of a circle, and a small correction was subsequently made to the algorithm by Tanaka [37] in order to rectify this. The group introduced a similar algorithm that helped the robots distribute themselves uniformly in a circle and a few more ways that needed manual intervention to help the system form a polygon, form a line and divide into groups. In the limited visibility case, they showed that the robots could converge to a point only when they were synchronous i.e. executed their tasks in unison.

In the Suzuki model, the different kinds of patterns can be formed only in the unlimited visibility case. When the application requires a large number of robots, the assumption that each robot can sense all other robots becomes unreasonable. A second group of researchers worked on a similar model of robots but explored the limited visibility case in detail. They showed that the robots could converge to a single point even in the asynchronous case, provided they had a sense of direction (possessed a compass) [25]. They analyzed the different shapes that are achievable by such a system, starting with the case when the robots have no common knowledge i.e. they do not share a common sense of direction or orientation, and increasing the knowledge step by step up to the case when they share everything [38, 27].

A refreshingly different approach to mobile multi-robot formation is provided by Gordon et al. [30, 31]. The robots here are actually Micro-Air Vehicles (MAVs) that need to distribute uniformly in a hexagon or a square in order to form a sensing grid that can operate like a radar. The motion of the MAVs is governed by laws that mimic natural gravitational forces, which the authors term *artificial physics* forces. It is obvious that, in order to compute the virtual forces underlying their control, such robots need to have capabilities greater than those assumed in the Suzuki model.

Defago and Konagaya [39] present a method for circle formation based on extensions of the Suzuki model. This method consists of two algorithms executed one after the other. The first algorithm places the robots along the circumference of a circle and the second uniformly distributes the robots along the circumference. In the circle formation algorithm, the robots are initially in arbitrary positions. The goal of these robots is to determine the *smallest enclosing circle (SEC)* and then move to occupy positions along this circle. The SEC can be defined often by two, and at most three, extremal robots. The robots can have their own local coordinate systems, and hence different views of the environment, but the smallest enclosing circle (SEC) every robot computes for a given configuration of robots will be the same. However, the configuration of the robots changes as they move and hence it becomes necessary to ensure that the robots move in such a way that the SEC remains the same. To achieve this, the robots move according to certain rules based on Voronoi tessellation of the space around the robots [39]. Once all the robots take positions along the circumference of the SEC, the second algorithm is executed to spread the robots along the circumference. Every robot tries to move half the distance towards the mid-point of its nearest left and right robots. This method of circle formation is computationally complex because of the calculation of the Voronoi tessellation [39].

In order to reduce the complexity of the above algorithm, another method was developed by Markou et al. [40]. In this method, once the SEC is computed, a given robot determines the point on the circumference of the SEC, that is closest to it and moves towards this point. When all the robots have taken positions along the circumference, they spread along the circumference as described earlier [40].

Yun et al. [41] present a novel method for circle formation called the *Merge then Circle* algorithm. All the robots move initially towards the midpoint between their nearest and farthest robots. This, when executed for a long period of time, brings all the robots together in a cluster. After this step, the robots sense the positions of other robots and move in a direction of empty space for a distance equal to the radius of the desired circle. Once this is done, each robot uses the positions of its two nearest robots to move towards their midpoint for a more uniform distribution.

3.4 A Communication-Based Multi-Robot System

Most of the methods presented in the previous section are based on the Suzuki model of robots, which makes the assumption that each robot can sense the positions of all other robots irrespective of their distance from the given robot. This assumption becomes unreasonable when there are more than a handful of robots. A more practical model based on wireless communication has recently been developed by us [42], and is described below. Two circle-formation algorithms — Batch Broadcast of Coordinates (BBC) and Individual Broadcast of Coordinates (IBC) — have been developed for this model based on the algorithms in [28, 35, 37].

In the communication-based model, the robots are not equipped with sensors, but are assumed to possess transmitters and receivers, using which they can communicate messages to each other. The messages are mostly coordinate positions of the robots. However, the robots are assumed to be able to broadcast messages only over a small distance, analogous to the limited visibility condition in the Suzuki model. In this way, a robot can transmit its position to its neighbors directly and to distant robots through these neighbors by propagation (hops). The robots are anonymous in the sense that they cannot be distinguished from each other either by appearance or by the programs they possess. They are autonomous in that they do not need intervention from other robots, humans or external controllers. They are assumed to be able to move small but definite distances called *steps*. There are (N) robots ($R_1,\ldots,R_N$) placed initially in random positions on a unit grid and loaded with the same programs which, when executed by each of them, lead to the formation of different patterns. Coordinates are shared through communication and the robots are assumed to follow a global coordinate system. This can be achieved by either pre-initializing the coordinates externally followed by path integration, or by the robots themselves setting up a self-organized coordinate system [43, 44, 38, 45]. The external initializer can be the same as the one that deploys the robots and is not needed during the system's pattern formation phase.

Using this model, and the circle formation algorithms (BBC and IBC) as the base, other algorithms were developed that exploit the arrangement of robots in a circle to solve other pattern formation and coordination problems in such a multi-robot system. This work is presented in the following sections.

3.5 Beyond Circle Formation - Formation of Other Patterns and Groups

A circle is a very simple and perfectly symmetric geometric pattern, which is precisely why its formation is easy with oblivious and anonymous robots. The formation of other shapes is not so simple because they require designated vertices, orientations, etc. However, the ability to form a circle can be exploited to form other shapes, and for achieving other coordination goals in multi-robot systems. Basically, forming a circle is a way of organizing an initially randomly scattered population of robots into a regular shape, which can then be used to localize and tag specific robots for special tasks. The goal, as always, is to do this in an entirely distributed manner. This section presents algorithms that allow robots to achieve non-circular formations without manual intervention, only using circle formation as the basic organizing principle. Robots can be pre-programmed so that they first form a circle and then switch to the rules for forming the desired shape. Alternatively, the robots could be switched to the algorithm for a particular shape through a one-time global message.

In this work, we consider the formation of a vertical line, a triangle, a square and a semi-circle. We also look at the canonical problem of splitting the robot population into groups of approximately equal size. The key problem in all these cases is to get groups of robots to follow *different* rules — corresponding to different roles — with constraints on the relative sizes of the groups. Since the robots are initially anonymous and randomly distributed, it is difficult to assign roles systematically. Distributed algorithms for limited role assignment (such as cluster formation, leader election or gateway designation in ad-hoc networks) do exist, but these typically assign roles relative to the *actual* distribution of agents — e.g., assigning clusters based on agent density — whereas in formation problems, the roles are assigned relative to a future *desired* distribution. Relaxation-style algorithms for these assignments can be developed, but it is difficult to guarantee their correctness, convergence, etc. The algorithms based on circle-formation, on the other hand, allow systematic role assignments in ways that are much easier to analyze. Thus, the proposed algorithms should not be seen as solving a previously unsolvable problem, but as an elegant general approach to solving a broad class of important problems.

The basic idea is to use the circle as an ordering basis to split the robots into nearly equal groups and/or identify robots that define boundaries between these groups. For the cases presented here, the robots are split into four groups by dividing the circle into four quadrants. The fact that the robots follow a global coordinate system is used for this. The robots that have the highest y coordinate, lowest y coordinate, highest x coordinate and lowest x coordinate are termed as the R^{north}, R^{south}, R^{east} and R^{west} robots, respectively (Refer to Figure 3.1). This

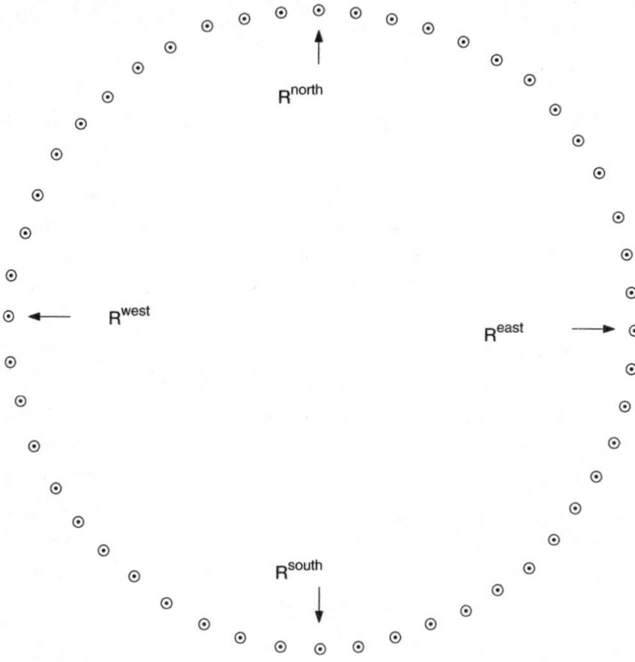

Fig. 3.1. Diagram depicting the North, South, East and West robots in a circle formation of robots

naming, however, is only for explanation purposes. Within the algorithms, the robots are still anonymous and autonomous until the algorithm below labels them. The identification of these four robots is common for all algorithms, and is done as follows. Every robot R_i executes the following steps:

Step 1: Broadcast own coordinates (x_i, y_i).

Step 2: Receive coordinates from other robots. Store the first two distinct coordinates received as (x_i^{n1}, y_i^{n1}) and (x_i^{n2}, y_i^{n2}).

Step 3: Calculate distances from (x_i, y_i) to (x_i^{n1}, y_i^{n1}) and (x_i^{n2}, y_i^{n2}) as D_i^{n1} and D_i^{n2}, respectively.
Step 4: If $(D_i^{n2} > D_i^{n1})$, swap (x_i^{n1}, y_i^{n1}) and (x_i^{n2}, y_i^{n2}).
Step 5: Broadcast own coordinates (x_i, y_i).
Step 6: Receive coordinates (x_j, y_j) from another robot. If these coordinates are the same as (x_i^{n1}, y_i^{n1}) OR (x_i^{n2}, y_i^{n2}), go to Step 11.
Step 7: Calculate distances from (x_i, y_i) to (x_i^{n1}, y_i^{n1}) and (x_i^{n2}, y_i^{n2}) as D_i^{n1} and D_i^{n2}, respectively.
Step 8: Calculate distance between (x_i, y_i) and (x_j, y_j) as D_i^{temp}.
Step 9: If $(D_i^{temp} < D_i^{n1})$, then set (x_i^{n2}, y_i^{n2}) as (x_i^{n1}, y_i^{n1}), and then set (x_i^{n1}, y_i^{n1}) as (x_j, y_j). Go to Step 11.
Step 10: If $(D_i^{temp} < D_i^{n2})$, then set (x_i^{n2}, y_i^{n2}) as (x_j, y_j).
Step 11: Repeat Steps 5 to 10 for N^{iter1} iterations (chosen empirically).
Step 12: If $(y_i > y_i^{n1})$ AND $(y_i > y_i^{n2})$, then identify oneself as R^{north}. STOP.
Step 13: If $(y_i < y_i^{n1})$ AND $(y_i < y_i^{n2})$, then identify oneself as R^{south}. STOP.
Step 14: If $(x_i > x_i^{n1})$ AND $(x_i > x_i^{n2})$, then identify oneself as R^{east}. STOP.
Step 15: If $(x_i < x_i^{n1})$ AND $(x_i < x_i^{n2})$, then identify oneself as R^{west}.

Each robot executes the above steps to determine if it is one of the four robots. Following this, the North, South, East and West robots alone broadcast their coordinates for a few iterations. These messages are received by the other robots and stored. The other robots also rebroadcast these messages. It should be noted that the robots are in a circle, so these messages literally travel around the circle. Also, the messages broadcast by the North, South, East and West robots need not be tagged. Once the other robots have four distinct coordinates stored, they can easily determine which corresponds to North, South, East and West (the North robot will have the highest y coordinate among the four and so on). The above algorithm is termed here as the *NEWS algorithm*. At the end of the NEWS algorithm the north, south, east and west robots are first correctly identified and labelled. Then their coordinates are made available to all other robots. Each robot R_i stores these coordinates as $(x_i^{north}, y_i^{north})$, $(x_i^{south}, y_i^{south})$, (x_i^{east}, y_i^{east}) and (x_i^{west}, y_i^{west}). Then the group can execute one of the following algorithms to form other shapes or dividing into groups, as described next.

3.5.1 Formation of a Vertical Line

This algorithm is for a vertical line but could be used easily for a horizontal line as well. The idea behind line formation is that the North robot remains fixed, while all the other robots move along the direction of the x-axis to align themselves below the North robot. Every robot R_i executes the following steps:

Step 1: Execute the NEWS algorithm.
Step 2: If (R^{north}), then do not move. STOP.
Step 3: Set the target coordinates (x_i^{tar}, y_i^{tar}) to be reached as (x_i^{north}, y_i).
Step 4: Take a step towards (x_i^{tar}, y_i^{tar}).
Step 5: Repeat Step 4 for N^{iter2} iterations (chosen empirically).

By replacing R^{north} (or R^{south}) by R^{east} (or R^{west}), one can form a horizontal line instead of a vertical line. The algorithm was tested using simulations and results are depicted in Figure 3.2, where the dots $(.)$ represent initial positions along a circle and the asterisks $(*)$ the final positions along a line.

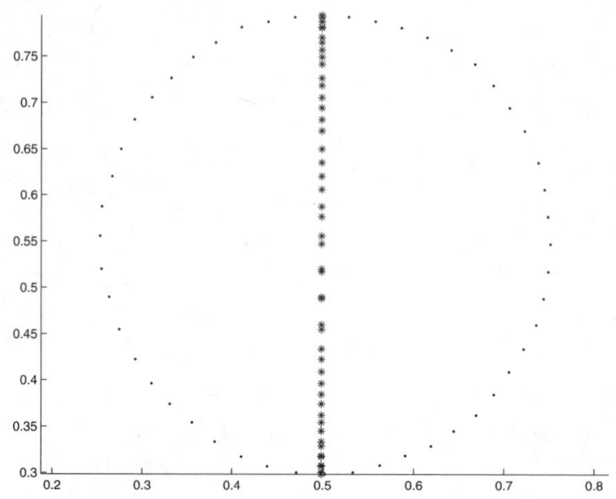

Fig. 3.2. Result from an actual simulation of robots having executed the line formation algorithm

3.5.2 Formation of a Triangle

The idea behind this algorithm is to form a triangle with the North, East and West robots as the vertices. The robots in the bottom two quadrants of the circle form the base, while the robots in each quadrant on top form one side each. This results in an isosceles triangle. Basically, each robot moves along the direction of the y-axis to reach a point on the line segment of its interest. The orientation of the triangle can be changed easily. Every robot R_i executes the following steps:

Step 1: Execute the NEWS algorithm.
Step 2: If (R^{north}) OR (R^{east}) OR (R^{west}), then do not move. STOP.
Step 3: If $(x_i > x_i^{north})$ AND $(y_i > y_i^{east})$, then set the target coordinates (x_i^{tar}, y_i^{tar}) as the point on the line connecting $(x_i^{north}, y_i^{north})$ and (x_i^{east}, y_i^{east}) with x-coordinate x_i. Go to Step 6.
Step 4: If $(x_i < x_i^{north})$ AND $(y_i > y_i^{west})$, then set the target coordinates (x_i^{tar}, y_i^{tar}) as the point on the line connecting $(x_i^{north}, y_i^{north})$ and (x_i^{west}, y_i^{west}) with x-coordinate x_i. Go to Step 6.

Step 5: Set the target coordinates (x_i^{tar}, y_i^{tar}) as the point on the line connecting (x_i^{west}, y_i^{west}) and (x_i^{east}, y_i^{east}) with x-coordinate x_i.
Step 6: Take a step towards (x_i^{tar}, y_i^{tar}).
Step 7: Repeat Step 6 for N^{iter3} iterations (chosen empirically).

By changing the choice of stationary robots (between R^{north}, R^{south}, R^{east} and R^{west}) the orientation of the triangle can be changed.

3.5.3 Formation of a Square

The algorithm for a square is nearly the same as the triangle algorithm described earlier. The only difference is that, once a robot determines that it is one of the cardinal (North, South, East or West) robots, it does not move. All the other robots try to form lines with these robots as end points forming a quadrilateral that is approximately a square. Every robot R_i executes the following steps:

Step 1: Execute the NEWS algorithm.
Step 2: If (R^{north}) OR (R^{south}) OR (R^{east}) OR (R^{west}), then do not move. STOP.
Step 3: If $(x_i > x_i^{north})$ AND $(y_i > y_i^{east})$, then set the target coordinates (x_i^{tar}, y_i^{tar}) as the point on the line connecting $(x_i^{north}, y_i^{north})$ and (x_i^{east}, y_i^{east}) with x-coordinate x_i. Go to Step 7.
Step 4: If $(x_i < x_i^{north})$ AND $(y_i > y_i^{west})$, then set the target coordinates (x_i^{tar}, y_i^{tar}) as the point on the line connecting $(x_i^{north}, y_i^{north})$ and (x_i^{west}, y_i^{west}) with x-coordinate x_i. Go to Step 7.
Step 5: If $(x_i < x_i^{south})$ AND $(y_i < y_i^{west})$, then set the target coordinates (x_i^{tar}, y_i^{tar}) as the point on the line connecting $(x_i^{south}, y_i^{south})$ and (x_i^{west}, y_i^{west}) with x-coordinate x_i. Go to Step 7.
Step 6: Set the target coordinates (x_i^{tar}, y_i^{tar}) as the point on the line connecting $(x_i^{south}, y_i^{south})$ and (x_i^{east}, y_i^{east}) with x-coordinate x_i.
Step 7: Take a step towards (x_i^{tar}, y_i^{tar}).
Step 8: Repeat Step 7 for N^{iter4} iterations (chosen empirically).

3.5.4 Formation of a Semi-circle

For a semicircle, the robots in the upper two quadrants remain in place, while the robots in the bottom half form a line. Every robot R_i executes the following steps:

Step 1: Execute the NEWS algorithm.
Step 2: If (R^{east}) OR (R^{west}), do not move. STOP.
Step 3: If $(y_i > y_i^{west})$ OR $(y_i > y_i^{east})$, do not move. S TOP.
Step 4: Set the target coordinates (x_i^{tar}, y_i^{tar}) as the point on the line connecting (x_i^{west}, y_i^{west}) and (x_i^{east}, y_i^{east}) with x-coordinate x_i.
Step 5: Take a step towards (x_i^{tar}, y_i^{tar}).
Step 6: Repeat Step 5 for N^{iter5} iterations (chosen empirically).

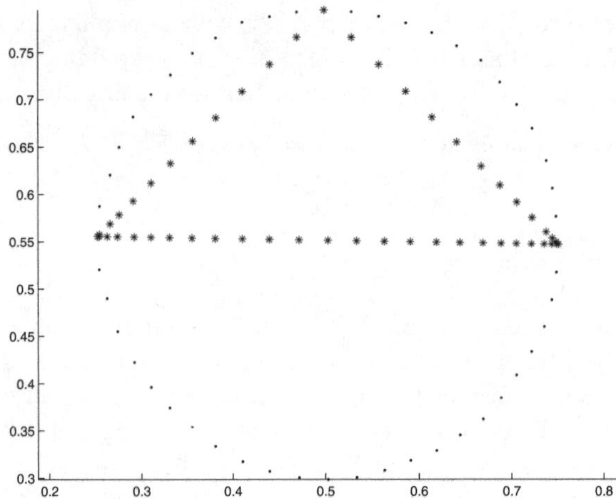

Fig. 3.3. Result from an actual simulation of robots having executed the triangle formation algorithm

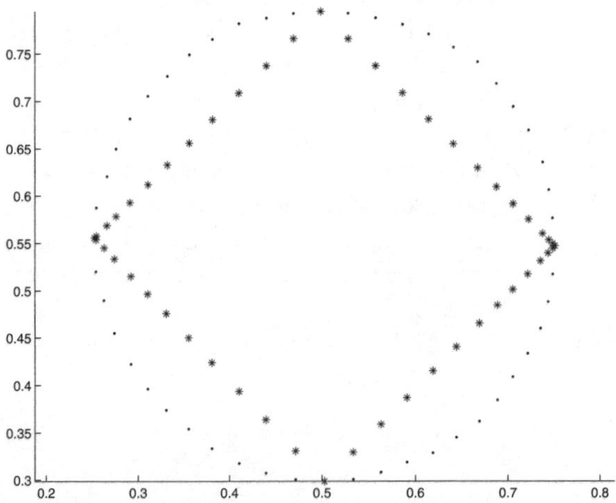

Fig. 3.4. Result from an actual simulation of robots having executed the square formation algorithm

The results for formation of triangle, square and semi-circle are presented in Figure 3.3, Figure 3.4 and Figure 3.5, respectively. It can be observed that the distribution of the robots along the edges of these patterns is not uniform. This was expected, since distribution was not addressed in the algorithms. The focus was only on achieving different patterns.

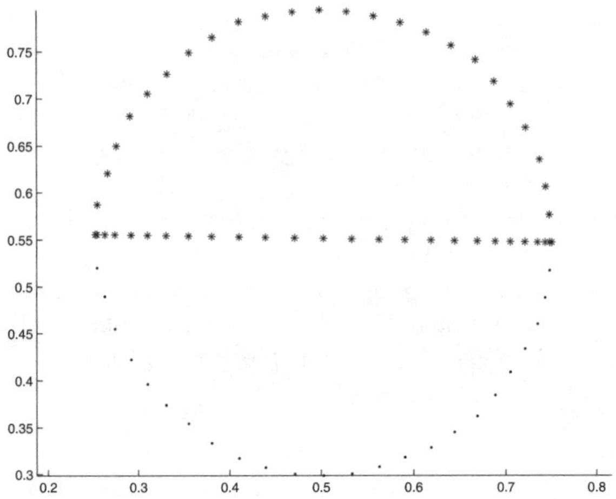

Fig. 3.5. Result from an actual simulation of robots having executed the semicircle formation algorithm

3.5.5 Splitting into Four Groups

This method is similar to the algorithms discussed earlier. The robots first determine the quadrant, and thus the group in which they belong. Following this, they make sure they coordinate only with the robots in their group for future tasks. To achieve this, they tag messages they broadcast with a group identification based on the quadrant in which they fall. Using this group identification, they make use of messages from their group and ignore messages from members of other groups. Each group can form any pattern based on all the algorithms explained in the earlier sections. For the group determination algorithm, every robot R_i executes the following steps (this is similar to the steps used in the square formation algorithm, but is presented here for sake of completeness):

Step 1: Execute the NEWS algorithm.
Step 2: If (R^{north}) OR (R^{west}) OR (R^{south}) OR (R^{east}), assign oneself a value of 1, 2, 3 and 4 as the group identification, respectively. STOP.

Step 3: If $(x_i > x_i^{north})$ AND $(y_i > y_i^{east})$, then assign oneself a group identification of 1. STOP.
Step 4: If $(x_i < x_i^{north})$ AND $(y_i > y_i^{west})$, then assign oneself a group identification of 2. STOP.
Step 5: If $(x_i < x_i^{south})$ AND $(y_i < y_i^{west})$, then assign oneself a group identification of 3. STOP.
Step 6: If $(x_i > x_i^{south})$ AND $(y_i < y_i^{east})$, then assign oneself a group identification of 4.

Now every robot has a group identification and hence can execute separate behaviors by coordinating only with robots which have the same group identification. It should be noted that execution of the *same* algorithm by each robot results in it assigning itself a group identification. This does not give an identity to the individual robots, which still remain anonymous. It only serves the purpose of helping the robots distinguish the messages coming from members in their group from those coming from other groups. The simulation results are presented in Figure 3.6 where the robots, after determining their groups, interact with only their group members and execute the circle formation algorithm (diameter of the smaller circles can be chosen to be a fraction of the larger circle). Circle formation by groups is just an example and they can perform other tasks as well.

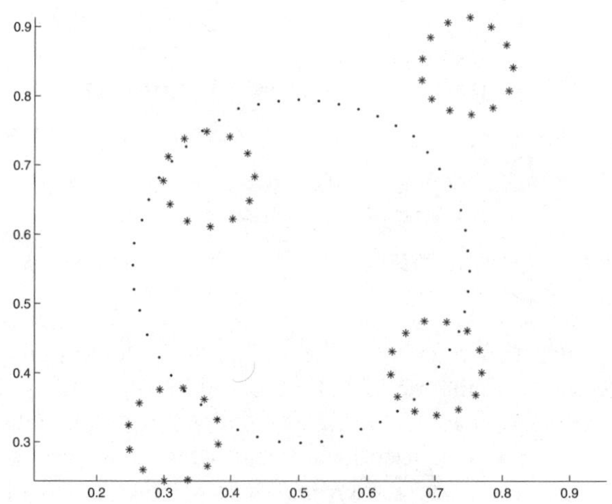

Fig. 3.6. Result from an actual simulation of robots having executed the splitting into groups and performing group tasks algorithm

3.5.6 Splitting into Groups — Generalized Case

While splitting robots into four equal groups is interesting, many applications require a different number of possibly unequal groups. In this section, we discuss how circle formation can be exploited to achieve this.

In a circle, the robots are distributed uniformly, so one can employ a splitting technique based on the angular position of the robots with respect to the center of the circle. The robots are aware of the positions of the cardinal (North, South, East and West) robots and using their position can determine their angle with respect to the line segment between the center of the circle and the East robot. (The center of the circle can be taken as the geometric center of the cardinal robot positions). Depending on the application, the robots can be pre-programmed with certain angles so as to divide the circle into sectors, and hence groups. Based on the sector in which a robot lies, it can assign itself a pre-specified group identification. Then as shown in the earlier subsections, the robots can coordinate only with its group members and perform tasks assigned to the group. Note again that, while robots do determine special labels in the field and act accordingly, no *specific* robot is *pre-tasked* with a particular behavior. All behavioral differentiation emerges through self-organization.

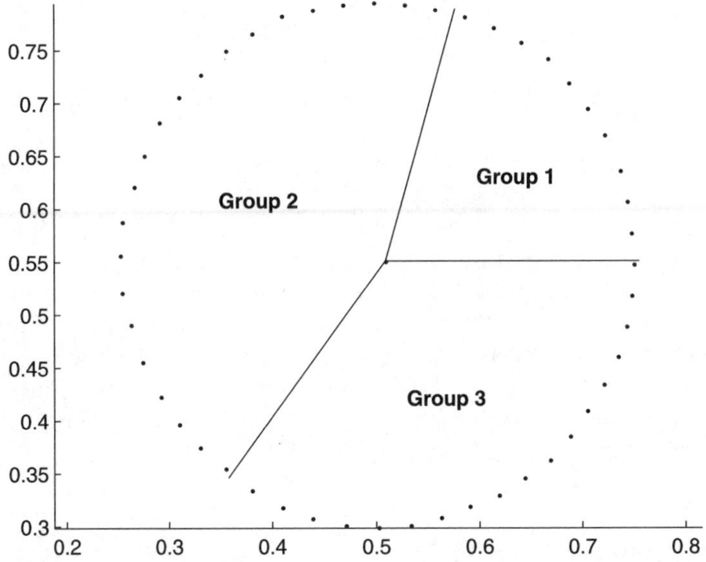

Fig. 3.7. Diagram depicting splitting of robots in a circluar formation based on angular constraints

Figure 3.7, depicts this technique. In this case, the robots are split into three groups. The robots need to given two angles for splitting into three groups. In the

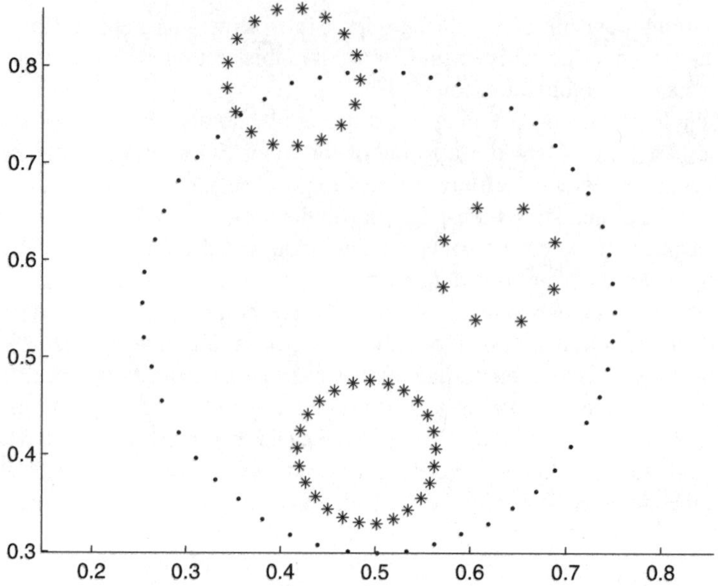

Fig. 3.8. Result from an actual simulation of robots having executed the splitting into 3 unequal groups algorithm

special case in which the angles of the sectors are specified such that they are equal, the resulting groups are also equal.

An example of such a splitting technique is presented in Figure 3.8. In this example, the angles specified were 60 and 120 degrees and hence resulted in sectors with angles 60, 120 and 180 degrees. The groups then performed circle formation algorithms (with one-fourth the original diameter). The distribution of the three circles can be seen to vary depending on the number of robots that ended up in each group, which is directly proportional to the magnitude of the angles. Since circle-formation is the basis of forming other shapes, robots in each small circle can then form different shapes, This is shown in Figure 3.9.

3.6 Programming the Robots

The real power of multi-robot systems lies in distributed control and it is essential that each robot be made entirely autonomous with no need for manual intervention. Thus, if the robots are to perform complex tasks comprising several subtasks such as circle formation, group formation, etc., there must be a *canonical procedure program* that can be pre-loaded into each robot such that, as each robot executes its program,

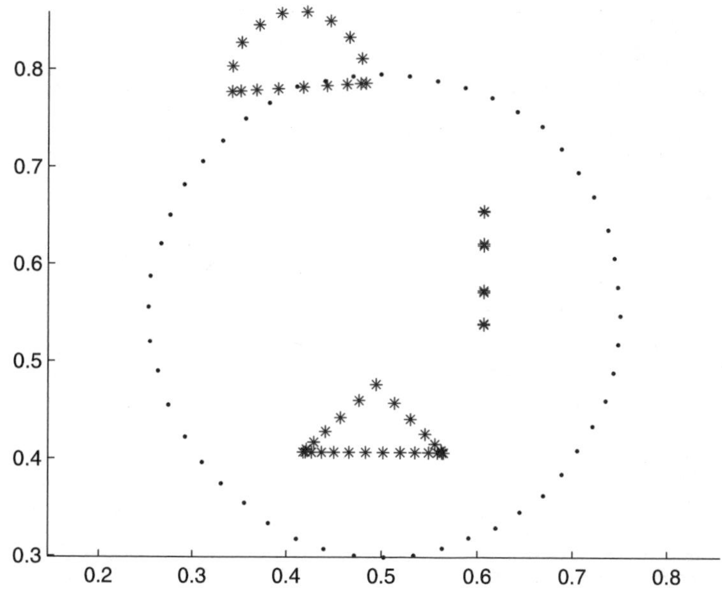

Fig. 3.9. Result from an actual simulation for splitting into 3 unequal groups and performing different group tasks

the result is the performance of the complex task. One possible approach to this is discussed here.

The general program followed by each robot can be described as follows:

- Perform a circle formation algorithm for a given number of iterations.
- Perform the NEWS algorithm for a given number of iterations.
- Determine the group identity.
- Perform the task assigned to the group.

Once, the parameters like, number of iterations and the task for each group are given prior to deploying the robots, there is no need for any manual intervention. Also, it is important to note that all robots are given *identical* programs. Differences in behavior arise only as a result of labelling generated during *execution*, and are *not* specifically assigned to individual robots a priori.

This idea is illustrated in Figure 3.10. Initially, the robots are in arbitrary positions. They then execute the circle formation algorithm, followed by the self-determination of groups. Then the robots execute (say) the circle formation algorithm as a group task and from there, they can form different shapes, which they know beforehand.

Similar programs can be written for other complex tasks. When a group of robots is to perform that task, the appropriate program can be loaded into all the

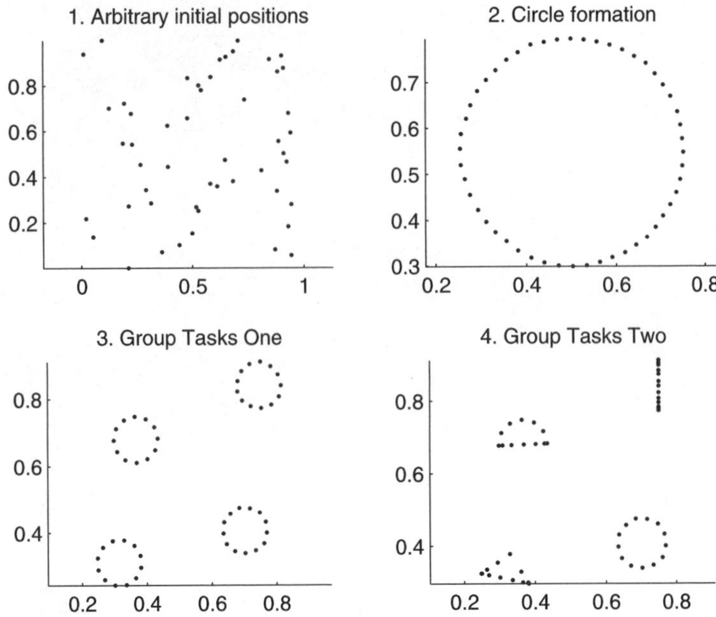

Fig. 3.10. Combination figure showing results from simulations performed successively. 1-Robots are randomly distributed, 2-Robots execute circle formation algorithm, 3-Robots split into four groups and each forms a circle, 4-Each group then forms other pattern s

robots or, more likely, triggered by circumstances. In the latter scenario, a large number of programs would be stored in each robot — the same ones for each — and would be triggered through an external instruction or observations made by the robots themselves. This will raise issues of consistency, i.e., how to ensure that all robots switch to the same program, but this is beyond the scope of the present work. Similar issues exist even with groups of human agents, which must be told to switch tasks. The focus in this work is on developing algorithms where these "external imperatives" can be made as independent of the detailed state of the robot group as possible (e.g., not requiring information on identities or positions of specific robots).

3.7 Discussion and Conclusion

The work described in this chapter has built upon circle-formation algorithms developed by other researchers to explore whether this can be used as a generic coordination mechanism for solving other, more complex organization problems. We have described methods to make robots form different shapes like lines, semicircles, triangles and squares, and to split into groups of specified size to perform specialized group tasks. A key point is that, using this approach, each robot for a specific organization task can be *pre-loaded* with a generic program whose execution by all individual robots will lead to the desired organization *without need for manual intervention*.

Most problems in formation or organization are solved by the assignment of *roles* or *tasks* to specific individuals. For example, a square can be formed if four robots are assigned to be the corners and the rest then use them to line up along the edges. However, in the absence of a central controller and with anonymous robots, there is no objective basis on which to make these assignments; all agents execute the same algorithm, and none is inherently disposed to behave in a special way. One possible way to get around this is to adapt *leader-election algorithms* [46, 47, 48, 49, 50, 25, 26] developed in the distributed systems literature for various purposes. These algorithms distinguish certain agents/robots for specific tasks (e.g., cluster-heads or gateways in ad-hoc networks) through a distributed process. Once these leaders are elected, they can perform functions such as indicating vertices or serving as beacons, and the rest can execute appropriate algorithms to fall in place using these indicators. However, different groups of agents must behave differently relative to the leaders/beacons to achieve the formation, e.g., four different groups of robots must choose a different edge to line up along in order to form a square. Thus, the decentralized organization process involves three steps: 1) Selection of specific leader/beacon agents; 2) Assignment of different behaviors to appropriate subgroups of agents; and 3) Execution of the designated behaviors by all agents. The assignments made in Steps 1 and 2 must satisfy certain criteria, e.g., number of beacons and relatively equal sizes of groups. Our work explores the use of circle formation as a generic mechanism integrating both leader-election and subsequent group assignment in a reliable and efficient way. The circle is especially appropriate for this function because of its uniquely homogeneous, arbitrarily symmetric shape. It provides a natural manifold for defining a problem-specific coordinate system for a wide class of problems. In work not presented here we have also studied the robustness of our algorithms and found them to be quite robust to many, though not all, types of errors [42] The results presented in this chapter demonstrate clearly that our approach is both powerful and generic. However, several issues, such as the positioning of groups, remain open and will be addressed in future work.

References

1. Brooks R (1991). Science, 253(5025):1227–1232.
2. Brooks R, Connell J (1987) SPIE Conference on Mobile Robots, Cambridge, MA (USA), pp. 77–84.
3. Cao Y, Fukunaga A, Kahng A (1997) Autonomous Robots. 4(1):1–23.
4. Parke LE (2000) Proceedings of the 5th International Symposium on Distributed Autonomous Robotic Sys tems, DARS, Springer Verlag, 4:3–12.
5. Yamaguchi H (1997) IEEE International Conference on Robotics and Automation, 3:2300–2305.
6. Mataric MJ (1995) Robotics and Autonomous Systems, 16:321–331.

7. Mataric MJ (1991) A Distributed Model for Mobile Robot Environment-Learning and Navigation. Technical Report TR-1228, Artificial Intelligence Laboratory, Massachusetts Institute of Technology.
8. Mataric MJ (1998) Journal of Experimental and Theoretical Artificial Intelligence, 10(3):357–369.
9. Mataric MJ (1998) In IEEE Intelligent Systems, 6–9.
10. Goldberg D, Mataric MJ (2000) Technical Report IRIS-00-387, USC Institute for Robotics and Intelligent Systems.
11. Ostergaard E, Mataric MJ, Sukhatme GS (2001) In IEEE/RSJ International Conference on Intelligent Robots and Systems(IROS), 2:821–826.
12. Mataric MJ, Fredslund J (2002) IEEE Transactions on Robotics and Automation, 18(5):837–846.
13. Mataric MJ, Jones C (2004) In Proceedings of the Hawaii International Conference on Computer Sciences, Waikiki, Hawaii, 27–32.
14. Nolfi S(1998) Connection Science, 10(3-4):167–184.
15. Unsal C, Bay JS(1994) IEEE International Symposium on Intelligent Control, Columbus, Ohio, 249–254.
16. Shen W, Chuong C, Will P (2002) IEEE/RSJ International Conference on Intelligent Robots and System, 3:2776–2781.
17. Baldassarre G, Nolfi S, Parisi D (2003) Artificial Life, 9(3):255-267.
18. Bonabeau E, Dorigo M, Theraulaz G (1999) Swarm Intelligence: From Natural to Artificial Systems, NY: Oxford University Press Inc.
19. Beni G Wang J (1989) Swarm intelligence in cellular robotics systems. Proceeding of NATO Advanced Workshop on Robots and Biological System, I1 Ciocco, Tuscany, Italy.
20. Arkin RC (1998) Behavior-Based Robotics, Cambridge, MA: MIT Press.
21. Millonas M (1994) Swarms, phase transitions, and collective intelligence. In: Palaniswami M, Attikiouzel Y, Marks R, Fogel D, Fukuda T (eds) Computational Intelligence: A Dynamic System Perspective, pp.137–151. IEEE Press, Piscataway, NJ.
22. Dudek G, Jenkin M, Milios E, Wilkes D (1993) A taxonomy for swarm robots, IEEE/RSJ Int. Conf. on Intelligent Robots and Systems, Japan, 1:441–447.
23. Sugihara K, Suzuki I (1990) Distributed Motion Coordination of Multiple Mobile Robots, In IEEE International Symposium on Intelligent Control, pp. 138–143.
24. Dudenhoeffer DD, Jones MP (2000) A Formation Behavior for Large-Scale Micro-Robot Force Deployment, In Proceedings of the 2000 Winter Simulation Conference, pp. 972–982.
25. Flocchini P, Prencipe G, Santoro N, Widmayer P (2001) Gathering of Asynchronous Mobile Robots with Limited Visibility. In 18th International Symposium on Theoretical Aspects of Computer Science (STACS), 331(1-3):147–168.
26. Desai J, Kumar V, Ostrowski J (1999) Control of changes in formation for a team of mobile robots, In Proceedings of 1999 International Conference on Robotics and Automation, pp.1556–1561.
27. Flocchini P, Prencipe G, Santoro N, Widmayer P. Pattern Formation by Autonomous Mobile Robots. Interjournal of Complex Systems, Article 395, (on line publication http://www.interjournal.org).
28. Sugihara K, Suzuki I (1996) Journal of Robotic Systems, 13:127–139.
29. Wang PKC (1989) Navigation Strategies For Multiple Autonomous Mobile Robots Moving In Formation, IEEE/RSJ International Workshop on Intelligent Robots and Systems,pp. 486–493.
30. Gordon FD, Spears MW (1999) Using Artificial Physics to Control Agents. In Proceedings of IEEE International Conference on Information, Intelligence and Systems.

31. Gordon FD, Spears MW, Sokolsky W, Lee I (1999) Distributed Spatial Control, Global Monitoring and Steering of Mobile Agents. In Proceedings of IEEE International Conference on Information, Intelligence and Systems (ICIIS).
32. Molnár P, Starke J (2001) Control of distributed autonomous robotic systems using principles of pattern formation in nature and pedestrian behavior. IEEE Transactions on Systems, Man, and Cybernetics, Part B, 31(3): 433–435.
33. Chen Q, Luh J (1994) Coordination and control of a group of small mobile robots. In Proceedings of the IEEE International Conference on Robotics and Automation, pp.2315–2320.
34. Trivedi M, Hall B, Kogut G, Roche S (2000) Web-based teleautonomy and telepresence, In 45th SPIE Optical Science and Technology Conference, Applications and Science of neural networks, fuzzy systems and evolutionary computation III, San Diego, v.4120.
35. Suzuki I, Yamashita M (1999) Distributed Anonymous Mobile Robots: Formation of Geometric Patterns, Siam J. Comput., 28(4):1347–1363.
36. Suzuki I, Yamashita M (1996) Distributed Anonymous Mobile Robots- Formation and Agreement Problems, In Proc. Third Colloq. On Struc. Information and Communication Complexity (SIROCCO), pp 313–330.
37. Tanaka O (1992) Forming a Circle by Distributed Anonymous Mobile Robots. Bachelor thesis, Department of Electrical Engineering, Hiroshima University, Hiroshima, Japan.
38. Flocchini P, Prencipe G, Santoro N, Widmayer P (1999) Hard Tasks for Weak Robots: The Role of Common Knowledge In Pattern Formation by Autonomous Mobile Robots. In 10th International Symposium on Algorithm and Computation (ISAAC), pp.93–102.
39. Defago X, Konagaya A (2002) Circle Formation for Oblivious Anonymous Mobile Robots with No Common Sense of Orientation. In Proceedings of the second ACM international workshop on Principles of mobile computing, pp.97–104.
40. Chatzigiannakis I, Markou M, Nikoletseas S (2004) Distributed Circle Formation for Anonymous Oblivious Robots. In 3rd Workshop on Efficient and Experimental Algorithms, Lecture Notes in Computer Science, 3059:159–174.
41. Yun X, Alptekin G, Albayrak O (1997) Line and circle formation of distributed physical mobile robots. Journal of Robotic Systems, 14(2):63–76.
42. Swaminathan K (2005) Self-organized formation of geometric patterns in multi-robot swarms using wireless communication. MS Thesis, University of Cincinnati, Cincinnati.
43. Ando H, Suzuki I, Yamashita M (1995) Formation and agreement problems for synchronous mobile robots with limited visibility. In Proceedings of the 1995 IEEE International Symposium on Intelligent Control,pp.453–460.
44. Gordon N, Wagner IA, Brucks AM (2003) Discrete Bee Dance Algorithms for Pattern Formation on a Grid. In IEEE/WIC International Conference on Intelligent Agent Technology, pp.545–549.
45. Dudek G, Jenkin M, Milios E, Wilkes D (1993) Robust Positioning with a Multi-Agent Robotic System. In Proceedings of the International Joint Conference of Artificial Intelligence (IJCAI) on Dynamically Interacting Robots, Chambery, France, pp.118–123.
46. Arora A, Kulkarni S (1995) Designing masking fault-tolerance via nonmasking fault-tolerance, 14TH Symposium on Reliable Distributed Systems, pp.174.
47. Arora A, Gouda M. Distributed Reset. IEEE Transactions on Computers 43(9):1026-1038.
48. Nagpal R. Programmable Self-Assembly Using Biologically -Inspired Multirobot Control (2002). ACM Joint Conference on Autonomous Agents and Multiagent Systems, Bologna.

49. Coore D, Nagpal R, Weiss R (1997) Paradigms for Structure in an Amorphous Computer. MIT Artificial Intelligence Laboratory Memo 1614.
50. Nagpal R, Coore D (1998) An algorithm for group formation in an amorphous computer. Proceedings of the Tenth International Conference on Parallel and Distributed Systems (PDCS).

4

Stigmergic Navigation for Multi-Agent Teams in Complex Environments

Alfred Wurr and John Anderson

Autonomous Agents Laboratory
University of Manitoba
Winnipeg, Manitoba, Canada R3T2N2
awurr,andersj@cs.umanitoba.ca

Summary. Robotic agents in dynamic environments must sometimes navigate using only their local perceptions: for example, in strongly dynamic domains where paths are outdated quickly. Reactive navigation alone has limited power: domain features such as terrain undulation, geometrically complex barriers, and similar obstacles form local maxima and minima that can trap and hinder agents that use it exclusively. Moreover, agents navigating in a purely reactive fashion forget their past discoveries quickly. Preserving this knowledge usually requires that each agent construct a detailed world model as it explores or be forced to rediscover desired goals each time, and share elements of this knowledge explicitly in group situations. The cost of explicit communication can be substantial, however, making it desirable to avoid its use in many domains. In this chapter we present the design and implementation of cooperative methods for reactive navigation: allowing a team of agents to assist one another in their explorations through implicit (*stigmergic*) communication. These methods range from simple solutions for avoiding specific problems such as individual local maxima, to the construction of sophisticated branching trails. We evaluate these methods individually and in combination using behaviour-based robots in a complex, three-dimensional software environment.

4.1 Introduction

Mobile robots are being used with increasing frequency for a wide range of applications, from rescue and security to exploration of remote areas. Simulated robotic agents are also employed as characters in the dynamic and complex simulated worlds of computer games. Whether the domain in which a robot is embodied is physical or exists only in software, navigation is an essential component: robots must not only move, but explore their environment in a coherent way in order to achieve their goals [1]. The challenge of navigation increases in tandem with the topographical complexity of the environment. Navigating in a static open area with uninterrupted sight lines and few obstacles is relatively straightforward. Few physical domains are this simplistic, however: even a basic building interior is a complex environment containing many corridors, doorways, walls and obstacles.

A. Wurr and J. Anderson: *Stigmergic Navigation for Multi-Agent Teams in Complex Environments*, Studies in Computational Intelligence (SCI) **31**, 85–116 (2006)
www.springerlink.com

At its most basic, navigating within buildings involves finding and negotiating relatively narrow doorways and hallways and possibly locating and climbing stairs (or controlling an elevator). This is still making an assumption of regular floor surfaces. In outdoor environments, or in situations such as robotic rescue where an indoor environment is severely disturbed, domains become much more complex. Agents may be faced with terrain undulation, irregularly shaped obstacles, and winding maze-like layouts of debris that make finding and navigating to a goal problematic. Intelligent navigation in complex environments typically involves dual processes of localization – determining an agent's position in the environment, and path planning – constructing a collision-free path to an agent's goal assuming its current position is known [2]. Path planning requires either pre-existing knowledge about the environment, such as a map and the locations of objects within it [3] or the ability to acquire this information while exploring the environment. Accurate localization is essential for path planning to be successful, because if the agent is in error about its current location, the path that is calculated to get it to its destination will be likewise incorrect [4].

A variety of localization techniques exist, including methods that use odometry, global-positioning, and radio-sonar positioning [5]. In odometry-based localization, agents begin in a known starting location and continuously update their locations as they travel [5]. Global positioning, on the other hand, works by dividing the environment into a grid where locations can be identified by precise coordinates. Radio-sonar methods triangulate position by sending out sonar pings and calculating position based on differences in ping times, and an analogous approach can be used with more accurate laser equipment. Unfortunately, odometry-based navigation works only over short distances [1], because cumulative errors in odometry can quickly result in discrepancies between the agent's actual and perceived location [6]. The remaining methods can require preparation of the environment ahead of time in some approaches, but more importantly require more complex sensory equipment [5]. In addition to accurate localization, path planning algorithms require and operate on symbolic representations of the environment. These may be provided *a priori*, constructed at runtime based on an agent's perceptions, or both in combination. Storing, constructing and maintaining these maps can demand significant storage and computational resources [7]. This is one reason that for even very complex agents it is helpful to work on a simple map based on previous travels (e.g. a topological map) rather than a map showing every detail of the world around the robot [8]. To make matters worse, if the environment is dynamic and subject to extensive change, any dynamically constructed map may have only a limited viable lifetime. In such situations, path planning becomes limited in its utility, as the path planner is reduced to planning based only on the agent's immediate perceptions and/or a partial map that the agent builds as it navigates. Consequently, embodied agents must often navigate in environments that are partially (if not wholly) unknown for periods of time [9]. Re-planning based on changes in the environment is certainly possible. However, path planning is computationally expensive, and continual re-planning is not always feasible. Indeed, in a fast-changing environment, even creating a single full path that is still useful upon completion may not be possible.

These drawbacks are even more severe when agents are planning and acting as a team, because of the greater complexity of group plans and uncertainty about the possible actions of teammates [10]. Agents must be prepared to adjust to changes in the environment caused by others, as opposed to simply the spontaneous changes offered by the world itself. For example, an agent's intended path may be abruptly cut off by objects placed by others or by the bodies of other agents themselves.

All of these factors contribute to making systems that rely exclusively on path-planning impractical in many real world scenarios [11]. Even when not relied upon exclusively, the difficulties described here still hinder the system merely because path planning is present. While it brings obvious benefits where it works well, it raises the issue of balancing cost and benefit: how much can be achieved without path planning? This is one facet of the overall issue of parsimony, a very important one in robotics. The more that sophisticated hardware is relied upon, the greater the likelihood that any unreliability in that hardware can cripple the system, irrespective of the benefit it brings. For example, a system relying on GPS for localization can currently be accurate enough for many outdoor environments, but is useless in situations where interference makes GPS contact unreliable. More sophisticated sensors also make robots significantly more expensive, which is an important factor in domains where equipment may be destroyed or damaged. This also becomes an issue when attempting to work with a team of robots, in that every component that makes an individual more expensive decreases the size of an affordable team. At some point, the benefit of additional team members may outweigh the benefit of increased sensors among all individuals. Adding more sophisticated hardware also means adding more control software and adding additional levels of complexity to the entire system along with the potential for unexpected interaction between components [12, 13]. Finally, being parsimonious in terms of specialized hardware also leads to a focus on more intelligent software, which is important from the perspective of advancing artificial intelligence [13]. A desire to avoid the disadvantages of an over-reliance on path planning, combined with a desire to achieve as much as possible with very parsimonious agents, has led us and others to explore the use of reactive navigation techniques in unknown domains [9, 14]. Instead of prescriptively navigating to a desired location via path planning, reactive navigation emerges from an agent's responses to its immediate local perceptions [10]. This allows agents to react quickly to changing circumstances, does not require a map of the environment, and since agents maintain at most a limited model of the world, localization is often not as crucial [15]. Reactive navigation is also referred to as *local* navigation, since the agent continually decides on the best direction in which to move based mainly on its local perceptions [9]. Reactive control mechanisms, such as those based on potential fields [16] are highly vulnerable to local minima. Local minima situations occur when all local perceptions lead an agent to a location that is not its goal; for example, when an agent can see its goal but for reasons of geographic topology must move *away* from the goal to ultimately reach it [17]. Typically, a certain amount of noise or randomness is added to the agent's chosen movement vector (desired velocity and orientation of travel) in order to partially deal with the problem by allowing an agent to move away from a local minimum and hopefully find new perceptions that lead it to more promising places [18, 12, 17]. Agents can similarly become trapped in local maxima where they lack the stimuli needed to choose a movement vector. For example, box-canyons (Figure 4.1) are a common local maxima problem, where agents are unable to see a goal of any type [9] and are reduced to wandering randomly before stumbling upon an exit [7]. The constricted zone between the two open areas is also an example of a problematic bottleneck, an area that is difficult to navigate due to the restricted space, especially when it is an area through which many agents must pass.

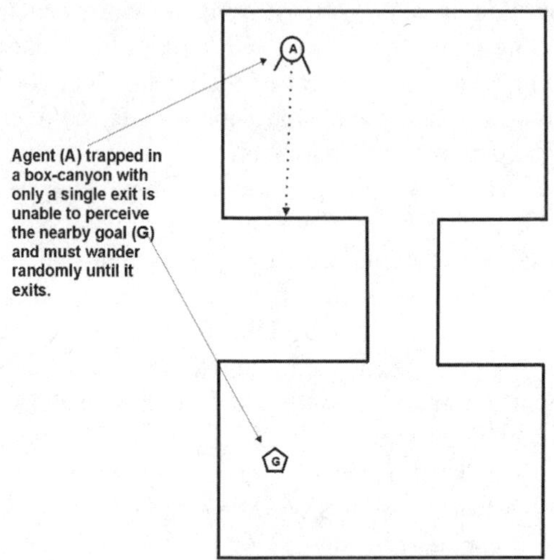

A
Agent (A) trapped in a box-canyon with only a single exit is unable to perceive the nearby goal (G) and must wander randomly until it exits.
G

Fig. 4.1. A box-canyon situation

The lack of goal-directed ability makes purely reactive navigation unsystematic, and agents (as individuals and as a team) do not benefit from the knowledge they have gained by exploring an environment (unless they are constructing a detailed world model as they explore) [19]. Thus, if an agent trapped in a box-canyon finds its way out and gets trapped in the same box-canyon again, it must repeat its search for an exit. In addition to bottlenecks and local minima and maxima, agents are also subject to cyclic behaviour, where they oscillate between multiple stimuli and never reach their goals (or at best waste time) [20, 21]. An ideal agent navigation system is one that has the responsiveness of reactive navigation and the proactivity of path planning methods. One partial solution to this problem is to allow agents to maintain state to remember experiences and places they have been [7]. However, this is contrary to a fundamental design goal of the purely reactive approach in that such agents are intended to maintain little or no state information [22]. Even though behaviour-based approaches relax this restriction to varying extents, it is still generally desirable to minimize state information in these agents for reasons of parsimony. As a compromise, both reactive and deliberative methods can be employed, resulting in a hybrid agent where local navigation is typically handled by a reactive subsystem while path planning methods are employed in a deliberative controller (e.g. [21, 8]). In systems that employ path planning, reactive navigation allows robots to respond to the unexpected in real time and handle aspects of the environment that have changed or were not known ahead of time [7]. Consequently, improving local navigation is desirable, whether it is used exclusively or as part of the reactive subsystem of a hybrid agent. Moreover, if reactive navigation can be made more effective, it can be employed more widely without resorting to more expensive path planning techniques. There are numerous potential improvements one can make to reactive navigation, including predicting what may occur in the environment and taking advantage of domain-specific phenomena. We are interested in ultimately supporting large teams of parsimonious agents, and so our efforts have been focused on improving reactive navigation by taking advantage of the experience of teammates in the environment. As agents explore the environment, they encounter a wealth of information that can be shared with others. With this comes a variety of decisions that must be made in any agent design: what mechanisms (both physical and linguistic) are to be used to communicate this information, what to communicate, when such information is to be shared, and with whom, for example. The principle of parsimony applies here as well: while one can develop agents that employ very elaborate communication schemes, these may not have a significant performance advantage over simpler agents once costs and benefits are weighed. Accordingly, we attempt to deal with the problems encountered in reactive navigation without the use of any explicit communication. Instead, agents assist one another in navigation by identifying desirable locations and sharing this knowledge with each other by modifying the environment. This form of implicit communication is properly referred to as *stigmergy*, a term used in biology to describe the influence that previous changes to the environment can have on an individual's current actions [23]. Naturally-occurring stigmergy is most commonly associated with the behaviours of social insects [24].

The remainder of this chapter overviews previous literature on reactive navigation and stigmergy, presents a number of increasingly sophisticated techniques for using stigmergy to assist in reactive navigation, and evaluates these in a complex three-dimensional software domain that embodies all of the classic problems associated with reactive navigation.

4.2 Related Literature

To date, most work with stigmergy has dealt with its effects on simple homogeneous robots collectively performing foraging or sorting tasks (e.g. [5, 23, 12, 25]). These differ from goal-directed navigation in that foraging has very general goals (e.g. find any food, which is satisfied simply by obtaining any instance) as opposed to the more specific nature of constructing a path to a single goal. Tasks such as foraging and sorting are representative of some robotic applications, and are also closely related to the behaviours of the natural creatures (e.g. ants and termites) that inspire the model.

Social insects such as ants are homogeneous, and redundant in the performance of group tasks. This redundancy ensures that group goals are still met, even when an individual performing a task fails to complete it, because the sub-tasks necessary to complete a larger task do not have to be completed by a single individual. The presence of environmental cues alone eventually ensures that a complete sequence of actions is executed, even if the steps in a sequence are performed by different individuals [23].

The potential benefits of modifying the environment to affect the actions of others can also be realized in more complex domains. Werger and Mataric [5, 26] use a form of stigmergy in which robot bodies are substituted for chemical pheromone. In their work, a team of robots searches an area without global positioning and only limited sensors by forming robot chains. The primary drawback with these robot chains is that close contact between robot bodies limits parallel action. Most of the team members are prevented from actually doing useful work by the requirement that they keep their position in the chain. In addition, the area of exploration is bounded by the maximum length of the robots' bodies laid out in a line. Balch and Arkin [7] have explored using limited local spatial memory to deal with navigating box-canyons in reactive schema-based robotic agents. Agents remember and avoid visited locations by recording this information internally in a 2-dimensional integer array. Although this improved navigation, they found that the robot could become surrounded by visited locations and find no direction in which to move. Also, since the stigmergy depends on an agent maintaining a rudimentary world model, this model occasionally became inconsistent with the world, causing a divergence between the agent's perceived and actual location.

Vaughan et al. [27] implemented a team of robots capable of locating and recovering a supply of resources within a complex and initially unknown environment. Their robots generate and share waypoint coordinates via radio communication, which they maintain as an internal list of *crumbs* and *places*. These lists form trails that the agents then follow. Like Balch and Arkin's work, stigmergic markers are used here as part of an overall internal world model rather than a set of perceptible markers in the environment. Simulating stigmergy within agents' world models as opposed to marking the physical world avoids the problem of marker clutter, which would be a significant issue here because crumbs are dropped by all agents at regular intervals. However, because agents must communicate their trails explicitly, this compromise means that all the disadvantages of explicit communication remain. The method depends on agents being able to localize themselves accurately, and was found to fail after the cumulative error between the robots' coordinate records and the real world became too great for crumb trails to be shared effectively.

Parunak, Sauter, et al., [25, 28] employ ant-inspired stigmergy in a software environment that uses synthetic pheromone to coordinate unmanned aircraft employing potential field-based navigation. Their method uses a distributed network of *place* agents that are used to record and control pheromone aggregation, propagation and evaporation. Thus while the pheromone is virtual, the trails are physical in that they are maintained by embodied computational entities. These trails allow *ghost* agents to move across place agents by following the stigmergic trail. This approach is interesting but requires extensive infrastructure: place agents must first be evenly distributed throughout a predetermined region, be able to communicate reliably with each other via a wireless network, and also have appropriate sensory apparatus to detect hostile targets that are of interest to the ghost agents that traverse the network. The question of maintaining this extensive infrastructure under the conditions a combat environment presents has not yet been adequately addressed.

Kube and Bonabeau [29] examine ant-inspired coordination in a transportation task using a team of robots with decentralized control. In their work a robot team cooperatively performs a coordinated movement task (pushing a box that is too heavy for one agent toward a goal marked by a spotlight) without explicit communication, using locally sensed information only [29, 30]. When robots involved in the task detect a lack of progress (i.e. the box is not moving), they re-orient their position until box motion is once again observed. Thus, if robots are pushing from opposite ends and hinder one another's efforts, they gradually realign themselves. Directed box-pushing to a position specified by a spotlight is also achieved without explicit communication. This is accomplished by designing robots to push the box only when they are in contact with it and when the goal spotlight is not detectable. This has the effect of aligning the robots on the side of the box furthest from the goal, causing them to push it in the desired direction. Though this work focuses on stagnation recovery behaviours for a team of reactive robots, it is relevant to work in stigmergy in that the robots sense progress (or lack thereof, caused by the incompatible movements of other robots) in the environment and adapt accordingly through performance of the task itself. This emphasizes the feasibility and validity of using stigmergy to improve agent performance as individuals and a team.

Rumeliotis et al. [31] suggest a somewhat different ant-based approach to navigation using landmarks rather than pheromone for navigation. They note that desert ants navigate using a combination of visual landmarks and path integration. Path integration is the process of keeping an accurate idea of the direction of one's starting point relative to one's current position by updating this global vector as one travels with angles steered and distances covered [31]. Even for humans, path integration in unfamiliar or confusing areas can be quite difficult and easily subject to error. Consequently, desert ants learn and integrate a series of local vectors between a number of landmarks to navigate [31], rather than relying exclusively on path integration. This allows them to follow meandering and complex paths to their nest or outward from it by breaking the journey into discrete vectors between recognized landmarks. Though the ants are not explicitly creating the trail, they are actively interpreting their environment to their advantage.

Howard et al. [32] use stigmergy in a similar manner. In their work, the bodies of a team of mobile robots are used as landmarks to allow robots to localize themselves based on the relative range, bearing and orientation of other agents in a dynamic and/or hostile climate. While this shows the potential for using stigmergy by actively creating landmarks when none exist, the problem with this approach is the use of the robots themselves. A robot serving as a landmark is not likely doing useful work in such an environment.

4.3 Stigmergic Navigation Techniques

In contrast to these works, we employ a number of different stigmergic techniques that involve the deployment and exploitation of physical markers (as opposed to markers used only internally and communicated explicitly) in order to assist in team navigation. We refer to the process of using these marking techniques collectively to make more purposeful reactive navigation decisions in an unknown environment as *stigmergic navigation*. These techniques are described and demonstrated here in the context of reactive agents navigating in an unknown three-dimensional indoor environment. The particular agent architecture was not a strong consideration, but we chose a schema-based approach [33], one of a range of behaviour-based approaches where overall intelligent behaviour results from the complex interaction between behaviours and the world. While these techniques should be equally applicable to other types of agents, such as hybrid agents, we chose this mechanism in order to employ a very simple set of agents, so that the value of the techniques themselves could be considered separately from the sophistication of the agents used.

For stigmergy to be possible, agents need to be capable of marking their environment in a manner recognizable to others. A convenient mechanism for accomplishing this, and the one used here, is to simply place perceivable markers (i.e. physical objects) on the ground at particular locations. Our techniques make use of two types of stigmergic markers: *homogeneous* and *heterogeneous* markers. In conjunction with the agent's behaviours, each of these can be designed to permanently or temporarily attract, repel, or induce other specific behaviour in an agent.

Homogeneous markers are those where there is no distinction made between individual markers: the presence and location of a marker alone is significant. While they have numerous applications, their main limitation is that agents have no ready mechanism for mediating between markers when more than one is visible, other than aspects of the environment such as marker distance that may have no meaning to the marker's context. Heterogeneous markers, in addition to imparting information by their location, encode a *value* (perceptible by an observing agent, and possibly unique) that can serve to identify a group of markers or uniquely identify a marker, allowing markers to be grouped to serve unique purposes. In addition to encoding such labels, the value associated with heterogeneous markers can also be used as an attractive value, allowing agents to differentiate markers to follow. While it is possible in theory to use homogeneous markers in a heterogeneous fashion by defining a sufficient number of different marker types (for example, rather than encoding a value in a painted marker, different colours of paint could be employed), this may prove difficult in practice due to the number of marker types that may be necessary (e.g. the limitation on the perceivable differences in shades of paint).

Stigmergic navigation involves selecting appropriate marker types, conditions for dropping markers, and conditions for following markers under particular situations. The following subsections describe how combinations of these elements can be used to solve problems traditionally associated with reactive navigation as outlined in Section 4.1. These range from very simple actions that require little coordination, to more complex activities that involve a team of agents cooperatively and dynamically constructing marker trails leading to discovered goal locations. As the simpler approaches can be used to provide additional support for more complex activities, we begin with these. Further details on all techniques may be found in [14].

4.3.1 Marking Bottlenecks

Bottlenecks are extremely common in indoor environments. For example, any path from a given room to another must pass through some number of doorways and/or hallways, each of which forms a zone (small in the case of a doorway, but likely of much greater extent in hallways) within which agent navigation is markedly more difficult because of a lack of movement space. Moreover, each of these zones is small from the perspective of perception, and therefore difficult for a reactive agent to be drawn into. Such bottlenecks are not only constrictive, but important to the agent's successful performance. Within the open area of an empty room, for example, any one coordinate position in the enclosure is no more critical than another in terms of choosing a path to get to another room (while some may be considered better in terms of being more direct, all will be adequate). The exception to this is the position corresponding to the doorway itself - the bottleneck which must ultimately be part of the path. One of the simplest uses of stigmergic markers is the placement of heterogeneous markers in such constricted areas, in order that agents can supply each other with an attractive stimulus in the absence of pre-existing naturally-occurring stimuli. Traversing such an area is an interesting problem in that in many cases attracting the agent is not enough. In the case of a doorway, perception of a new area is immediate once one is present in the constricted area. However, in a hallway such as that forming the box-canyon shown in Figure 4.1, once an agent is drawn into the constricted area, measures must also be in place to ensure that the agent continues through it, rather than moving back to the area that it was exploring previously.

The technique we employ requires the use of heterogeneous markers with unique identifiers, as well as designing agents to be able to perceive markers and record recently visited markers in a short-term memory. To achieve a balance between exploring an area and moving to a new area, an agent is only attracted to markers when it has not followed a marker trail for a given period of time. In this state, the agent is considered to have *marker-affinity* and will move toward any marker of this type that it sees. As soon as the agent is successful in reaching a marker, it enters into a *marker-following* state. In this state the agent continues to be attracted to bottleneck markers that it has not visited earlier (as long as it detects that it is in a constricted area such as a doorway or hallway), but is not attracted to those markers it has already visited.

By recording and ignoring markers that have been visited while in marker following mode, the agent is drawn along by the trail of unvisited markers until it enters another open area. As soon as the agent detects that it is no longer in a narrow space (doorway or hallway) it transitions to a *marker-neutral* state during which it is no longer attracted to bottleneck markers. Each computational cycle, any markers that have not been visited within a set period of time are purged from the agent's list of visited markers. This is necessary to allow the agent to potentially traverse a previously travelled marker trail at some future time. This purging also serves to keep the agent's memory requirements very minimal. The marker-neutral state persists for a predetermined period of time, before the agent once again becomes marker-affinitive.

These bottleneck markers act to draw agents through tight spaces that are difficult for agents using local navigation to find and move through. If the room has only one entrance/exit, an agent can escape it more quickly and continue its explorations via the trail it followed to get in. Conversely, if the room has several unexplored portals the agent still has the opportunity to discover these alternate exits during the period in which it is not attracted to markers. While in this case the agents are told the perceptual features that distinguish a bottleneck, in future we intend to employ an information-theoretic measure of entropy as a criteria for dropping these markers, in order that the measurable need for a marker can be directly reflected in the likelihood of an agent placing one.

4.3.2 Marking Local Maxima

In addition to using markers to draw agents through constricted spaces in the environment, agents can be assisted in negotiating local maxima through stigmergic means. Our philosophy here is that agents will be attracted to local maxima already through the basic principles of reactive navigation, and thus it is worthwhile to move the agent through the environment faster by having these be more attractive. However, once an agent is drawn to a local maximum, it must be moved along through the environment in order that it does not simply stay there. Having an agent be repulsed by a local maximum would allow it to move away from the local maximum, where it would be likely to find some exit or perceive another local maximum marker (or some other stigmergic marker if multiple techniques are being combined) that would allow it to be coherently moved through the environment. We achieve this by using homogeneous markers that are initially attractive to an agent, but cause agents to be repulsed once the agent reaches a specific proximity. This avoidance state persists for a predefined period of time, allowing an agent time to exit the local maximum situation.

An agent marks a local maximum when no goals are visible from the agent's current position (the essence of a local maximum), no local maxima markers are perceivable within a given range (to minimize the number of markers), and no walls or other obstructions are nearby. The last constraint is intended to ensure that these markers are placed in the open, so that they force agents to the edges of local maxima areas when they are repelled, where the agents are more likely to happen upon an exit. This also has the effect of drawing the agent further into the open when they are attracted to the markers, where the agent is likely able to see a larger portion of its environment.

Unlike the previous marker type, in most domains these markers must vanish (or the ability to perceive them be altered) after some period of time. This is because unlike hallways, local maxima are likely due to the interaction of the environment and the agents' current goals, and must change when these goals do.

This method of marking can be likened to the use of spatial grid maps by Balch and Arkin [12]. Unlike their method, however, our agents are storing information about visited locations in the environment, rather than in an internal data structure. In addition, this information is less fine-grained, making it less verbose, and can be generated and used by multiple agents simultaneously.

4.3.3 Marking Local Minima

Another application of stigmergic markers is to allow agents to deal with local minima. These can be especially difficult in environments where three dimensions must be considered. For example, low lying obstacles such as a stair railing can allow an agent to perceive its goal while physically preventing the agent from moving directly toward it (as opposed to many two dimensional environments, where any obstacle is assumed to completely occlude objects beyond it).

In order to deal with this particular situation, we introduce another homogeneous marker type - *local minima* markers. Rather than attracting or repelling agents, these markers are unique in that they prescribe an *action to undertake* in response to a specific situation at a particular place in the environment.

In the indoor environment we use for our implementation, for example, these markers can be used to compel agents to jump when sufficiently close to them, allowing an agent to leap over a low lying obstacle. In order for this prescribed action to be a useful one, agents only drop these local minima markers when they perceive an attractor and a low-lying barrier in their path to it and when they are sufficiently close to the barrier itself. Unlike other markers, these are domain dependent by default, a natural consequence of encoding a particular action to take.

In our implementation, these markers are designed to disappear after an interval in the same fashion as local maxima markers. We deemed this necessary in that a local minimum in our implementation arises not solely from the environment, but from its interaction with the changing goals of agents. In domains where this is not the case, local minimum markers can be permanent.

4.3.4 Stigmergic Trail-Making

Markers in the basic techniques described above have indirect relationships to one another. For example, marking the start of a bottleneck may lead an agent further into a corridor, where it may continue to follow other markers placed by different agents at an earlier time or place markers itself. However, there are no explicitly defined relationships between these markers. While markers may even be uniquely labelled, there are no relationships between the labels used. In contrast, the markers in a real trail do not coincidentally mark individual phenomena, but have an explicit ordering to them. Analogously, *stigmergic trail-making* is the process of constructing and interpreting a purposeful trail of markers *cooperatively* to a useful location, by any number of agents over time as they explore the environment. A stigmergic trail emerges naturally out of agents' collective desire to improve the navigability of the environment. By following a stigmergic trail, agents should be able to repeatedly locate a discovered goal without internal world modelling, internal maps, or internal path planning.

The process of stigmergic trail-making is best illustrated by way of an example. Consider the situation in Figure 4.2. A primitive stigmergic trail is created when an agent perceives a goal and drops a marker on the ground at its own current location (Figure 4.2a). The dropped marker identifies a vantage point from which the goal should be visible to an agent standing at the marker's location, and on its own, would allow an agent not within perceptual distance of the marker to be able to find the goal faster. By itself, a trail this rudimentary is of limited utility, since it only serves as a visual cue to agents that have an unobscured view of the marker, up to the limits of their perception. To make the trail more sophisticated, it is necessary to extend the trail, forming a chain of markers that will ultimately end at the goal. We must be able to know the placement of each marker in the trail, and we use a numeric value associated with a heterogeneous marker to represent the marker's position in the trail (in this case, 1) to do so.

Building a trail from a goal marker involves agents successively recognizing situations where the trail can usefully be extended from what can currently be perceived. Figure 4.2b illustrates this extension over time as an agent (possibly the same agent that started the trail, possibly not) sees an end-point of the trail (the 1 marker) and drops a second marker. The second marker dropped is assigned a value one higher than the marker that the agent has observed (in this case, 2), indicating an extension of the trail to that point. As the process repeats, possibly through the actions of many different agents, the trail gradually lengthens (Figure 4.2c, d).

To minimize clutter, agents drop markers only under constrained conditions. When the agent perceives the goal (or a marker already on the trail), it checks for the presence of stigmergic markers already serving as attractors to this target. If no markers are visible, the agent drops a marker with a value one greater than that of the marker it perceives. Ultimately this can create multiple useful trails due to the limitations of agents' perception: when a new branch is created, it will be because the remainder of the trail is out of sight, and will serve to guide agents from a different location from which the trail was not previously useful. This is illustrated in Figure 4.2e, where an agent sees the end of a trail, by perceiving a 3 marker but no 4 marker, and so extends the trail in another direction. A 4 marker exists, but is not within the agent's perception, and the trail is now useful from two incoming directions. The increased usability of this trail arises directly out of its construction by multiple agents.

In keeping with the objective of limiting marker clutter, an agent only extends the trail when it perceives what appears to be its true end. If the agent perceives other markers from its location, it only drops a marker if the goal or a higher valued marker is not also visible. There are never any truly redundant markers, since a marker is not dropped if one serving the same purpose is perceived, and if such a marker cannot be perceived, it is necessary. For example, in Figure 4.2f: the agent perceives marker 3 and two marker 4's; it knows that marker 3 is not the end of the trail (because it

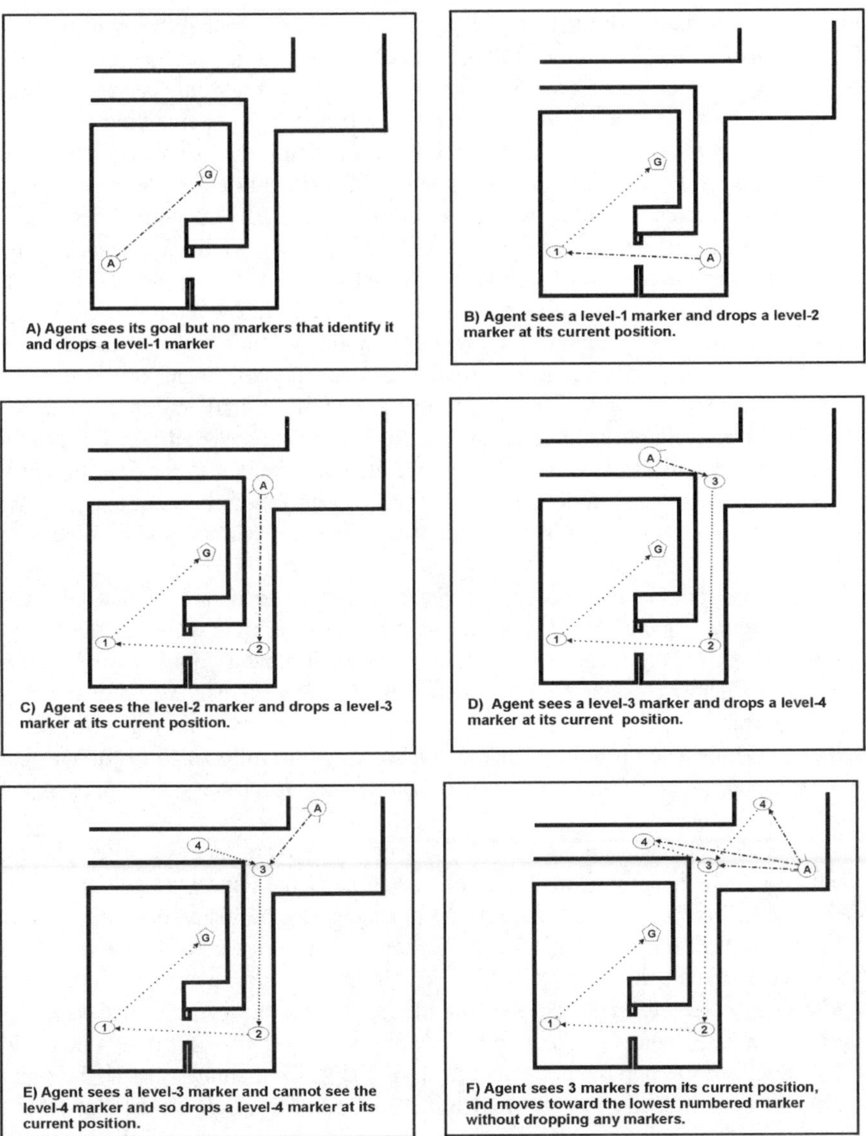

Fig. 4.2. Stigmergic trail making

perceives a 4 marker), and that its vantage point is not far enough away to warrant dropping another marker (since existing markers already supply pertinent navigation information from the agent's current location, or the 4 marker would not be present).

In the case where occlusion or some other factor makes perceiving a marker difficult, and another marker gets dropped because of this, this is entirely desirable: another agent in the same position is likely to suffer similar occlusion or perceptual difficulties and such a marker would be of assistance. To deal with cases where occlusion can occur due to the movement of other robots, we do wait a short time before dropping a marker to take into account the dynamic nature of the environment.

These conditions allow an agent to simply follow a trail without dropping new markers when there is enough navigation information available to do so, and assist in improving the trail when enough information is not available). Agents utilize the trail to locate the goal at the other end by following the trail of markers, always moving toward the lowest numbered marker visible. Since the markers are dropped based on visibility, the agents are able to consistently follow one marker to the next to locate the goal. This also helps minimize the length of a path followed by an agent - if an agent perceives a 3 and a 4 marker, for example, the agent will not bother moving to the 4 marker.

The two important features to emphasize in this approach are a) that this trail-building occurs collaboratively; and b) that it occurs over time while agents follow existing trails and otherwise navigate through the environment. In the case of several goals in reasonably close proximity, the trails may connect. This may cause an agent to perceive more than one lowest valued marker: in this case an agent will be attracted to the marker in closest proximity at any time, so may start following toward one goal and end up at another. Either way, however, the agent reaches a goal, which is the point of the process.

In our implementation, we did not have these markers fade, in order that we could explore the maximum advantage of being able to make and use trails. However, such markers could be built to fade where goals change. In this case, since markers are dropped at different times, with higher-numbered values extended from lower-numbered markers, the useful lifetime of a marker would have to be higher for markers that were closer to the goal (i.e., lifespan in inverse proportion to marker number). In cases where trails were broken due to marker fade, partial value could still be obtained through the portion of the trail that was remaining (that is, the agent could still be coherently led partway to the goal), and if a trail were to be rebuilt from the goal it would eventually intersect the still-existing portion. Like ant trails, mechanisms could also be put in place to allow agents to extend the lifespan of these markers as they traversed them.

The four methodologies for dropping and responding to markers described in these subsections form two complementary halves of stigmergic navigation. Bottleneck and local maxima markers serve to promote exploration and allow agents to more quickly locate an undiscovered goal somewhere in the agents' environment, while stigmergic trail and local minima markers facilitate subsequent trips to a goal once it has been discovered. Having described the basic principles followed by these techniques, we now turn to more detailed information about implementing these in a software environment.

4.4 Implementation

Agents in this work were implemented using a schema-based control approach. Schema-based behaviours divide control into *perceptual schemas* and *motor schemas* that are tailored to specific aspects of an agent's environment [34, 17]. Motor schemas are responsible for manipulating the agent's effectors to accomplish well-defined tasks, such as moving toward a visible attractor. When activated, a motor schema produces a vector of parameters to physically control a mechanical agent (e.g. desired direction and velocity with which the agent should move). The output of a motor schema can be weighted by the overall strength of the schema's activation, allowing a blended response where several different motor schemas may be applicable at the same time.

Motor schemas are activated either by being invoked by motor schemas that are already active (a collection of schemas properly termed an *assemblage* allows a series of activities to unfold, with each motor schema activating the next, for example, and can also act in a subroutine-like fashion), or by perceptual schemas. Perceptual schemas act as filters on information coming from sensors, looking for perceptions that will be of interest to particular motor schemas (or assemblages of schemas) to which they are connected. In our implementation, for example, the *percept-marker* perceptual schema is responsible for using an agent's sensory apparatus to detect when a marker is visible and activating interested motor schemas. Information provided about the marker, such as its location, provide an activation strength to the motor schema or assemblage, allowing it to affect the robot's effectors.

More sophisticated behaviours such as marker dropping are handled in our implementation by assemblages. For example, a *drop-marker* assemblage consists of motor schemas most relevant to placing a marker of some type on the ground, and is activated when the *percept-drop-marker* perceptual schema returns a value of *true*, and deactivated otherwise. Our implementation allows as many motor schemas as necessary to be active at any given time, and their vectors are normalized according to the activation strength of the motor schemas and combined into a single normalized vector for orientation and velocity that is then followed by the robot.

The agents operated by these control mechanisms inhabit a simulated world provided by the computer game Half-Life. This simulation provides a large-scale, three-dimensional environment with realistic physics. In addition, the environment has a first-person view that is useful for observing the agents' behaviour during experiments. To allow agents to mark their environment, a pre-existing game object (a small satchel) was adapted to function as a basic marker type. Agents were given an unlimited supply of these items to drop on the ground as required. The agents and their constituent schemas were implemented in Visual C++ as part of an overall agent architectural framework and designed to allow for dynamic reconfiguration of agent assemblages, schemas and associated gain values [14]. This object-oriented class framework allows for new motor and perceptual schemas to be developed that leverage existing code as much as possible.

An agent's decision making method is invoked by the Half-Life simulator a variable number of times per second (depending on the system load) to respond to changes based on perceptions provided by the simulator. Our implementation hooks this decision-making method to each perceptual schema, which regard the environment, invoke motor schemas and assemblages, whose outputs are combined into an overall response.

A number of perceptual schemas were developed to detect the conditions under which various marker types should be dropped. In addition, motor schemas were implemented to cause agents to react appropriately to the presence of the various marker types employed, according to the descriptions provided in Section 4.3. For example, the *avoid-marker-local-maxima*, alternates between moving the agent toward and away from a local maxima marker.

4.5 Evaluation

To examine the improvement in navigation by a team of agents using the stigmergic techniques described here, a series of experiments were conducted. In each of these, a team of six agents all employing the same combination of marking techniques were placed in a three-dimensional environment, with each agent's objective being to travel to an initially unknown goal while using its marking technique(s). Upon reaching the goal, an agent was moved back to a starting position to repeat this process, gaining the advantage of markers that had been placed by itself and others in the environment in the interim. The number of times the goal was reached in a 30 minute period was tracked, along with the time required to initially discover the goal, as well as other statistics. In this situation, a more helpful marking technique (or combination of techniques) should support a higher number of traversals to the goal over a particular time period.

The specific three-dimensional domain chosen was *Crossfire*, one of the standard maps packaged with Half-Life. A simplified two-dimensional overhead view of this domain is depicted in Figure 4.3. This environment was chosen because it displayed all the difficulties associated with most indoor environments. Beyond providing significant path complexity and ample opportunity for the production of incorrect paths, it includes open areas, box-canyons, stairwells, ramps and elevation changes. Since the map is made for challenging human multi-player interaction, it is designed to be non-trivial to navigate. As such it contains many areas where limited visibility makes reactive navigation by agents particularly difficult.

The size of the world defined by a Half-Life map is 8192 units in height, width, and depth. Space in this environment is measured in terms of an abstract *unit* that is 1/8192 the length of a dimension. There is thus no direct relationship to any measurement in the real world. However, an agent in Half-Life is intended to be roughly human-sized, and is defined by a bounding rectangle 72 units high, 32 units wide, and 32 units deep. Common-sense approximation to the real-world equivalent

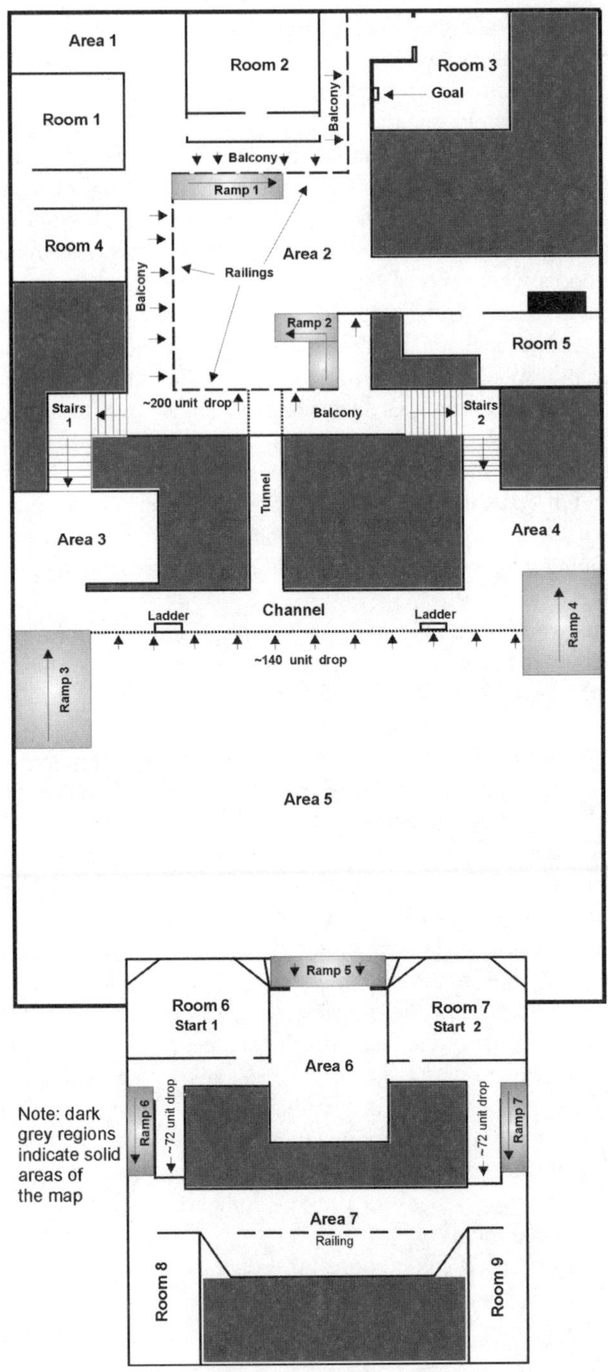

Fig. 4.3. Map of Experimental Environment

of a unit would thus be about an inch, assuming a human is six feet tall. In referring to the environment, however, we use the same unit term used in Half-Life in order to be consistent with other research employing this domain.

In order to accurately measure the impact of each marking method, the environment, goal location and agent starting positions were consistent throughout all experimental trials, and all trials were run on the same machine (a Pentium III - 933 MHZ computer, running Windows 2000 Pro Service Pack 3 with 512 MB RAM). Agents began each trial in either Room 6 or Room 7 (noted on Figure 4.3). Two starting rooms were used so that agents would not interfere with one another excessively due to overcrowding at the start of a trial. Since the two starting positions are positioned symmetrically, navigating to the goal from either should have no effect on results. The goal was located in a room on the far side of the environment (Room 3). As can be seen from Figure 4.3, agents had to navigate over ramps and stairs, and negotiate a number of doorways, open areas and box-canyons in order to find the goal. After locating and moving to within 40 units of the goal, a point was added to the running total of goals discovered in the trial. The agent discovering the goal was then transported back randomly to either Room 6 or Room 7 to attempt to find the goal again (gaining the advantage of any additional markers that had been placed in the meantime).

Experiments were performed using the simple stigmergic techniques described in Section 4.3, both individually and in combination, followed by more sophisticated stigmergic trail-making, and finally supplementing stigmergic trail-making with combinations of these simpler techniques.

The stigmergic techniques described in Section 4.3 employ a number of parameters: for example, marker lifespan and range of influence. To select appropriate values for this evaluation, we ran numerous initial trials varying these parameters. We chose the parameter values that appeared to be the most promising based on these initial trials. The specific values for these parameters depend on the marker types employed, and are detailed in the subsections below.

Each experiment involved 40 contiguous trials of 30 minutes each. In each trial, we tracked not only the real-world time involved for events such as the initial discovery of the goal, but also the number of agent decision making invocations - that is, the number of times the simulator invoked each agent to make a decision on an action. This was used as a secondary measure of time passing in the system, because in practice we found that the computational overhead of simulating some of the more complex marker types in combination had a side-effect of slowing down the entire simulation, resulting in fewer agent decision-making invocations over the same time period. This additional measure allows a comparison with issues of variability in simulation overhead removed.

The details of the specific conditions for each experiment and associated results are described in the following subsections. A table of all experimental results follows in Section 4.5.4.

4.5.1 Simple Marking Schemes

We began by experimenting with simple marking schemes and combinations of these. To establish a baseline, we ran a set of trials with a team of reactive agents using no markers and relying entirely on random search to locate the goal. In this case agents discovered the goal 161 times in 40 trials for an average of 4.03 goals per trial (with a standard deviation for goals reached in a trial of 1.67). It took them on average 592.35 seconds (with a standard deviation of 362.25 seconds) and an average of 57982.88 agent decision-making invocations to initially discover the goal. The agents' best performance in locating the goal repeatedly in a single trial was 8. The best time to initial discovery was 89.98 seconds over 8820 decision cycles.

We then performed experiments examining the simple marking techniques, and combinations of these, in order to compare them to this baseline. The results of these experiments along with detailed experimental conditions are presented in the subsections that follow, and are summarized in Figure 4.4.

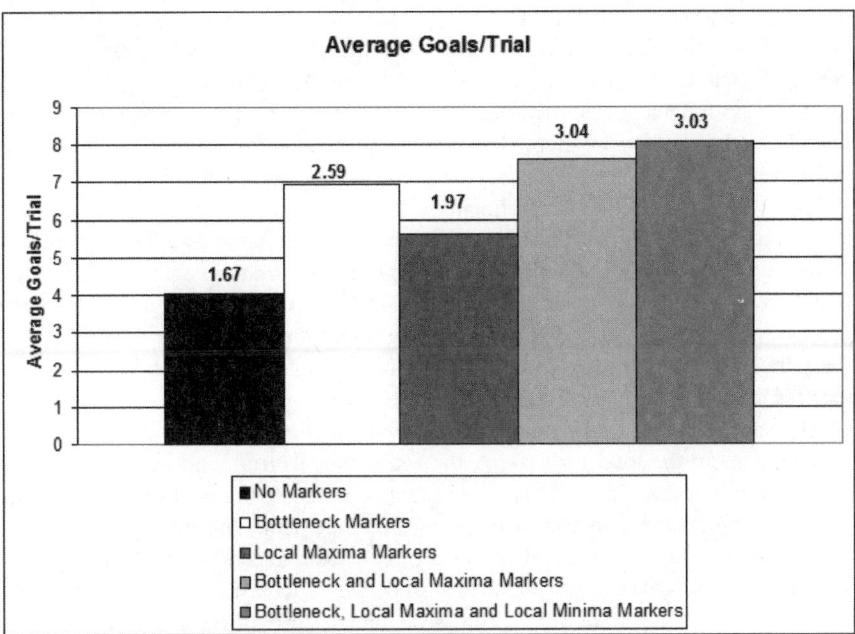

Fig. 4.4. Comparison of average goals/trial (the standard deviation for the number of goals/trial is indicated atop each bar)

Bottleneck Markers

Having established a baseline, we then explored the effects of agents that employed bottleneck markers (Section 4.3.1), to lead agents out of box-canyons (Section

4.1). The *percept-drop-marker-bottleneck* perceptual schema was configured in this experiment to return a true value whenever the agent was standing in a doorway or a hallway, unless a marker could already be seen within 60 units of the agent's current position. This perceptual schema was used to activate the *motor-drop-marker-bottleneck* motor schema, which caused the agent to drop a marker on the ground at its current position.

Another perceptual schema, *percept-marker-bottleneck*, was configured to maintain a dynamic list of markers visited by the agent. A marker was considered to be visited when the agent was standing within 40 units of the marker. These visited markers were purged from the short-term list after 30 seconds had elapsed between the agent's visit to the marker. This was intended to discourage cyclic behaviour and cause the agent to proceed along a trail of markers more deliberately (as described in Section 4.3.1). Otherwise, an agent might move back and forth between markers if it happened to turn around and see a previously visited marker again.

The *motor-move-to-marker-bottleneck* motor schema was also configured to cause the agent to move toward a visible bottleneck marker until it was within 40 units of the marker. At the point that the first marker was reached, the motor schema continued to move the agent toward any subsequent markers as long as the agent detected that it was still in a doorway or hallway. As soon as the agent detected that it was no longer in a narrow passageway (after reaching a marker), the *motor-move-to-marker-bottleneck* motor schema was suspended for a period of 30 seconds (during which any markers in its list of visited markers are gradually purged).

The agents in this set of experiments located the goal 278 times in 40 test runs for an average of 6.95 goals per trial, with a standard deviation of 2.59 (Figure 4.4). On average these agents discovered the goal for the first time in a trial within 463.67 seconds (with a standard deviation of 241.87 seconds), taking 43917.58 agent decision-making invocations to do so. The minimum time to initially locate the goal in any trial was 211.86 seconds and 20074 decision-making invocations. The most goals reached in a single trial was 12.

As can be seen, just marking bottlenecks alone resulted in significant improvements in the total goal count, the average goals/trial, and even the average time to initially locate the goal. The average time to locate the goal for the first time by any agent in the group was reduced by 21.72 percent. The average number of decision-making invocations to discover the goal was reduced by 24.26 percent.

The average goals discovered per trial was increased by 72.67 percent. Accordingly, the total goals for all trials was increased by the same percentage. In addition, the team of agents using bottleneck markers located the goal more often in the majority of trials. The second bar from the left in Figure 4.4 illustrates these findings in comparison to other techniques.

The standard deviation for goals per trial, at 2.59, was wider than that of agents not employing markers. The increased variability in goals is perhaps due in part to the fact that these markers are dropped without the agents knowing where the goal is located in the environment. As a result, the possibility exists that many (or all) of the agents might be initially led in a completely incorrect direction, slowing their time to find the goal. However, since a marker no longer affects an agent once the agent has visited it (that is, once the agent comes within a few units), the effect of being misled would be temporary. In any case, it is equally probable that the marker would in fact lead the agent in a better direction than a worse one (accounting for the variability in goals found).

Local Maxima Markers

We then examined the effect of solely using local maxima markers as described in Section 4.3.2. Agents were compelled to drop a local maxima marker at their current position when no goal was visible, no walls were nearby (within 160 units), and no local maxima markers were already present within a range of 500 units. A dropped marker was set to have a lifespan of 150 seconds.

Agents alternated between moving toward local maxima markers and avoiding them. When first perceiving a local maxima marker, while in a local maxima marker-affinitive state, an agent moved up to the marker's position. Once the agent physically reached the marker (moved within 40 units), the agent transitioned to avoiding local maxima markers. The local maxima marker *avoidance* state persisted for 30 seconds, during which the agent had an opportunity to find an exit from its current area (since it had nothing interesting in it). To allow for a certain freedom of movement, agents are only repelled by local maxima markers when they are in within 120 units. This has the effect of keeping them to the edge of the area they are in and closer to any exits from it, without trapping them against one particular side. If the agent does not find an exit, it is eventually drawn back into the open to try again upon becoming, once again, local maxima marker-affinitive.

The use of local maxima markers improved performance over using no markers in the majority of trials. The agents in this experiment discovered the goal a total of 224 times in 40 trials, for an average of 5.6 goals per trial with a standard deviation of 1.97 (the middle bar in Figure 4.4). The highest number goals found in a single trial was 11, and the average time taken to discover the goal for the first time was 533.48 seconds (with a standard deviation of 242.33) over 52135 decision-making cycles. The best time to reach the goal over all trials was 173.77 seconds over 17029 decision-making invocations.

In addition, the time to initially locate the goal was improved on average by 9.94 percent over using no markers. The average number of agent decision-making invocations to discover the goal was reduced by 10.09 percent. The number of goals discovered in total was increased by 39.13 percent over no markers. This was not as strong an improvement as gained by marking bottlenecks. However, this is not surprising, since local maxima markers work more indirectly, by pushing agents away from uninteresting places rather than actively drawing agents to new regions.

Bottleneck and Local Maxima Markers

We then examined the combined effect of bottleneck and local maxima markers, by allowing agents to employ both as per the conditions of the previous two experiments. Agents followed the additional restriction of not dropping local maxima markers within 150 units of bottleneck markers, and were configured to value bottleneck markers more strongly than local maxima markers via a higher gain value, so that the presence of a local maxima marker would not prevent them from moving to a simultaneously visible bottleneck marker.

A team of such agents located the goal 305 times in 40 trials for an average of 7.63 goals per trial, with a standard deviation of 3.04 (second bar from the right in Figure 4.4). The highest number of goals in a single trial was 13, and the best time to initial discovery was 244.12 seconds with an average time taken to discover the goal for the first time at 567.02 seconds (with a standard deviation of 222.61 seconds) or 51697.6 agent decision-making invocations.

The average goals per trial was increased over no markers by 89.44 percent using both marker types. This is compared to 72.67 percent for bottleneck markers and 39.13 percent for local maxima markers.

Adding Local Minima Markers

To this point, we have not yet introduced the use of local minima markers. This is because in the placement of the goal in the environment (Figure 4.3), there is no immediately occurring local minimum, making it unlikely that such markers would be helpful alone. Other marker types, however, form multiple intermediate goals that dynamically create local minimum situations, and so local minimum markers were expected to be more suited to augmenting these other types. To study this, we added local minima markers to the previous combination, allowing all of the simple techniques to be examined in tandem. Like local maxima markers, the fade period for local minimum markers was set to 150 seconds.

Unlike the other types of markers, local minima markers do not just generally attract or repel agents. Instead they instruct the agent to jump in the air upon moving sufficiently close to these markers. This is because the major local minima condition occurring in this domain were railings and other forms of low obstacles. For example, the balconies overlooking Area 2 of Figure 4.3, are lined by railings that can block agents attempting to move toward markers lying in Area 2. Agents coming up the stairs from Area 3 can sometimes see an attractive marker placed by another agent in Area 2 below them. However, when the agent attempts to move toward the attractive marker, the railing blocks the agent's path. In order to handle this and other low obstacle situations, it is possible for an agent to jump over the railing rather than remaining stuck or having to navigate the long way around the railing.

Using bottleneck, local maxima, and local minima markers in combination resulted in 323 goals in 40 trials for an average of 8.08 goals per trial (with a standard deviation of 3.03). As the rightmost bar in Figure 4.4 shows, this was a significant improvement over agents using no markers, but only a small improvement over that of the other marking techniques.

4.5.2 Stigmergic Trail-Making

Having examined the effects of the simplest techniques, we can now compare these to the more sophisticated trail-making technique explained in Section 4.3.4. As illustrated by the fourth bar from the right in Figure 4.5 (the bars before this are those from Figure 4.4 reproduced proportionally), stigmergic trail markers increased performance substantially over agents using no markers and that of all marking techniques examined previously. The average goals per trial for a team of agents using these markers resulted in more than 10 times the number of goals than a team of agents using no markers at all.

A team of agents using stigmergic trail markers located the goal 1828 times in 40 trials for an average of 45.7 goals per trial, with a standard deviation of 48.14 goals. On average they located the goal for the first time in 462.19 seconds of first appearing in the environment, with a standard deviation of 229.9 seconds. The most goals reached in a single trial was 182, and the fastest time to discover the goal was 103.62 seconds or 10213 decision making cycles for the agent that discovered the goal.

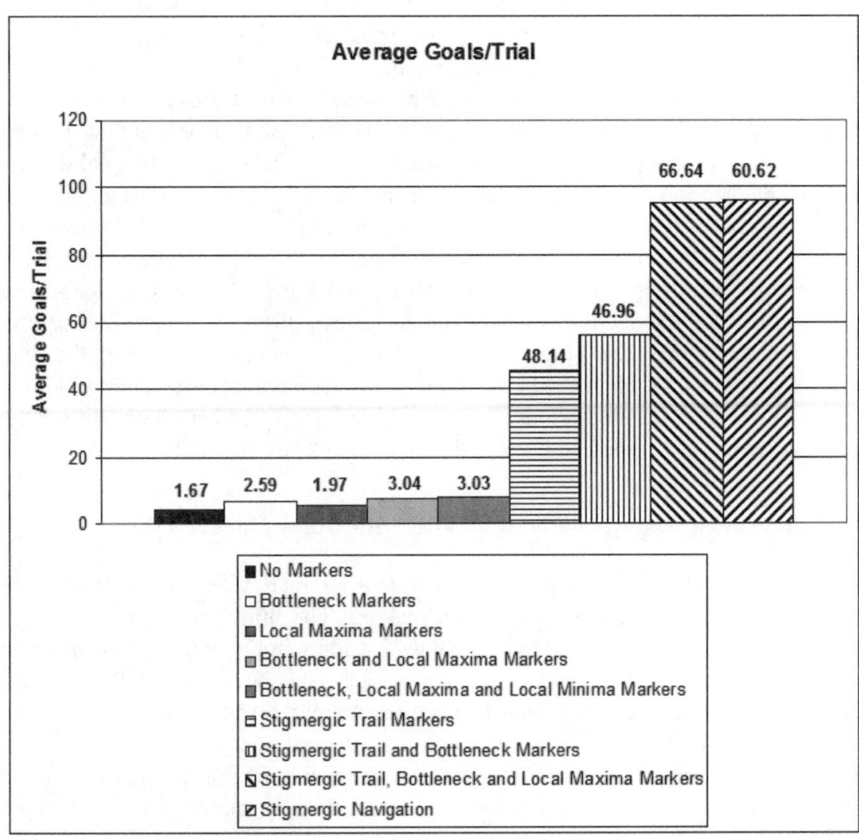

Fig. 4.5. Comparison of average goals/trial (the standard deviation for the number of goals/trial is indicated atop each bar)

During the course of the experiment, a trail of markers was gradually built up, working backward from the goal after it was first discovered, and the trail branched appropriately where its end could not be perceived. Consequently, subsequent agents were eventually able to follow a trail of markers directly to the goal almost from the moment they appeared at the starting point (and improve the trail where they came upon its end from a novel location). The eventual result was a parade of agents heading (more or less) directly to the goal. This method exceeded the performance of agents using no markers in all trials and that of the combination of bottleneck and local maxima markers in 36 out of 40 trials.

As one would expect, the greatest factor affecting performance in these trials was how quickly the marker trail gets created. Once the marker trail was in place, the agents were able to follow it repeatedly and consistently until time expired. The poorer performances only occurred when this trail was not constructed early enough for the agents to benefit. When the agents failed to locate the goal early enough or frequently enough, their performance was comparable to using no markers at all. However, once the trail extended far enough back from the goal, the agents rapidly outpaced all others methods by a wide margin.

The average time to initial goal discovery was also improved by almost 22 percent over agents using no markers. One possible reason for this is that the first stigmergic trail marker dropped ensured that the agent discovering the goal did not wander inadvertently back out of the room in which the goal is situated. Without the use of these markers, this counterproductive activity was observed repeatedly (usually due to the influence of the *motor-noise* motor schema, which adds some randomness to an agent's movements but also on occasion when obstacle avoidance schemas directed the agent away from the goal when obstacles were present) With stigmergic trails, the agent drops a stigmergic trail marker upon first seeing the goal, allowing it to recover from even counterproductive behaviour more quickly. This is because even after wandering out of the goal room, the agent can often still see the trail marker it just dropped and is thus drawn back into the room directly.

4.5.3 Improving Stigmergic Trail Making with Other Marker Types

We then examined the integration of the previous stigmergic techniques to examine how they could improve stigmergic trail-making. The third bar from the right in Figure 4.5 illustrates the improvement gained by adding only bottleneck markers to the stigmergic trail-making process. Experimental data showed that this combined approach yielded better results than agents using no markers in 39 of 40 trials. At the same time, it resulted in a higher number of goals than stigmergic trails alone in 25 of 40 trials. This is also shown in that a total of 2253 goals were reached across all trials for an average of 56.33 goals per trial. The standard deviation for goals reached per trial was 46.96. The average time to discover the goal for the first time was 508.15 (with a standard deviation of 259.44) over an average of 46896 decision-making invocations. This is a 14.21 percent decrease in time and 19.12 percent decrease in decision-making invocations to discover the goal compared to using no markers.

Using both markers in combination resulted in 13 times (1299 percent) more goals being located on average per game versus agents not using markers, compared to 10 times more on average using stigmergic trail markers by themselves. The improvements gained here over using stigmergic trail markers alone are due to the bottleneck markers promoting greater coverage of the environment by drawing agents to different areas periodically. Once the stigmergic trail has lengthened a certain distance back from the goal, the faster other agents notice the trail, the better. Bottleneck markers help in this regard by drawing agents back and forth from room to room more regularly than they do when wandering randomly. Thus, once the trail is sufficiently long, other agents see its end sooner and the complete trail gets built earlier as a result. Since the trail gets built earlier, the agents have a longer period of time in which to benefit from it, resulting in a higher performance overall. The second bar from the right in Figure 4.5 illustrates the very dramatic improvement that results from adding local maxima markers to stigmergic trails and bottleneck markers. The combination of these three marker types resulted in 3817 goals being reached in total over all 40 trials: 22.7 times more goals than a team of agents using no markers and an average of 95.43 goals per game (with a standard deviation of 66.64 goals). The average time to initially discover the goal in a trial was 525.3 seconds (with a standard deviation of 291.27 seconds) over an average of 30869.03 decision-making invocations. This is an 11.32 percent reduction in average time to locate the goal in a trial. More significantly, it is a 46.76 percent reduction on average in the number of decision-making invocations required to locate the goal. The large difference in these percentages is primarily due to the additional processor load incurred by using all three marking methods in combination, slowing down the amount of simulation time that is performed in the real time units we measure.

As with the previous experiment, the improved performance using the three marker types in combination is due to the stigmergic trail being constructed and extended faster than would otherwise occur. The addition of the local maxima markers promotes even wider coverage of the environment than bottleneck markers by themselves. Local maxima markers have the additional effect of drawing agents into the open initially where they can see a wider portion of their surroundings, and increasing the likelihood that one or more agents will perceive a trail end sooner.

These effects in combination contribute to the stigmergic trail being established earlier in the trial, allowing the team to benefit from its use for a longer period of time. This is also evident in that this combination of marker types resulted in a best performance in a single trial: reaching the goal 247 times. This is significantly higher than any previous marker combination. Finally, we added local minima markers, resulting in the complete combination we have termed Stigmergic Navigation. As can be seen in Figure 4.5, the increase in performance achieved by adding local minima markers was again a small one relative to other marker types. Using all marker types resulted in 3856 goals being reached in 40 trials for an average of 96.4 goals per trial (with a standard deviation of 60.62 goals). The average time to locate the goal for the first time in a particular trial was 490.11 (with a standard deviation of 262.24) in 24814.38 decision-making invocations. The former is a 17.26 percent decrease (compared to agents using no markers) in the average time to initially locate the goal, and a 57.2 percent decrease in the average number of decision-making invocations required to initially locate the goal, compared to agents using no markers.

There are several likely reasons that local minima markers did not result in a substantial improvement. First, the situation that local minima markers are intended to handle appears in only a small portion of the environment. Consequently, the opportunity for them to have a positive effect on the agents' performance is limited. Moreover, the agents themselves are fairly well-equipped to deal with local minima via introduction of noise into their movements as part of their behaviour-based design. Finally, the other marker types can themselves reduce the frequency of local minima in this experiment.

Observations of the agents in a number of experiments showed that the stigmergic trail leading to the goal was typically built along a path that bypassed the upper balcony where local minima are known to appear. Instead, the trail typically takes the agents across Area 5 (see Figure 4.3) to either Ramp 3 or 4, down the channel near the center of the map, down the tunnel connecting it to Area 2 before proceeding down the last hallway before the goal. As a result, the occasional positive impact of the local minima markers did not have a substantial effect on overall performance.

4.5.4 Summary of Results

Table 4.1 illustrates the total number of goals achieved across all 40 trials, illustrating the efficacy of employing no markers (*None*), Bottleneck Markers (*BTl*), Local Maxima Markers (*LocMax*), Stigmergic Trail Markers (*StigTrail*), all of these together (Stigmergic Navigation - *StigNav*), and the various combinations described in the previous subsections. This table shows the average real-world time (in seconds) and the average number of agent decision-making invocations required to find the goal the first time in a trial for the same combinations of marker types.

The standard deviations for the time taken to reach a goal the first time and the number of goals reached were both large. The former is due to the nature of reactive navigation itself - while areas that are covered can benefit from the effects of markers, the areas that have not yet been covered are still slow to navigate. While making a stigmergic trail should not intuitively increase the speed at which a goal is first found, there are some specific phenomena in this domain to account for improvement in cases where this does occur (Section 4.5.2). The standard deviation in the number of goals reached is directly connected to the variation in finding the goal the first time. In cases where stigmergic trails are built, the team is essentially following this trail repeatedly until time expires. If that trail is not built early, performance suffers correspondingly.

4.6 Discussion

In this chapter we have described a series of increasingly sophisticated stigmergic techniques for navigation in a three-dimensional domain, and have examined the efficacy of these in a software environment. By marking bottlenecks and local

Table 4.1. Summary of Results: Total number of goals, average initial goal discovery time, and average number of agent decision-making invocations before initial goal discovery, by combination of marker types.

Marking Method	Total Goals	Average Time	Average AgentInv
None	161	592.35	57982.88
BTl	278	463.67	43917.58
LocMax	224	533.48	52135.00
BTl/LocMax	305	567.02	51697.60
BTl/LocMax/LocMin	323	467.67	24068.98
StgTrl	1828	462.19	45043.30
StigTrl/BTl	2253	508.15	46896.00
StigTrl/BTl/LocMax	3817	525.3	30869.03
StigNav	3856	490.11	24814.38

None=No Markers, BTl=Bottleneck Markers, LocMax=Local Maxima Markers, LocMin=Local Minima Markers, StigTrl=Stigmergic Trails, StigNav=Stigmergic Navigation

maxima, agents were able to more frequently discover an initially unknown goal. At the same time, through stigmergic trail-making, agents were able to greatly increase the frequency and ease with which they subsequently located a previously discovered goal, and this improvement was increased substantially by combining this with simpler techniques. All these methods are noteworthy in that the team of agents cooperatively marks the environment over time and shares knowledge of their explorations with one another, resulting in significantly better performance.

Despite having been examined in a software domain, the principles behind each of the techniques described here are all translatable to the physical world. For example, the heuristic of marking bottlenecks to deal with navigational difficulties such as doorways and hallways was useful because in this environment, it is difficult (if not impossible) for an agent to detect the existence of a doorway at a distance. Here, the markers allow the agents to place a more easily detectable object at the location of interest, lightening the sensory load on other agents. In a physical environment this presents similar difficulties, since it is likely that a robot would have to actually be at the doorway already in order to detect it (without a map and accurate localization). Using markers, the agent can "see" the doorway at a distance and utilize that information more readily.

The work presented here bears some relationship to the literature cited in section 4.2. The main difference between this work and those previously described [27, 28, 25] is that here markers are physically dropped in an environment (albeit a simulated one for the purposes of implementation), as opposed to being maintained and communicated between agents as part of larger world models. Moreover, these other methods require explicit communication to work. Agents therefore require communication hardware and a reliable network to share information. This makes such approaches unsuitable for many domains where direct communication might be intermittent or blocked, making stigmergy an attractive alternative.

The work of Sauter, Parunak, et al. [28, 25] has the added requirement of special place agents that must be evenly distributed throughout the environment ahead of any navigating agents. While having an advantage in that the approach can easily model the natural decay of pheromone trails, this additional factor makes this approach impractical for many real world environments. Although outside the scope of this work, it is worth noting that others are working on physical stigmergic mechanisms that allow for physical fading [35].

Finally, though the methods of Vaughan et al. [27] allow robotic agents to navigate to a goal location more quickly and reliably than otherwise, they do nothing to reduce the time it takes for the goal to be initially discovered. In contrast, our method to promote exploration, described in Section 4.3.1, seeks to allow the team to discover an unknown goal for the first time faster than otherwise. This is especially advantageous in applications where quickly locating a particular unknown goal is critical (such as rescue scenarios).

4.7 Challenges and Future Work

While the results presented here are interesting, there is much future work to be done in this area. Deciding when to drop and when to follow markers is critical to the effectiveness of stigmergic navigation. If agents drop markers indiscriminately, they introduce noise and distraction that can hinder rather than help agent performance. Here, we have concentrated on showing the advantages to be gained by being discriminate in choosing when to mark the environment in order to avoid marker clutter in the first place. However, several of the marker types fade in simulation to achieve the results here, which limits the effect of clutter in these marker types.

While in the real world, such fading would have to be implemented using physical mechanisms (e.g. chemical pheromones perceivable by a robot [35]), we are interested in both emphasizing the issue of reducing marker clutter as well as the use of physical markers that require less specialized equipment to perceive and manipulate. Our future work will thus entail the use of teams of agents that can remove as well as place markers, something that is more easily adaptable to the physical world given current technology. Adding the ability to remove a placed marker when it does not match the conditions for its placement is relatively simple on its own, and deals with the issue of marker clutter without the need for fading. This becomes a more involved strategy, however, when a marker being moved or removed has a relationship to others, as it would in stigmergic trail making. We are currently exploring a number of strategies for marker removal under changing goals that are intended to limit the damage done to a trail when markers are removed simply due to misperception or misinterpretation. The ability to remove markers also adds one more crucial behaviour to the set that make up a marking approach: an agent must decide how the removal of a marker alters its reaction to other markers. For example, if an agent has removed a bottleneck marker, there are likely additional markers upcoming that bear an indirect relationship to this and could also be candidates for removal - but at the same time, other upcoming markers may be completely independent.

Related to the issue of marker fading and clutter is the supply of markers themselves - our evaluation does not deal with a finite supply of markers in an agent. Instead, we focus on attempting to limit the need for an overly-large number of markers, both from the standpoint of clutter and the standpoint of a finite marking supply. Being able to pick up markers implies being able to re-use them, and thus the work described above should also support working with a finite marker supply.

In addition to exploring issues of when to drop markers (or remove them), the issue of *when to follow* markers is an important one to explore further. Agents need to balance exploitation of markers with performance of other tasks. If an agent simply follows the markers left by others, without additional exploration, the added value of using these mechanisms is limited, since the markers are not being deployed as extensively as they could. When using bottleneck markers, for example, we employed the simple mechanism of having agents ignore markers for a period of time after reaching one. This allowed agents to wander around the area that they presumably entered upon reaching a doorway for a time, and perhaps find another different doorway. Analogous mechanisms exist for other marker types, and these can certainly be improved.

Adapting these techniques to the physical world also introduces new problems in control: behavior-based robots must have a supply of marker types (or some substance used for marking), and must have the ability to drop (and remove) these items. Dropping a marker is a high-level control command in our approach, but would have to be expanded to physically implement this operation at a motor control level. Similarly, the agents would also have to be physically able to perceive these markers and distinguish between marker types. In simulation, perception has a bounded distance for accuracy, but this is still more accurate than perception in the real world.

More immediately, we intend to use the existing system to examine the effect of the number of agents on the utility of these techniques, as well as changing conditions in the environment. For example, it would be useful to study the performance of local minima markers in an environment that *forces* agents to negotiate them to get to the goal (rather than bypassing them, as sometimes occurred in these experiments). We also intend to experiment with changing conditions in the agents' behaviours (e.g. how long to ignore a bottleneck marker).

A natural extension of this work will be to adapt the markers described here to systematically explore an environment as opposed to locating an isolated goal. By giving markers unique identifiers, agents can pass marker lists that indicate the order in which they should be traversed to reach a goal. This approach is likely to be more resistant to localization errors than that of Vaughan et al. [27], since markers exist externally.

Ultimately, we would like to build agents that *learn* when to place and follow stigmergic markers, in order to minimize the placement of markers and maximize discrimination in exploiting markers. One approach to accomplishing this could include categorizing perceptual features into exemplar states to allow agents to generalize and learn the characteristics of desirable locations and update them dynamically. Agents could then use stigmergic markers to identify valuable locations more generally as they travel.

Another application of stigmergic markers that warrants exploration is using markers to assist agents in localizing cooperatively in conjunction with path planning-based navigation. This could be accomplished by having agents dynamically create landmarks in environments that lack easily identifiable signposts. In this approach, each marker dropped would encode the agent's beliefs about the coordinates of the location that the agent is marking along with a confidence factor. This would allow other navigating agents to use the marker as a potential landmark and as a means to localize.

While stigmergic techniques, in nature and otherwise, are not new, we believe that both they and the central concept they embody – storing knowledge in a distributed fashion throughout the world as opposed to inside an agent – is very under-appreciated in artificial intelligence. Modern research is only now beginning to truly recognize the power of storing knowledge in the world, and techniques embodying this concept, such as stigmergic navigation, will be seen more extensively in future.

References

1. Nehmzow U, Gelder D, Duckett T (2000) Automatic selection of landmarks for mobile robot navigation. Technical report UMCS-00-7-1, Department of Computer Science, University of Manchester
2. Baltes J, Anderson J (2003) Flexible binary space partitioning for robotic rescue. In: Proceedings of the IEEE international conference on intelligent robots and systems, Volume 3. Las Vegas, pp. 3144–3149
3. Reece, D, Kraus M (2000) Tactical movement planning for individual combatants. In: Proceedings of the 9th conference on computer generated forces and behavioral representation. Orlando, FL
4. Verth J, Brueggeman V, Owen J, McMurry P (2000) Formation-based pathfinding with real-world vehicles. In: Proceedings of the game developers conference. CMP Game Group, San Jose, CA
5. Werger B, Mataric M (1999) Exploiting embodiment in multi-robot teams. Technical report IRIS-99-378, Institute for Robotics and Intelligent Systems, University of Southern California
6. Brooks R (1991): The role of learning in autonomous robots. In: Proceedings of the 4th annual workshop on computational learning theory. Santa Cruz, CA, pp. 5–10
7. Balch T, Arkin R (1993) Avoiding the past: A simple but effective strategy for reactive navigation. In: Proceedings of the IEEE international conference on robotics and automation, Volume 1. Atlanta, GA, pp. 678–685
8. Fu D, Hammond K, Swain M (1996) Navigation for everyday life. Technical Report TR-96-03, Department of Computer Science, University of Chicago

9. Sgorbissa A, Arkin R (2001) Local navigation strategies for a team of robots. Technical report, Georgia Tech Robotics Laboratory, Georgia Institute of Technology
10. Mataric M (1997) Behavior-based control: Examples from navigation, learning and group behavior. Journal of Exp and Theor AI 9:323–336
11. Brooks R (1991) Intelligence without reason. In: Mylopoulous J, Reiter R (eds) Proceedings of the 12th international joint conference on artificial intelligence. Sydney, Australia, pp. 569–595
12. Balch T, Arkin R (1994) Communication in reactive multiagent robotic systems. Autonomous Robots 1:27–52
13. Baltes J, Anderson J (2002) A pragmatic approach to robot rescue: The keystone fire brigade. In: Smart W, Balch T, Yanco H (eds) Proceedings of the AAAI mobile robot competition. Edmonton, Canada, pp. 38–43
14. Wurr A (2003) Robotic team navigation in complex environments using stigmergic clues. MSc thesis, Department of Computer Science, University of Manitoba. Winnipeg, Canada
15. Groner T, Anderson J (2001) Efficient multi-robot localization and navigation through passive cooperation. In: Proceedings of the 2001 international conference on artificial intelligence. Las Vegas, pp. 84–89.
16. Arkin R (1998) Behavior-based robotics. MIT Press, Cambridge, MA
17. Pirjanian P (1999) Behavior coordination mechanisms: State-of-the-art. Technical Report IRIS-99-375, Institute for Robotics and Intelligent Systems, University of Southern California, Los Angeles
18. Balch T, Arkin R (1995) Motor schema-based formation control for multiagent robot teams. In: Lesser V, Gasser L (eds) Proceedings of the 1st international conference on multiagent systems. AAAI Press, Menlo Park, CA, pp. 10–16
19. Moorman K, Ram A (1992) A case-based approach to reactive control for autonomous robots. In: Proceedings of the AAAI fall symposium on ai for real-world autonomous mobile robots. AAAI Press, Menlo Park, CA
20. Balch T, Arkin R (1997) Clay: Integrating motor schemas and reinforcement learning. Technical report, College of Computing, Georgia Institute of Technology
21. Arkin R., Balch T (1997) Aura: Principles and practice in review. Journal of Exp and Theor AI 9:175–189
22. Brooks R (1986) A robust layered control system for a mobile robot. IEEE Journal of Robotics and Automation RA2:14–23
23. Holland O, Melhuish C (1999) Stigmergy, self-organisation, and sorting in collective robotics. Artificial Life 5:173–202
24. Perez-Uribe A, Hirsbrunner B (2000) Learning and foraging in robot-bees. In: SAB2000 proceedings supplement. International Society for Adaptive Behavior. Honolulu, HI, pp. 185–194
25. Parunak H, Brueckner S, Sauter J, Posdamer J (2001) Mechanisms and military applications for synthetic pheromones. In: Proceedings of the workshop on autonomy oriented computation. Montreal, Canada
26. Werger B, Mataric M (1996) Robotic food chains: Externalization of state and program for minimal-agent foraging. In: Proceedings of the 4th international conference on the simulation of adaptive behavior: From animals to animats 4. MIT Press, Cambridge, MA, pp. 625–634
27. Vaughan R, Stoy K, Sukhatme G, Mataric M (2002) Lost: Localization-space trails for robot teams. IEEE Trans on Robotics and Automation 18:796–812
28. Sauter J, Matthews R, Parunak H, Brueckner S (2002) Evolving adaptive pheromone path planning mechanisms. In: Proceedings of the 1st international joint conference on autonomous agents and multi-agent systems. Bologna, Italy, pp. 434–440

29. Kube C, Bonabeau E (2000) Cooperative transport by ants and robots. Int Journal of Robotics and Autonomous Systems 30:85–101
30. Kube C, Zhang H (1994) Stagnation recovery behaviors for collective robotics. In: Proceedings of the IEEE international conference on intelligent robots and systems. Munich, pp. 1883–1890
31. Rumeliotis S, Pirjanian P, Mataric M (2000) Ant-inspired navigation in unknown environments. In: Sierra C, Gini M, Rosenschein J (eds) Proceedings of the 4th international conference on autonomous agents. Barcelona, pp. 25–26
32. Howard A, Mataric M, Sukhatme G (2002) Localization for mobile robot teams using maximum likelihood estimation. In: Proceedings of the IEEE international conference on intelligent robots and systems. Lausanne, Switzerland, pp. 434–459
33. Arkin R, Balch T (1998) Cooperative multiagent robotic systems. In: Bonasso R, Murphy R (eds) Artificial intelligence and mobile robots. MIT/AAAI Press, Cambridge, MA
34. Balch T, Arkin R (1998) Behavior-based formation control for multi-robot teams. IEEE Trans on Robotics and Automation 14:926–939
35. Kuwana Y, Nagasawa S, Shimoyama I, Kanzaki R (1999) Synthesis of silkworm moth's pheromone-oriented behavior by a mobile robot with moth's antennae as pheromone sensors. Biosensors and Bioelectronics 14:195–202

5

Physically Realistic Self-assembly Simulation System

Vadim Gerasimov, Ying Guo, Geoff James, and Geoff Poulton

CSIRO ICT Centre, Autonomous Systems Laboratory, Australia first.last@csiro.au

Summary. The self-assembly research bridges the area of swarm optimization with practical applications in chemistry, nanotechnology, and biology. The swarm optimization often deals with idealized geometric and logical environments. It is practically impossible, however, to constrain any real environment to exclude phenomena that break the rules of such idealized environments. Our research explores how to bring swarm optimization methodology into the real-world self-assembly applications. We focus on experimental exploration of group behaviour of physical entities with programmable logical properties in various physical environments. Setting up such experiments in the real world is rarely feasible. Software simulations not only can achieve sufficient accuracy to predict self-assembly behaviour in the real world, but also offer more flexibility in manipulating the physical and logical parameters of the experimental model. We describe a software system to model and visualize 3D or 2D self-assembly of groups of autonomous agents. The system makes a physically accurate estimate of the interaction of agents represented as rigid cubic or tetrahedral structures with variable electrostatic charges on the faces and vertices. Local events cause the agents' charges to change according to user-defined rules or rules generated by genetic algorithms. The system is used as an experimental environment for theoretical and practical study of self-assembly. In particular, the system is used to further develop and test self-assembly properties of meso-blocks.

We also describe the architecture of the system and a set of experiments which explore passive aggregation and active directed self-assembly of meso-blocks. The experiments demonstrate sensitivity of self-assembly results not only to the logical programming of the agents and initial configuration, but also to physical parameters of the system.

The software system can be applied to the analysis, prediction and design of self-assembly behaviour of agents from atomic- to macro-scales. In particular, it may become a platform for developing design techniques that can be implemented in real nano-scale systems to achieve useful structures.

V. Gerasimov et al.: *Physically Realistic Self-assembly Simulation System*, Studies in Computational Intelligence (SCI) **31**, 117–130 (2006)
www.springerlink.com

5.1 Introduction

Behaviour of a multiagent system stems not only from intrinsic properties of the agents, but also from the properties of the environment in which the agents interact [7, 13]. Our purpose is to create a simulated environment mimicking laws of classical mechanics that allows us to engineer and explore self-assembly behaviours of artificial and real multiagent systems. In addition to allowing the logical properties of the agents to be altered during self-assembly, the system makes it possible to vary simulated physical features of both the agents and the whole environment. For example, the system allows the user to change viscosity and boundary properties of the simulated world as well as mass, moment of inertia, shape, and frictional properties of the agents. This chapter is focused on our work on meso-block self-assembly. However, the simulation environment can be applied to a broader range of problems and theoretical models.

The research on molecular self-assembly [16, 17] demonstrates the possibility of designing molecular behaviour that assembles pre-defined structures from disordered building blocks. The research also shows the difficulty of performing experiments in this area in the real world. We, therefore, propose to conduct the major part of the behaviour design using the simulated environment and to conduct real-world experiments only to verify the results.

A strong link between swarm optimization and self-assembly already exists in macro-scale robotic self-assembly research [19, 14, 15, 18]. This demonstrates applicability of swarm approach in complex environments. The agents in that research are more complex than the ones we are dealing with. They also have more processing power than we could fit into molecular-scale entities. One of the goals of our research is to understand the minimal complexity of the agents in a particular environment necessary to build a desired range of structures.

5.2 Implementation

The simulation environment Fig. 5.1 expands the capabilities of the 2D self-assembly environment used in the earlier meso-block research [1, 2, 3, 4, 9, 10, 11, 12, 13]. The new system replaces schematic idealized interaction rules with rules that realistically reflect the complexity of physical interaction between objects. The simulation environment is designed for 3D modelling and visualization, but can easily be applied to 2D scenarios. The system is developed in Borland Delphi™7. The primary focus of the system design is to create a flexible tool that demonstrates multiagent behaviour in complex physically-realistic environment. The user can adjust the system parameters and observe the dynamics of the behaviour in real time. The general approach to the physical simulation is similar to that of the Open Dynamic Engine (ODE) www.ode.org. However, as opposed to ODE the physical model is designed and optimized to particle interaction and extended to program the behaviour of the agents. Our system can achieve sufficient performance to calculate and visualize behaviour of up to several hundred agents in real time.

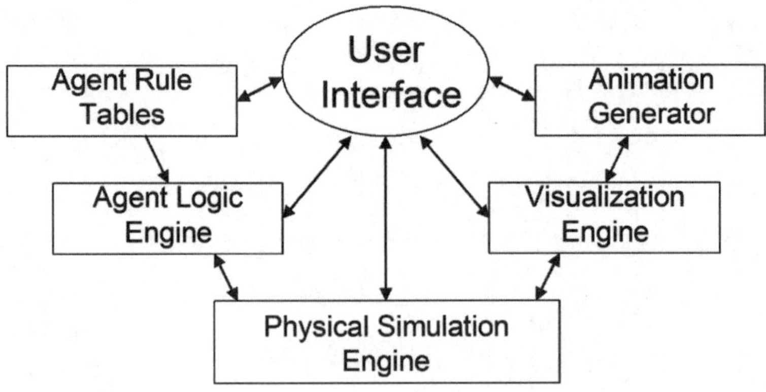

Fig. 5.1. System architecture diagram.

5.2.1 Agents

A representation of each simulated agent includes its position, momentum, and spin vectors as well as its rotation matrix (representing the current orientation) and mass. Since the agents may represent meso-blocks they also have a charge state associated with each face and/or vertex of an agent. The charge can be "+" - positive, "-" - negative, or "0" - neutral. In the current implementation all agents are rigid either cubic or tetrahedral structures with the charges located in the centre of each of the 6 cube faces or each of the 4 tetrahedron vertices. In this chapter we focus on the cubic meso-block agents. The state of meso-block cube charges can be described as a vector of 6 symbols:

$$Q = (a_1, a_2, a_3, a_4, a_5, a_6), \quad a_i \in \{-, 0, +\} \tag{5.1}$$

The cube face numbering is compatible with the 2D meso-block numbering used in earlier research [3] and is shown in Fig. 5.2. For 2D modelling the charge state can also be abbreviated as

$$Q = (a_1, a_2, a_3, a_4) \tag{5.2}$$

In this case a_5 and a_6 are 0.

The charges may affect motion of the agents by making them attract or repel each other. In the current implementation the forces obey the classical electrostatic laws: opposite charges (+1 and -1) attract and like charges repel. The neutral charge does not interact with other charges and does not affect the agent motion. A repulsive force operating at small inter-agent distances is also used to limit overlap between agents.

Each agent has a sensor at each charge location. The sensors detect proximity to other agent's charges and trigger changes of the agent state according to the rules outlined in the next section. The sensor can detect several different situations: detached (0), detaching (D), attached (1), attaching (A), and touching (T). The

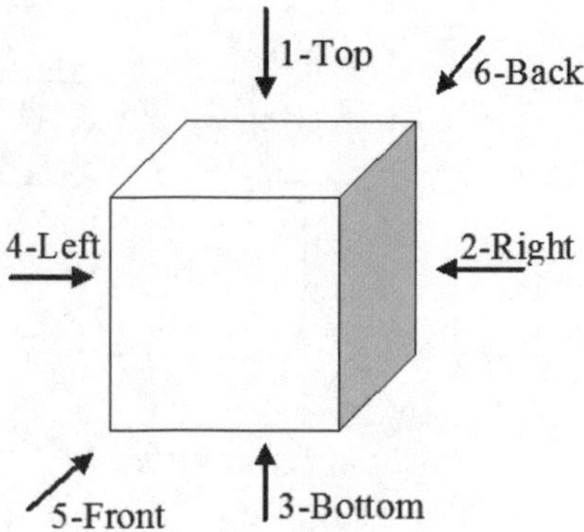

Fig. 5.2. Cube face numbering.

detached state is sensed when no other charges are within the sensor range. The attached state corresponds to the situation when an opposite charge of another agent is within the sensor range and the relative motion of the two agents is small enough to assume that the agents are going to stay connected if not disturbed. The sensor detects the touch state when another charge is within the sensor range, but the attached conditions are not satisfied. For example, this state may be detected when two agents bump into each other at high speed or when at least one of the charges is neutral and there is no attractive force between the charges.

Detaching is a transition from attached or attaching to detached or touch; similarly attaching is a transition from detached or touch to attached. The symbolic representation of the sensor states (0, D, 1, A, and T) is compatible with the 0 (unstick) and 1 (stick) symbols used in the meso-block rule definitions in the earlier work [3, 4]. The state of meso-block cube sensors can be described as a vector of 6 symbols:

$$S = (s_1, s_2, s_3, s_4, s_5, s_6), \quad s_i \in \{0, D, 1, A, T\} \tag{5.3}$$

The sensor states in 2D modelling can be abbreviated as

$$S = (s_1, s_2, s_3, s_4) \tag{5.4}$$

In this case s_5 and s_6 are 0.

The transitions to and from the attached state may depend on the relative velocity of the agents. To avoid noisy transitions while two attaching agents stabilize, the system implements a hysteretic gap between the speeds required to enter and to leave the attached state.

5.2.2 Rules

The logic of interaction between the meso-block agents is a superset of the one defined in the previous research [3]. The new system adds the ability to define charges and sensor states for 6 cube faces in the 3D case and expands the set of sensor states.

The charges of agents may change according to a set of rules. The rules define how the agents change their charges depending on their current charge configuration and the sensor states.

In general a rule can be written as

$$Q_s S \Rightarrow Q_r \tag{5.5}$$

where $Q_s$is the initial charge configuration, S is the sensor state, and Q_r is the resulting charge configuration. For example the rule

$$(-,-,-,-,-,-)(1,0,0,0,0,0) \Rightarrow (-,0,0,0,0,0) \tag{5.6}$$

changes a block with all negative face charges attached on side 1 and detached on all other sides to a block with the first side negative and all others neutral.

The rules can be added, deleted, or edited before or during the simulation. A complete set of rules is internally represented as a table that maps all possible charge and sensor states to resulting new charge states. The total number of entries in such a table is $N = 3^6 \times 5^6$. However, the number of rules used in most test scenarios is much lower. To make manual input of rules easier, the rule editor in the system allows the user to use a wildcard symbol (*) for sensor states as well as to define symmetrical rules. The symmetrical rules in the rule editor generate 24 entries in the internal rule table for the rotationally-symmetric situations. Symbol @ in front of a rule definition denotes a symmetrical rule.

5.2.3 Simulation process

The program has three principal processes: simulation of physical interaction among the agents, updates of the meso-block states according to the rules, and visualization.

The simulation of physical interaction is the most computationally intensive part of the system. The simulation runs in a loop recalculating the dynamic physical parameters of every object. At each step the system calculates forces acting between each pair of objects and then updates velocity and spin as well as rotational and linear position of every object. One of the most important parameters of the simulation is the time step used in the calculations. It corresponds to the time period that passes in the simulated physical system with each calculation cycle. Decreasing the time step improves the calculation precision, but makes the simulation run slower relative to real time. Increasing the time step makes the simulation faster, but less precise. If the time step is increased beyond a certain limit the simulation of physical system becomes unstable.

During the force calculation phase the system finds pairs of agents that are interacting with each other. In a real physical system electrostatic forces have no "cut-off limit" so that charged objects affect each other at any distance. However, the forces are inversely proportional to the square of the distance and may be negligibly small for objects that are far apart. The system includes an interaction cut-off parameter that can substantially increase the performance by limiting the force calculations only to the agents that are close together. This parameter defines the minimum distance between objects that enables force calculations.

The force between two face charges is calculated as

$$\mathbf{f} = \frac{kq_1q_2\mathbf{r}}{|\mathbf{r}|^3 + \varepsilon_c} \tag{5.7}$$

where $\mathbf{r}$ is the distance vector, q_1 and q_2 are the charges, k is the Coulomb's law coefficient, and ε_c is a small constant value that signifies that the charges do not interact as ideal points and allows the system to avoid infinite forces when the charges approach each other. Depending on the sign of the charges the force can be either attracting or repelling.

In addition to the electrostatic interactions the agents also repel each other with a force equal to

$$\mathbf{f_r} = \begin{cases} 0, & |\mathbf{b}| \geq d \\ \dfrac{cd\mathbf{b}}{|\mathbf{b}|^3 + \varepsilon_r}, & |\mathbf{b}| < d \end{cases} \tag{5.8}$$

$\mathbf{b}$ is the vector between the centres of mass of the agents, d is the size of the agent, and c is the repelling force coefficient. ε_r is similar to ε_c, but may have a different value. Similarly to the charge interaction force, the repelling force obeys the inverse-square law, but acts on the centres of mass of the agents and drops to 0 if the distance is greater than the diameter of the agents.

The system calculates a first order approximation of the solution for the motion differential equations.

The change of momentum for each agent at every time step is calculated as the product of the time step and the sum of all forces applied to the agent

$$\Delta\mathbf{p} = \Delta t \sum_i \mathbf{f_i} \tag{5.9}$$

The change of angular momentum for each agent at every time step is calculated as the product of the time step and the net torque vector

$$\Delta\mathbf{L} = \Delta t \sum_i \mathbf{f_i} \times \mathbf{l_i} \tag{5.10}$$

$\mathbf{l_i}$ is the force shoulder vector between the centre of mass and the point where the force is applied.

The position vector of every agent is then updated with

$$\Delta\mathbf{r} = \Delta t \frac{\mathbf{p}}{m} \tag{5.11}$$

The rotational orientation of every agent is updated as an absolute rotation around the current rotational momentum vector by the angle

$$\alpha = \Delta t \frac{|\mathbf{L}|}{I} \tag{5.12}$$

I is the moment of inertia. In the meso-block modelling the system uses the solid cube moment of inertia

$$I = \frac{md^2}{6} \tag{5.13}$$

To calculate the interaction forces among the agents the program has to perform an order of N^2 (where N is the number of agents) operations. The momentum, position, and rule updates require an order of N operations.

The physical interactions among the agents represent a balancing between potential and kinetic energy. The regular Newtonian calculations conserve the energy in the system. However, to make the system self-assemble energy has to be taken out of the system. The simulation system implements two ways of reducing the energy: gradual scaling down of the rotational and translational momenta of each agent and faster reduction of momenta of the agents that are close to each other. The former mimics viscosity of the intra-agent environment. The latter mimics friction between the agents.

All calculations performed in the simulation are based on Euclidian 3D math. The 2D simulations are also performed with 3D calculations where one of the x, y, or z coordinate components of all parameters of all agents is set to 0. When all forces, momenta, and positions have 0 as one of the coordinates the interaction calculations preserve the 0 coordinate and never deviate from the plane.

The user can choose one of the 3 possible boundaries for the simulated world: cube, sphere, or periodic. The cube boundary has cubic shape and bounces agents back inside preserving the kinetic energy. The sphere boundary has spherical shape and also bounces agents back preserving kinetic energy. The periodic boundary has a cubic visual space, but creates a toroidal space where opposite walls of the cube are linked. For example, agents moving through a wall reappear on the opposite side.

The meso-block rules can be executed either after each physical interaction phase or less frequently. In the current implementation the rules are executed synchronously with the physical interaction phases at every step.

The rules dictate transitions of an agent charge configuration depending on the sensor values.

5.2.4 Visualization

The visualization is performed in parallel and independently of the physical interaction and state update calculations. The visualization is currently implemented using MS DirectX API and uses 3D visualization hardware of the computer if present.

The user has control over the camera position in the visualization space and can choose how the agents are visualized. For example, the meso-block agents may be shown as coloured cubes that reveal the agent states or as monochromatic spheres that hide the charge structure of the system.

The camera position can be controlled by the mouse or keyboard in an intuitive way. The camera can be moved around as well as towards or away from the rendered scene.

In addition to observing the agent behaviour in real time the system can save a sequence of frames into an AVI file. Realtime observations are possible for groups of up to the order of 100 agents using an average 2003 PC. For larger numbers of agents or finer time steps the system can run in the background generating an animation of the simulation process over a long period of time.

5.2.5 User Interface

In addition to visualization controls the system includes a simulation control panel, the rule editor, and a menu system. The control panel includes "pause/resume", "cooler", "hotter", and "save movie" buttons. The "pause/resume" button controls the physical interaction simulation process, but does not affect the visualization, so that the scene can be dynamically observed from different angles while the agent motion is stopped. The "cooler" and "hotter" buttons are used to evenly add or remove kinetic energy from the system. The "save movie" button is used to initiate the AVI file saving process. The control panel also includes controls to dynamically switch between cubic or spherical visualization of meso-blocks as well as between spherical, cubic, or periodic boundaries. The control panel also allows the user to change the size of the enclosing boundary, cooling effect coefficient, friction coefficient, meso-block size and mass as well as other physical parameters of the system.

The rule editor can be used to dynamically update the meso-block rules of the simulated system as well as the initial configuration of the meso-block system (including the number and position of the blocks generated at the system reset).

The menu system duplicates the features available in the control panel and allows the user to save and restore sets of rules and sets of system parameters.

5.3 Experimental Results

We conducted a series of tests with a variety of simple meso-block configurations to debug the system and make sure that the block interaction is an acceptable approximation of the laws of physics. For example, one of the tests verified that the system preserves translational and rotational momenta with expected accuracy.

The objective of the preliminary experiments is to explore the range of possible self-assembly scenarios that can be modelled by the system and to replicate the result obtained earlier with the stochastic 2D simulator.

In all experiments the meso-block mass is 0.0003kg, the size is 0.002m. A typical simulation time step is in the range of $1 - 50\,\mu s$.

5.3.1 Aggregation

We observe a process we call aggregation when the meso-block rules are not used and blocks do not change their charge states. The most interesting structures assemble with blocks that have a combination of +, -, and 0 charges. We conducted a set of experiments with mixes of (+,-,0,0) and (+,0,-,0) 2D blocks. In this case all generated structures are either closed loops or elongated structures that have + on one end and - on the other.

The forward self-assembly simulation results (Fig. 5.3 left) are generated from a 50/50 – half-and-half – mix of (+,-,0,0) and (+,0,-,0) meso-blocks that are initially randomly placed and rotated within a cubic boundary. As opposed to previously used simulation environments the blocks do not have to connect at right angles only. The system can generate a broad variety of shapes. Slight changes in the initial conditions may lead to different sets of similar shapes. The structures generated in 3D are usually larger and geometrically more complicated.

Another example is aggregation of blocks that have a "checker" charge configuration: (+,-,+,-) in 2D or (+,-,+,-,+,-) in 3D. In this case the end result is usually a single fractal-like structure (Fig. 5.3 right).

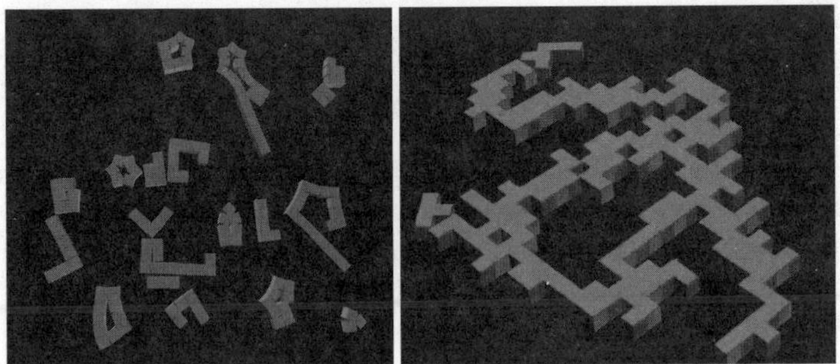

Fig. 5.3. Aggregation: half-and-half (left) and checker (right).

5.3.2 Snake

Another meso-block configuration we call "snake" includes a single rule and generates long chains of meso-blocks. Initially, one of the blocks (seed) has (+,0,0,0) charge configuration and all others ("sea" blocks) have (-,-,-,-). Any sea block can attach to the positive side of the seed. The rule

$$@(-,-,-,-)(1,0,0,0) \Rightarrow (-,0,+,0) \tag{5.14}$$

converts the attached block so that another sea block can be attached to the end of the chain.

We conducted experiments with the "snake" configuration using different friction, boundary, and moment of inertia parameters to see the range of possible behaviours.

In most cases the resulting structure is a single long chain of blocks (Fig. 5.4 left). Even though in many cases the chain may break apart in the middle, the pieces tend to eventually reattach restoring a single long chain.

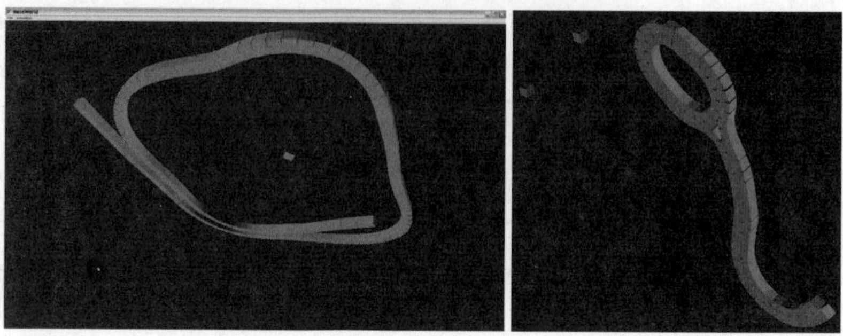

Fig. 5.4. Snake configuration.

In some cases in a high friction environment with a spherical or cubic boundary, a piece of the chain may break away and form a loop (Fig. 5.4 right).

We also discovered an interesting phenomenon of snake acceleration which is a side effect of setting an artificial limit to the range of action of the electrostatic field. To increase the update rate the simulation environment has an option of restricting electrostatic interaction calculations to the charges within a certain distance from each other. The electrostatic force effectively drops to 0 beyond this distance. The result of interaction of the charges of the snake with the edge of field from the sea blocks is to cause the snake to accelerate forward and the sea blocks to accelerate in the opposite direction.

5.3.3 Self-replicating enzymes

Our previous work explored how the meso-blocks could be programmed to assemble stable structures. A stable structure is an aggregate of meso-blocks attached to each other electrostatically in such a way that:

1. the structure does not have any free uncoupled charges that can interact with meso-blocks that do not belong to the structure;
2. the structure does not fall apart spontaneously during the self-assembly process.

The key concept for these structures is the artificial "enzyme". In the real world an enzyme is an organic protein which catalyses a specific reaction. Here, we use it in the context of self-assembling blocks to characterize an assembly of blocks with the capacity to produce another block assembly whilst itself remaining unchanged. The aim is for the product structures to serve as primitive building blocks for the self-assembly of more complex objects. As we have discovered, using enzymes leads to the more efficient production of a wider variety of structures than was possible with the earlier approach of [3]. A Genetic Algorithm is the main tool that we have used to generate enzymes.

The "enzyme" concept is introduced to develop a more flexible and directed method of producing basic building objects compared with the approach in [3], where the resulting structures are defined only by the rule set common to all "sea" blocks. By introducing an enzyme into the multi-agent environment, the structures which self-assemble depend also on the rule set and physical structure of the enzyme. Different enzymes will generate different final objects from the same "sea" blocks. Enzymes can also give control over where and when structures self-assemble, by appropriate placement and assuming that enzymes can be switched between active and inactive states.

The properties of an enzyme are defined as:

1. Each enzyme is a stable structure comprising K blocks.
2. Each block in the same enzyme has the same rule set.
3. An enzyme must remain unchanged after the self-assembly process. During the process it may change in size and shape, but it must cyclically return to its initial state.
4. Blocks in the environment ("sea" blocks), whilst identical, can have a different rule set from the enzyme blocks.

An example of an enzyme enabling a self-assembly process is shown in Figure 3. It is a 3-block structure sitting in a "sea" of blocks with all negative sides and an identical rule set. In addition to self-replication this enzyme produces 3x2-block stable structures that we call "6-packs". The enzyme has two positive charges on the inside that attracts a free sea block to initiate production (with 50/50% chance) of either a new enzyme or a new 6-pack.

We repeated some of the experiments from our earlier work [2, 4] on self-replicating enzymes. The results were similar to the ones obtained with the 2D stochastic simulation environment despite the substantial differences in the way the simulation systems work. However, the newer environment helped us to understand the range of physical parameters that would make the meso-block system stable. We could also see a few side effects and irregularities that are impossible in the previous simulation system, but would happen in any physical implementation of meso-blocks. For example, the blocks may attach at an unexpected angle or some connection can break apart because of the free interactions with the environment.

Fig. 5.5 shows an example of directed self-assembly of meso-blocks using top-down/bottom-up design. The meso-block rules are generated using a genetic algorithm to achieve a combination of self-replicating and neutral structures. The process starts with sea blocks and a single 3-block "enzyme" structure and ends with a mixture of self-replicated enzymes and "6-pack" neutral structures.

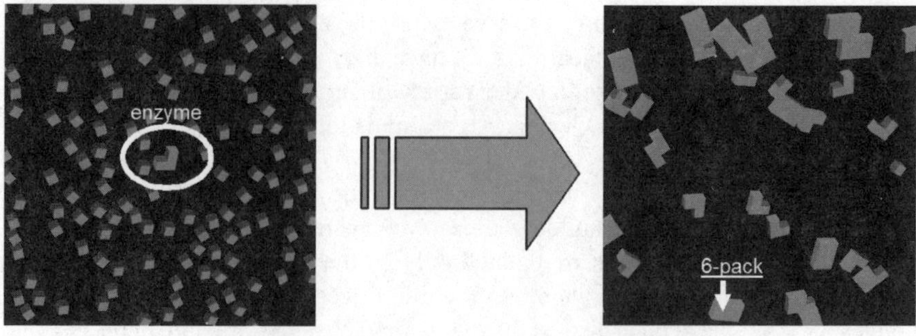

Fig. 5.5. Self-replicating enzyme.

5.3.4 Dynamic structures

The meso-block environment can also be used to explore behaviour of dynamic multi-agent structures. An example is a meso-block motor (Fig. 5.6). The motor consists of a preassembled frame with a sliding block inside. The frame has negative charges on two sides. The sliding block has a negative charge on one side, and positive on the opposite side. The sliding block changes its charges according to the following rule:

$$@(-,-,-,-)(1,0,0,0) \Rightarrow (-,0,+,0) \tag{5.15}$$

Frame blocks have no rules and therefore do not change state.

The sliding block should move towards one of the sides of the frame, then flip polarity and move towards the opposite side.

In the simulated environment the motor behaves as expected. However, under low-friction conditions the block may accelerate substantially deforming the frame. In some cases, the sliding block may even bounce one of the side blocks out of the frame effectively destroying the motor system.

Such dynamic structures can potentially mimic complex biochemical molecules such as proteins. They may also be used to create prototypes of nanomechanical systems.

5.4 Conclusion

The simulation system is a feature-rich tool for modelling and exploration of self-assembly and other phenomena in multi-agent systems.

The simulation environment bridges theoretical work in the area of multiagent systems with practical development in biology, chemical engineering, and nanotechnology.

The experiments conducted with the system have led to better understanding of self-assembly behaviour of meso-blocks and its dependence on the physical properties of the environment. We have explored the self-assembly possibilities

Fig. 5.6. Meso-block motor.

of our theoretical meso-block model and conducted a preliminary analysis of the presented self-assembly cases. The current meso-block model does not define means to store or transfer information among the agents. This somewhat limits the ability to assemble predefined shapes. The simulation system helps us to specify and test new additional features of the meso-block model to overcome the limitations.

The simulation environment has been used in combination with genetic algorithms to develop enzymes [4] and other static and dynamic structures with desired properties.

We expect to further develop the system to simulate other physical interaction models, such as quantum molecular interaction. For example, we would like to assess the possibility of simplified modelling of carbon atom interactions using the tetrahedral structures.

Another goal is to advance the theory of self-assembly in multi-agent systems, further explore swarm optimization approach in this context, and to find a robust physical implementation of self-assembly models.

References

1. Gerasimov V, Guo Y, James G, Poulton G, and Valencia P (2004) Multiagent Self-Assembly Simulation Environment. AAMAS 2004: 1382-1383
2. Poulton G, Guo Y, James G, Valencia P, Gerasimov V, and Li J (2004) Directed Self-Assembly of 2-Dimensional Mesoblocks using Top-down/Bottom-up Design. ESOA'04 Workshop AAMAS 2004, New York
3. Guo Y, Poulton G, Valencia P, and James G (2003) Designing Self-Assembly for 2-Dimensional Building Blocks. ESOA'03 Workshop AAMAS 2003, Melbourne, Australia
4. Poulton G, Guo Y, Valencia P, James G, Prokopenko M, and Wang P (2004) Designing Enzymes in a Multi-Agent System based on a Genetic Algorithm. 8th Conference on Intelligent Autonomous Systems, Amsterdam
5. Goldberg D (1989) Genetic algorithms in search, optimisation, and machine learning. Addison-Wesley Publishing Company, Inc., Reading, MA

6. Wolfram S ed. (1986) Theory and Application of Cellular Automata. World Scientific, Singapore
7. Haken H (1983) Synergetics. Springer-Verlag, Berlin
8. Garis H (1992) Artificial Embryology: The Genetic Programming of an Artificial Embryo. Chapter 14 in Dynamic, Genetic, and Chaotic Programming. ed. Soucek B and the IRIS Group, WILEY
9. Whitesides G, et al. (1999) Mesoscale Self-Assembly of Hexagonal Plates Using Lateral Capillary Forces: Synthesis Using the 'Capillary Bond'. J. Am. Chem. Soc., 121: 5373-5391
10. Whitesides G, et al. (1998) Three-Dimensional Mesoscale Self-Assembly. J. Am. Chem. Soc., 120: 8267-8268
11. Whitesides G, et al. (1999) Design and Self-Assembly of Open, Regular, 3D Mesostructures. Science, 284: 948-951
12. Raguse B (2002) Self-assembly of Hydrogel Mesoblocks. Personal Communication, CSIRO
13. Bojinov H, Casal A, Hogg T (2000) Multiagent Control of Self-reconfigurable Robots. Fourth International Conference on Multi-Agent Systems, Boston, MA
14. Gross R, Bonani M, Mondada F, and Dorigo M (2006) Autonomous Self-assembly in a Swarm-bot, Proc. of the 3rd Int. Symp. on Autonomous Minirobots for Research and Edutainment (AMiRE 2005), Springer Verlag, Berlin, Germany, 314-322
15. O'Grady R, Gross R, Bonani M, Mondada F, and Dorigo M (2005) Self-assembly on demand in a group of physical autonomous mobile robots navigating rough terrain, Proc. of the 8th European Conf. on Artifical Life, ECAL 2005, volume 3630 of Lecture Notes in Artificial Intelligence, Springer Verlag, Berlin, Germany, 272-281
16. Griffith S (2004) Growing machines. Ph.D. Thesis, Massachusetts Institute of Technology, Cambridge, MA
17. Griffith S, Goldwater D, and Jacobson J (2005) Self-replication from random parts. Nature 437: 636
18. Zykov V, Mytilinaios E, Adams B, and Lipson H (2005) Nature 435: 163-164
19. Dorigo M and Sahin E (2004) Swarm robotics - special issue editorial. Autonomous Robots, 17(2): 111-113

6

Gliders and Riders: A Particle Swarm Selects for Coherent Space-Time Structures in Evolving Cellular Automata

JJ Ventrella

Ventrella@earthlink.net
http://www.ventrella.com/Alife/Cells/GlidersAndRiders/Cells.html

Summary. Cellular Automata (CA) which are capable of universal computation exhibit complex dynamics, and they generate gliders: coherent space-time structures that "move" information quanta through the CA lattice. The technique described here is able to evolve CA transition functions that generate glider-rich dynamics. Pattern-recognition algorithms for detecting gliders can be used as a way to determine the relative fitness among a population of CA rules, for use in a genetic algorithm (GA). However, digital image-based techniques for this purpose can be computationally expensive. In contrast, this technique is inspired by particle swarm optimization: the particles guide evolution within a single, heterogeneous 2D CA lattice having unique, evolvable transitions rules at each site. The particles reward local areas which give them a "good ride", by performing genetic operators on the CA's transition functions while the CA is evolving. It is shown that the resulting dynamics converge numerically to the *edge of chaos*, using a measure of the lambda value. This technique is not only efficient in evolving glider-rich CA of great variety, but it also models a kind of symbiosis: the swarm and the CA dynamics engage in a mutual evolutionary dance. Using a swarm to evolve gliders places the search for universal CA into a new context. A proposal for detecting types of glider collisions, aided by swarm communication, is offered, which could be used as a tool to better understand emergent computation.

6.1 Introduction

From its earliest beginnings, discrete dynamical systems known as *cellular automata* (CA) have been used to explore the frontiers of theoretical biology. John von Neumann's self-replicating automata is one of the earliest examples, created at the dawn of the computer age. John Conway's *Game of Life* popularized CA and created a generation of casual enthusiasts, after a series of Scientific American articles by Gardner [4] were published on the subject. Much research has been devoted to identifying the numerous structures that emerge as a result of applying the relatively

JJ Ventrella: *Gliders and Riders: A Particle Swarm Selects for Coherent Space-Time Structures in Evolving Cellular Automata*, Studies in Computational Intelligence (SCI) **31**, 131–154 (2006)
www.springerlink.com

simple rules for *Game-of-Life* [13]. In particular, the 5-cell, diagonally-moving structure (the "glider"), has become a familiar emblem in the CA community.

Wolfram, in identifying four classes of CA dynamics (1. homogeneous, 2. periodic, 3. chaotic, and 4. complex), suggested that universal computation ("universality") was possible in class 4 automata [23]. This is the class that supports gliders (of many varieties). It has been shown that the interactions of multiple gliders, and other structures, in controlled arrangements in Conway's *Life* Universe, can represent the logical operators necessary for universality [1, 16]. Smith [19] and others have developed other CA rules which are capable of universality. An overview of CA's used in computation is given by Mitchell [10]. The existence of gliders is an indication that a CA is poised at the "edge of chaos", where universality is likely to occur. Are gliders in fact *necessary* in order for universality to exist? Some form of propagating pattern at least would have to be, in order to carry signals across the CA lattice.

Langton [9] introduced the lambda (λ) parameter, which places Wolfram's classes in a logical order whereby the rules are organized along a parametric continuum, and demonstrated that class 4 CA are actually located between Wolfram's classes 2 and 3 (between order and chaos). He places this observation in the context of the origin of life, as the spontaneous emergence of an information-dynamic, at this critical edge.

Genetic algorithms (GA's) have been used to evolve complex CA. Building upon earlier experiments by Packard [12], Mitchell et al. [11] have developed GA's to evolve populations of CA rules. The resulting dynamics are able to perform simple computational tasks. Bilotta et al. [2] employ a GA to evolve self-replication in CA. Sapin et al. [17] used evolutionary algorithms to evolve a universal CA.

The method for evolving glider-rich CA dynamics described here employs a biologically-inspired searching technique, based on the behavior of many simple agents (a particle swarm). Using this technique, a single non-uniform CA evolves - each cell having its own unique transition function. Starting from a chaotic, heterogeneous state, it evolves through various stages, having clumps of competing homogenous regions, and eventually gives way to a single complex global dynamic. This does not use the standard GA, as in most of the research to date, but instead applies genetic operators asynchronously over time in a single simulation run. This is explained in the next section.

6.1.1 Particle Swarm as Catalyst for Evolution

Particle Swarm Optimization (PSO) [7] has proven to be not only effective in many problem domains but it is also evocative as a paradigm for modeling adaptation and intelligence - "intelligence" being a term that applies not only to brains, but to all adaptive systems comprised of many parts - human societies and bee hives included.

This paper introduces a technique which has similarities with PSO, whereby a swarm of 2D particles selects for evolution of coherent space-time structures in 2D CA dynamics. The swarm is superimposed upon the space of the CA. The particles are affected by the CA dynamics, and also directly affect the evolution of the CA. They do this by manipulating the transition functions (genetics) of the cells in the lattice, applying the GA operators of selection, reproduction, mutation, and crossover.

Each particle moves freely over the CA lattice. At every time step, a particle occupies the area corresponding to a unique cell in the CA lattice (imagine a marble rolling across a square-tiled floor). Particles are attracted to "live" cells, and can detect them if they exist within a local neighborhood. These live cells generate an attraction force which is applied to the particle's velocity. The quality of the particle's velocity is then subject to interpretation by the particle itself, based on a selection criterion. When a superior quality is found, the particle stores the selection criterion value along with the genes from its associated cell. (analogous to *pBest* in PSO). Copies of the selected genes are continually deposited back into the CA over time as the particle continues to move about. When two particles come in contact with each other, the particle with the highest value gives a copy of its genes to the other particle. This local, social exchange reinforces the effect of selection. There is no concept of a global best (analogous to *gBest* in PSO).

Soon after initialization, local regions of relatively homogeneous dynamics emerge. Some of these are dense, highly chaotic, and effectively "trap" particles - a behavior that has the effect of preserving genetic identity in these regions. Some of these local regions give birth to short-lived propagating structures, or "proto-gliders". These proto-gliders have the effect of transporting the particles across greater distances, thus spreading further the genes that gave birth to them. These regions are rewarded by way of particles selecting and reproducing their genes. Often in a simulation run, after an initial phase characterized by localized homogeneous areas, a secondary phase takes over in which new gliders begin to expand the genetically homogeneous regions from which they originated. This phase can often result in rapid evolutionary convergence, and a quick take-over by a transition function supporting a dynamics characterized by distinct gliders propagating against a quiescent background.

To personify the goal of a particle, it is always "looking for a good ride", and so the particles have been given the nickname "riders". In turn they reward local areas of the CA for a good ride by distributing the genes associated with that ride.

This extends the basic concept of a previous simulation [21] which includes a population of physically-based organisms that evolve locomotion through pressure to mate. Similarly, in this simulation local gene pools emerge from within a heterogeneous initial state, and the gene pools which give birth to agents with superior locomotion have the competitive edge, and thus spread. In this case the agents of locomotion are *gliders*.

6.1.2 Interactive Swarm vs. Image-Based Analysis

The human eye-brain system is well suited for detecting coherent space-time structures, and the subtle differences between CA dynamics are readily appreciated while watching a CA in real-time on a computer screen. In a previous exploration in evolving gliders, an interactive evolution interface was developed and published online as an application, called "Breeding Gliders with Cellular Automata", [20]. With repeated viewings of the dynamics of a population of CA rules, the user is asked to judge them, and is encouraged to respond according to aesthetics. The users choices become the selection operator for a modified GA. Favored dynamics tend to be the most complex and have the most gliders, perhaps due to the *biophilia* [22] of life-like motion. The question of how to automatically select for *gliderhood* was considered, and this lead to the current investigation.

Wuensche [26] has developed a suite of techniques for finding gliders in CA dynamics. This includes a visualization tool for filtering out background patterns to detect coherent space-time structures. Similar techniques have been used by Das et al. [3] for detecting boundaries within space-time domains for their experiments.

Digital image filtering of CA dynamics has the potential to draw from a long tradition of image processing and pattern recognition techniques, for the purpose of determining a fitness metric to apply to a GA. It's likely that a high level of sophistication can be achieved for classifying CA dynamics qualitatively, using convolution filters, etc., especially when applied as a post-process: after the dynamics have had sufficient time to generate space-time structures.

An earlier stage of this investigation incorporated a simple convolution filter for detecting structures as they emerged, but it proved to be computationally expensive, particularly since it was executed continuously in simulation time. Kazakov and Sweet [6] developed a GA to evolve CA with stable "creatures". They used entropy-variance as a base fitness function. But a secondary fitness function was used for identifying coherent structures which requires storing patterns in memory as they emerge – a process which the authors admit is very expensive to compute.

After articulating the motivations for using pattern-recognition while the dynamics emerged, a conclusion was drawn: such image-based techniques are not only expensive, but they also model high-level visual systems, and are thus not particularly germane to the subject of interest.

In contrast, an interactive swarm such as the one described here is characterized by a synergistic relationship with the problem domain (exploring the spontaneous emergence of complexity in primordial life – a scenario which almost certainly did not include the gaze of eyes or brains).

The work of Ramos and Almeida [14], and Ramos and Abraham [15] in modeling artificial ant colonies to detect features in digital images, using pheromone modeling, provided further motivation for taking a swarm approach. Such techniques use collaborative filtering, and demonstrate *stigmergy*, a principle introduced by Grassé [5], to explain some of his observations of termite nest-building behaviors. Stigmergy recognizes the environment as a stimulus factor in the behavior of an organic collective, which in turn, affects that environment.

6.1.3 Motion

The smooth pursuit system of mammalian eyes for tracking objects may have evolved as a consequence of, or in tandem with, the evolution of flying insects, slithering snakes, and the movements of herds in a field. Eye-brain systems are good at sensing motion. This particle swarm may be seen as modeling, in a loose sense, the smooth-tracking of a visual system.

Life itself could be described in terms of motion. Langton asked: "In living systems, a dynamics of information has gained control over the dynamics of energy, which determines the behavior of most non-living systems. How has this domestication of the brawn of energy to the will of information come to pass?" [9] Life is the motion of matter and it is also the motion of information. The technique described here is founded on such a perspective on the emergence of life, as moving objects (particles) and moving information (gliders) interact in a rain-dance of emergent complexity. If Fredkin, Wolfram, and others are right, everything after all can be reduced to discrete interactions in a vast cellular space [24], in which case, the swarm particle and the glider represent two ways of modeling motion – just at different levels.

6.1.4 Purpose

The technique described here serves two primary purposes:

1. It demonstrates a way to efficiently search for CA rules that support dynamics of Wolfram's class 4. These CA are based on a large set of possible transition functions, and an arbitrary number of states. Having tools for finding glider-rich CA, as Wuensche suggests, can be an aid in the search for universality [26]. This is an alternative to image-based pattern-recognition schemes, which can be computationally expensive, and are modeled on high-level vision.

2. This technique sets up an environment for asking questions about symbiosis and interaction between multiple agents in a dynamical system. It is a model of emergent complexity and co-adaptation.

6.2 A Description of the CA

The computer simulation incorporates a 2D lattice of automata (cells) arranged in a square grid. The domain has periodic boundary conditions, and so opposing boundaries wrap around, forming a torus topology. Time is measured in discrete steps t. A cell can assume any state in the range from 0 to the number of possible states K. State 0 is called the *quiescent* state, denoted by Q. At each time step the state of each cell can be changed to another state according to a transition function.

6.2.1 The Transition Function

Each cell has a unique transition function. The 9-cell Moore neighborhood is used as a local environment to determine the new state of the cell at each time step. The transition function consists of R counting rules which are applied in sequence. (So as not to confuse these rules with the transition function itself, which is often called the "transition rule" in CA language, these will be referred to as "sub-rules").

A sub-rule is expressed as follows: the cell's current state is compared to a reference state (the *referenceState*). If the cell's state does not match *referenceState*, then the sub-rule is not applied. Otherwise, the sub-rule compares the number of neighbors having a specific state (the *neighborState*) to a specific number (the *neighborCount*). If there is a match, then the sub-rule sets the cell's new state to a specific result state (the *resultState*).

Thus, four parameters are used for each sub-rule, which are *genetic*, i.e., they can have a range of possible values, and can be changed by a genetic operator. The ranges are:

1. *referenceState* can be any state in K
2. *neighborState* can be any state in K except for Q
3. *neighborCount* can be any number from 1 to M
4. *resultState* can be any state in K

K has been tested with values as high as 20 and the number of sub-rules R has been tested with values as high as 80 – with these values, the number of genes per transition function, $4R$, can be as high as 320. The set of all genes for a transition function is referred to as the *gene array*.

Algorithm 1. shows how the sub-rules are applied to determine the new cell state.

Algorithm 6.1 Cell Transition Function

```
Set the value of new cell state to quiescent
For each subRule
    If current cell state EQUALS subRule.referenceState THEN
        IF number of neighbors of state subRule.neighborState
        EQUALS subRule.neighborCount THEN
            SET value of new cell state to subRule.resultState
        End If
    End If
End For
```

To place this in a familiar context: the transition function would require a minimum of three separate sub-rules to define *Game-of-Life*, and the gene values would be as shown in Table 1.

Table 6.1. The minimal sub-rule genes for *Game-of-Life*

sub-rule#	referenceState	neighborState	neighborCount	resultState
1.	0	1	3	1
2.	1	1	2	1
3.	1	1	3	1

Or:
1. A dead cell (of state 0) with 3 live neighbors (of state 1) is born (1).
2. A live cell with 2 live neighbors remains alive.
3. A live cell with 3 live neighbors remains alive.
(In every other case, the cell is dead).

The minimal gene array representation is thus (013111211131). Of course, the essence of *Game-of-Life* can be expressed with fewer parameters, using more elegant notations, but the purpose of this experiment is not elegance – it is to evolve potentially complex and flexible CA rules from random initial conditions.

Thus, each cell's unique transition function is represented as a gene array, consisting of $4R$ integer values (genes), whose ranges are given above.

6.2.2 Sub-rule Redundancy and Over-Writing

Given a large number of sub-rules, and the usual set of initial random values, it is often the case that some of these sub-rules are redundant – having the exact same values. Furthermore, it is likely that many of these sub-rules, as they are applied in sequence, will "over-write" the results of previously applied sub-rules. This may seem inefficient – however, it allows for better evolvability – a more elastic genetic space within which experimentation can occur. Throughout the course of evolution, there can exist "latent" sub-rules that can be activated by way of crossover or mutation. Having some sub-rules exist in this latent form allows the genotype to more freely adapt.

Implicit in this representation is the logical operator, OR: if any two sub-rules have the same value as their *referenceState*, and the same value as their *resultState*, then the algorithm could conceivably be re-written to combine the two associated comparisons using an OR in one expression. The choice to leave out this structure allows more flexibility in terms of how the sub-rules are applied.

6.3 The Particle Swarm

The swarm consists of n (typically 500) particles. Unlike the cells of the CA, the particles occupy a real number space (using double-precision variable types), and can thus move continuously. The square domain which the particles occupy is normalized to size 1, and maps to the CA lattice, as illustrated in Figure 6.1.

Particles have position and velocity, and no representation of direction or orientation. Similar to the CA, particles have periodic boundary conditions, and so if a particle falls off an edge of the domain, it re-appears at the opposite edge, and maintains its velocity.

Velocity and position are updated every time step t as follows:
for the ith particle...

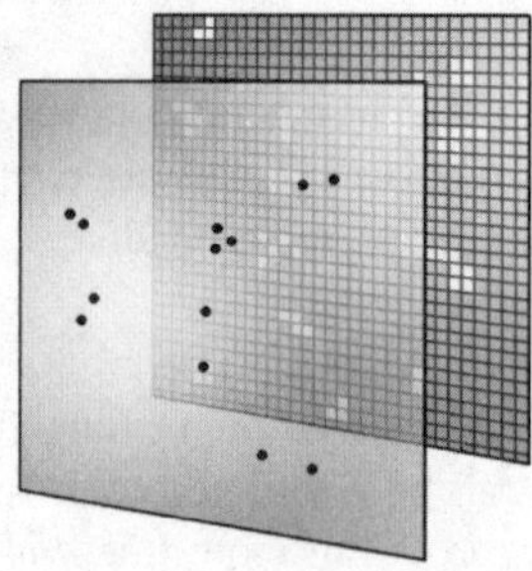

Fig. 6.1. The particle swarm maps to the CA lattice

$$\mathbf{v}_i(t) = \mathbf{v}_i(t\text{-}1)\omega + \mathbf{a}_i + \mathbf{r}_i$$

$$\mathbf{p}_i(t) = \mathbf{p}_i(t\text{-}1) + \mathbf{v}_i(t)$$

$i = \{1, 2...n\}$, where n is the number of particles. **p** is position, **v** is velocity, and ω is the inertia weight ($0 \leq \omega \leq 1$). Two forces: **a** and **r**, are added to the velocity at every time step, and are associated with the modes of riding, and *not* riding, respectively.

To keep the particles from going out of control as forces **a** and **r** are added to **v** over time, **v** is scaled at each time step by ω. Shi and Eberhart [18] introduced an inertia weight to the standard PSO technique, which affects the nature of a particle's "flying" behavior. The particles in this scheme perform different tasks than in PSO (they don't fly, they *ride*). In either case, the precise tuning of the ω is important for optimizing swarm performance.

6.3.1 The Particle Cell Neighborhood

Due to the mapping of particle space to CA space, at every time step a particle occupies a unique cell in the CA lattice. The cell that the particle occupies at any given time is the called its "reference cell". The particle continually reads the contents within a Moore neighborhood of radius 2 (size = *PN* = 5X5 = 25 cells) which surrounds, and includes, the reference cell, as shown in Figure 6.2. (Note that this neighborhood is larger than the neighborhood for the CA transition function). If one or more of the cells in the neighborhood are non-quiescent, the velocity of the particle is accelerated by an attraction force **a**, which is the sum of all direction vectors from the reference cell to each non-quiescent cell in the neighborhood (it is then normalized, and scaled by the attraction weight *aw*).

The attraction weight *aw* is an important value, just like ω. Both of these in combination affect the quality of particle motion, and the ability to track propagating non-quiescent structures.

When a particle's velocity **v** is being accelerated by force **a**, the particle is said to be "riding", and the "quality" of the ride q ($0 \leq q \leq 1$) is determined by the selection

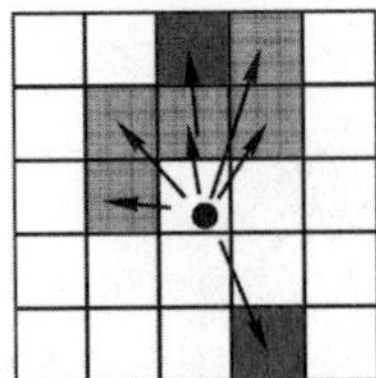 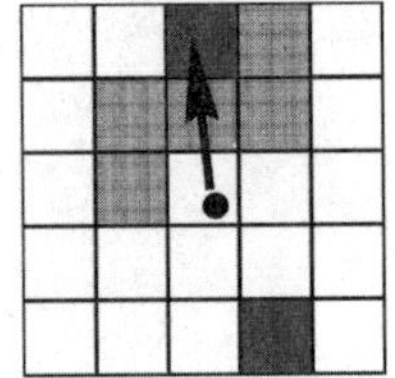

Fig. 6.2. An illustration of a particle cell-neighborhood containing 7 non-quiescent cells. Left side shows the 7 resulting attraction vectors. Right side shows the vector which is the result of summing these vectors and then normalizing the sum

criterion *S*. In most experiments, *S* is configured so as to favor a constant velocity and a maximum speed. This is explained below.

When the particle is not riding, a random force vector, **r**, is added at each time step. It has magnitude *rw* (random weight), and one of four possible orthogonal directions, randomly chosen at each time step. This force causes the particle to meander, Brownian-style, when it finds itself in a quiescent background. This random force allows momentary dense clumps of particles to dissipate. Higher values of *rw* cause faster dissipation.

6.3.2 Particle Re-birth

Every particle has a random chance of being re-born, at a rate of b ($0 \leq b \leq 1$) per time step. When a particle is re-born, its position is reset to a new random location, it's memory *bq* is set to zero, and it selects the genes from its new reference cell. This is explained below.

6.3.3 The Genetic Operators

Particles collect, carry, and exchange gene arrays originating from the CA, and deposit them back into the CA. But, while the particles are very active in manipulating genetic information, it is important to point out that they do not actually use these genes for their own purposes – they are entirely at the service of the CA and its evolution.

Particles perform four genetic operators, illustrated in Figure 6.3 and explained below.

1. Selection

A particle will store the genes from its reference cell when it experiences "the best ride in its life" – the best q found so far (stored as *bq*). If at any point a better q is found, it again stores the genes from its reference cell, and the new q as *bq*. This value is analogous to the "pBest" value used in PSO. In GA terms, the value of *bq* and the gene array are effectively a fitness metric and the genotype associated with that fitness.

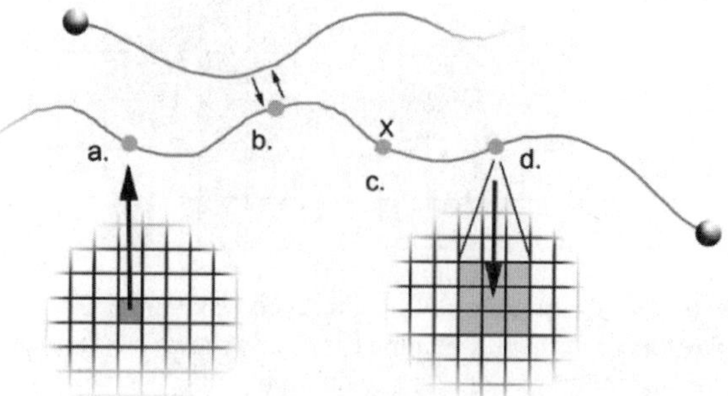

Fig. 6.3. The four genetic operators: (a) selection, (b) exchange, (c) mutation, and (d) reproduction

2. Reproduction

Particles deposit random portions of their gene arrays back into the CA. The rate of depositing is d ($0 \leq d \leq 1$) random chance per time step. This operator uses a variation on the crossover from the standard GA except that it's one-directional: genetic information originates from the particle only, and is transferred to a local Moore neighborhood of radius 1 (size = 9 cells) centered on the particle's reference cell. This is illustrated in Figure 6.4.

A Boolean *crossover switch* value is randomly set to true or false. Then, for each of the nine cells of the reference cell's Moore neighborhood, the particle gene array values are read off one-by-one, and either copied or not copied into the corresponding values of the cell gene array. During this process, *crossover switch* may be randomly switched at a rate of c (the crossover rate), determining whether of not the gene will be copied.

The effect of this operator is that approximately half of the genetic information originating from the particle is spread out into a local region of the CA. This also effectively introduces a form of sexual reproduction, and allows potentially successful new genotypes to emerge from the combination of gene array building blocks. Repeated crossover operations in the same region with similar genetic information increases genetic homogeneity in that region.

3. Exchanging Genes (reproduction via social communication)

Every particle has a random chance per time step e ($0 \leq e \leq 1$) of exchanging genes with another particle (if it comes in contact with another particle, i.e., the distance between the two particles is less than one cell width). When an exchange occurs, a copy of the genes of the particle with the higher bq value replaces the genes of the particle with the lower bq value. This exchange provides an extra level of reproduction, taking advantage of the swarm as a collective. In PSO, the best value within a local particle neighborhood is typically called *lBest*. While no such value

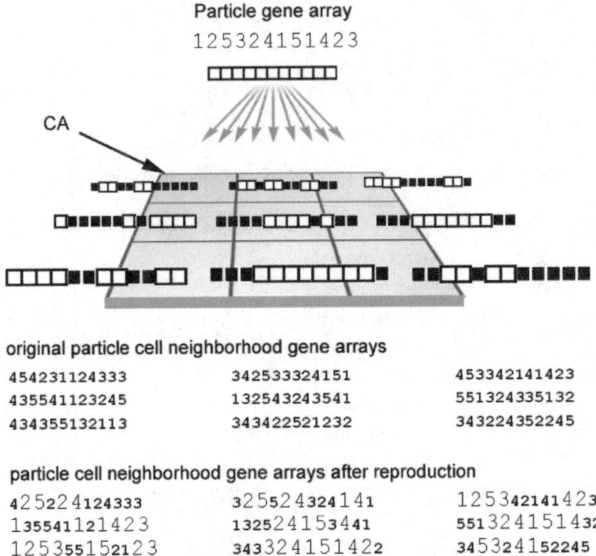

Fig. 6.4. An illustration of how a particle deposits portions of its gene array into its cell neighborhood, using a "one-way" version of crossover, resulting in random portions of each of the cell's gene arrays (shown in **small bold**) being replaced with corresponding portions of the particle's gene array

exists in this method, this social exchange effectively causes a local best to emerge as a result.

4. Mutation

Each particle has a random chance of genetic mutation, at a rate of m per time step ($0 \leq m \leq 1$). When a mutation occurs, exactly one of the genes in the particle's gene array is reset to a random value within its full range.

6.4 Experiments

All experiments were organized as controlled variations from a base experiment. The configuration for the base experiment is shown in Table 6.2.

At initialization, all cell gene arrays are randomized, and the particles are evenly distributed in random positions. The simulation begins with total quiescence.

6.4.1 The Periodic Noise Wave – Stirring the Primordial Soup

The one and only *stimulus* used in this simulation is a linear wave of random non-quiescent cell states which sweeps across the entire CA domain at a constant rate.

Table 6.2. Base Experiment Configuration.

Domain	Symbol	Description	Value
CA	*resolution*	number of cells in both dimensions	128
	K	number of possible cell states	4
	R	number of transition function sub-rules	16
	M	maximum neighbor count	5
Swarm (Motion)	*PN*	particle cell neighborhood size	(5X5 = 25)
	ω	inertia weight	0.7
	aw	attraction weight	0.4 / *resolution*
	rw	random weight	0.06 / *resolution*
Swarm (Genetics)	*m*	mutation rate	0.0001
	d	reproduction rate	0.05
	c	crossover rate	0.3
	e	exchange rate	0.1
	b	re-birth rate	0.001

The width of the wave is equal to the width of 1 cell. Density $w = 0.2$ in most simulations (a density of 0 has no effect, and a density of 1 creates a solid wave of non-quiescent cells).

The wave repeats its sweep every $W = 500$ steps. The wave moves twice as fast as the speed of light *C* (the sweep lasts *resolution*/2 time steps: it effectively leap-frogs over alternating cells). Moving faster than *C* reduces interference with the resulting dynamics – which can never have features that propagate faster than *C*. From a local Moore neighborhood vantage point, the noise wave appears as a sudden storm lasting no longer than 2 time steps, in which a subset of cells may acquire random non-quiescent states. The noise wave remains dormant between sweeps. This period of dormancy allows the dynamics to settle to its native behavior.

The CA can sometimes be in a highly chaotic state – either because *R* is large and the transition functions are generating a high proportion of non-quiescent states, or a chaotic transition function has temporarily dominated the CA. In this case, the noise wave is not necessary, as the soup is effectively stirring itself. However, the dynamics inevitably shifts towards the *edge of chaos*: it approaches that critical boundary in which order is just over the edge, and so the dynamics can easily slip into quiescence in large regions of the CA. The periodic noise wave in this case helps to provide some momentarily stimulation. In evolved CA that have strong quiescence and a semi-stable population of gliders, the noise wave is no longer required.

6.4.2 The Selection Criterion

A preliminary simulation was run in which the value from the selection criterion *S* was intercepted by an arbitrary random weight (i.e., it was not based on any property of the CA dynamics). When seeded with initially random genes, the result, as expected, was nothing more than a patchwork of mostly chaotic regimes formlessly changing boundaries. This scenario did not change after many tens of thousands of time steps. This reveals that the formation of homogenous regimes occurs simply as a function of locality, as the particles select genes (indiscriminately) and deposit them nearby.

In all other experiments S is set to “constant and fast”, favoring non-quiescent structures that move at a constant velocity, as close to the speed of light as possible. Its measurement, the ride quality per time step, q, is defined as...

$$q_i = (|\mathbf{v}_i| - |\mathbf{d}_i|) / C$$

where **v** equals velocity; **d** equals the change in velocity since the previous time step; and C equals the speed of light (equal to the width of one cell). With this selection criterion, if particle i is riding a glider that is moving steadily at the speed of light, then q_i will approach 1. Due to the usual settings for ω and aw, $|\mathbf{v}_i|$ is rarely $> C$, and so q is rarely > 1. Also, $|\mathbf{d}_i|$ is rarely $> |\mathbf{v}_i|$, and so q is rarely < 0. Still, q is clamped ($0 \leq q \leq 1$).

6.4.3 Simulation Output

Figure 6.5 shows four stages of a typical simulation run: (1) initial state; (2) clustering at 4,000 time steps; (3) gliders emerge at 30,000 time steps, and (4) convergence at 50,000 time steps. Particles are barely visible as small white dots. Non-quiescent cells are seen as shades of gray against a dark background of quiescent cells.

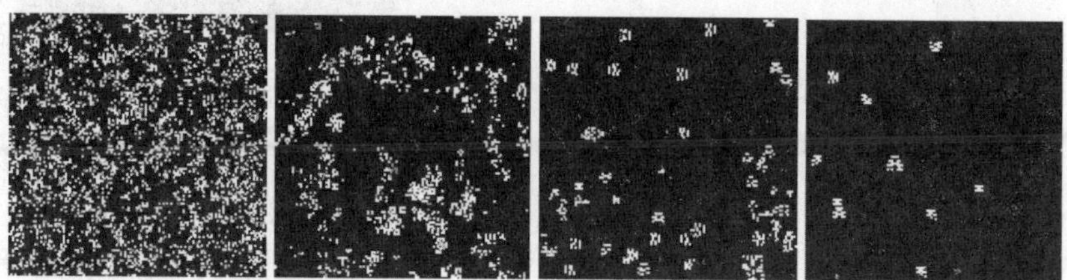

Fig. 6.5. Four stages of a typical simulation run

Figure 6.6 shows the output of a simulation run for approximately 100,000 time steps. At the left is a graph plotting the average ride quality of all particles *which are riding* (aq). At the right is a snapshot of the resulting CA dynamics.

Figure 6.7 shows the output of a simulation run for approximately 70,000 time steps. A sudden jump in aq can be seen after 30,000 time steps. This is attributed to the emergence of a local gene pool of transition functions that produced superior gliders.

Figure 6.8 shows the results of a simulation in which R was increased to 32 to encourage more complex transition functions. At around time 25,000 (Figure 6.8a) solid, long, and narrow structures become dominant. At around time 65,000 (Figure 6.8b) the solid structures have broken into individual gliders that are emanating rapidly from small localized chaotic regions, in opposite directions (indicated by

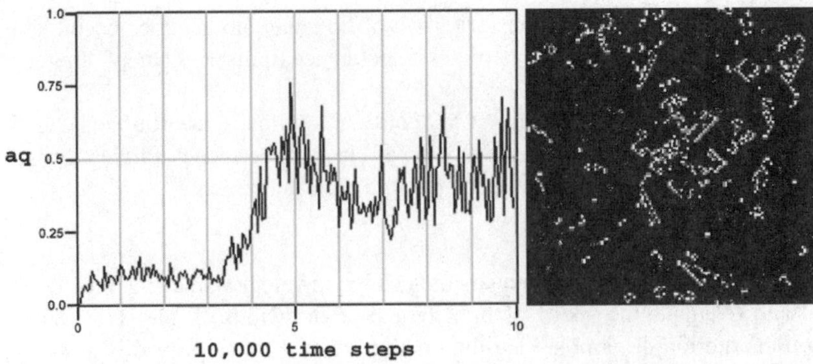

Fig. 6.6. A graph of average ride quality for a simulation – image of final CA dynamics at right showing "Brain-like" structures.

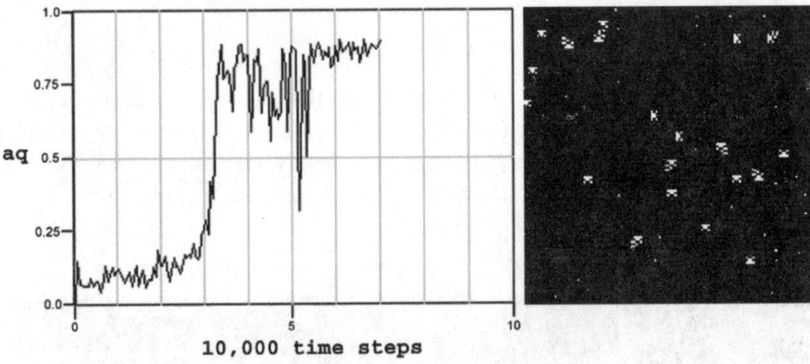

Fig. 6.7. Another simulation run, with resulting glider dynamics

the arrows). These chaotic regions may have different genetic signatures than the surrounding areas, causing a sort of nucleation for growth, and acting as *glider guns*.

Because the structures seen in 6.8b are separated by quiescent gaps, the particles are able to detect a better ride, due to more coherent attraction forces originating from variance in their cell neighborhoods. Note the increase in *aq* at around 60,000 in the graph (Figure 6.8d). By time 100,000 (Figure 6.8c), a uniform global dynamic has taken over – and the final scenario consists of small, sparsely-distributed gliders.

Figure 6.8e shows a close up of the box drawn in Figure 6.8c. A glider can be seen moving upward. It has recently passed through a cluster of particles, many of which have "caught a ride" on the glider because they were located in its path. Some of these particles were only temporarily propelled by the passing glider, and are consequently left trailing behind, while others can be seen riding on the back end of the glider. The gliders in this CA are smaller than the average glider. Larger gliders more easily collect particles because they occupy more non-quiescent cells in the particle's cell neighborhood.

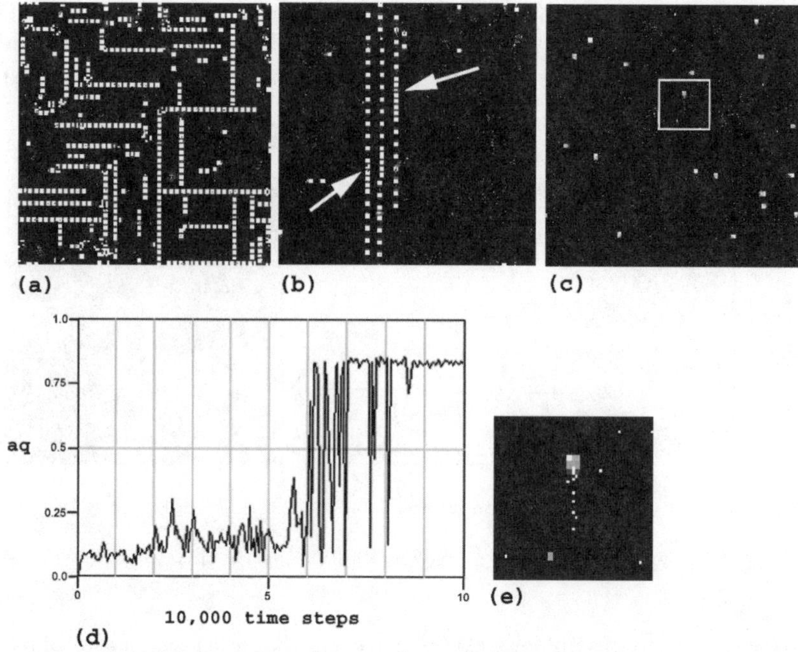

Fig. 6.8. Three stages in the evolution of a glider-rich CA

The erratic jumps in *aq* in the graph between time 60,000 and 90,000 are most likely the result of sparse data sampling since there is so much quiescence and fewer riding particles, and the CA has not yet settled into its final state. By around time 90,000, the gliders are few and sampling is still sparse, but the gliders are now moving at a constant rate, and collisions are infrequent.

6.4.4 Evolving Game-of-Life at $\lambda = 0.273$

The lambda value λ ($0 \leq \lambda \leq 1$) of a cell is the amount of non-quiescence resulting from its transition function. It is calculated using a function which generates every possible neighborhood state given K and R, and applies the cell's transition function on each of these neighborhood states. The sum of all non-quiescent values of *resultState* from applying the transition function is determined, and this is divided by the total number of possible neighborhood states to get λ.

A series of simulations were run which included this calculation. Lambda values are shown in Figure 6.8 as a scatter plot in the same graph as the value of *aq*. (Since both values map linearly within the range (0...1), they can be plotted in the same graphical window). The scatter plot is generated by measuring λ from 10 randomly selected cells at each time step, and then plotting these 10 values as dots.

Langton [9] found λ for *Game-of-Life* to be approximately 0.273. As an experiment, λ was plotted in a number of simulation runs with $K = 2$, and R ranging from 3 to 10. As hoped, *Game-of-Life* did emerge. While *Game-of-Life* requires only $R = 3$, it was found that it more easily evolves when $R > 3$. Figure 6.9 shows a simulation run with $R = 10$. As expected, lambda converges to 0.273 (indicated by the arrow).

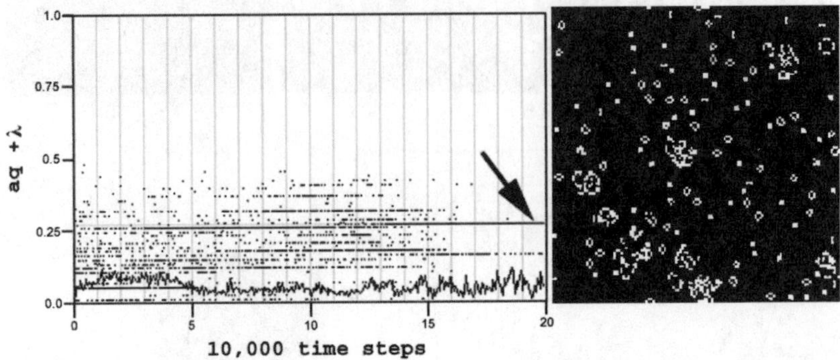

Fig. 6.9. *Game-of-Life* evolves with λ converging to 0.273

This simulation was run for 200,000 time steps (the particles are not displayed in the image of the resulting CA dynamics). A few observations are worth mentioning:

1. In simulation runs, *Game-of-Life* converges more slowly in general than CA with $K > 2$. As the graph shows, *aq* never gets very high, possibly because *Game-of-Life* is less glider-rich than CA with large *K* values Also, the gliders move more slowly (*C*/4 to be exact). Thus, *Game-of-Life* offers less opportunity for particles to find a good ride, at least given the selection criterion used.
2. The multiple horizontal lines in the λ plotting reveal the presence of competing regions in the CA domain, each gravitating toward different λ. At the end of the simulation run can be seen a band of lower λ values which competes with the λ for *Game-of-Life*, but begins to weaken at the very end.
3. It was discovered that *Game-of-Life* evolves more easily when the density of the noise wave *w* is higher, and so in these simulations *w* was raised from 0.2 to 0.5. The reason *Game-of-Life* prefers higher *w* was not determined.

6.4.5 Plotting λ with Other Values of *K*

The edge of chaos changes as *K* varies. A simulation was run with K=4 and R=8 for 200,000 time steps. Figure 6.10 shows the results. The graph shows that λ converges even though *aq* has not increased significantly, indicating genetic convergence. After around 160,000 time steps, *aq* increases, and λ converges more. In this particular CA, λ is slightly higher than 0.273.

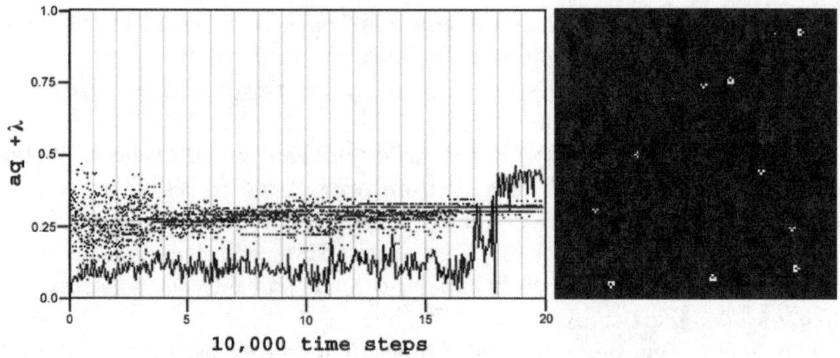

Fig. 6.10. Plotting λ over time for a CA with $K = 4$

6.4.6 Parameter Ranges For Optimal Swarm Performance

Changing the parameters associated with the CA has little effect on particle performance. Higher values of K and R cause more complex dynamics, often with exotic gliders and waves. As long as the CA is capable of generating coherent space-time configurations, the particles will perform in a predictable manner.

The parameters ω and *aw* are important for a particle's ability to generally track coherent motion. If *aw* is set too low, the particle will not experience enough attraction to live cells. If it is too high, particle motion becomes too erratic. Setting ω higher has the effect of smoothing out a trajectory that would otherwise be erratic due to small changes in the particle cell neighborhood. Gliders with large periods may change their shapes as they migrate across the CA lattice – larger ω allows the particle to respond less to these changes and more to the overall position and velocity of the glider. Low *aw* values and high ω values cause the effect of a low-pass filter on the trajectory. In general, keeping ω within the range of (0.6 to 0.8) and *aw* within the range of (0.3 to 0.5) allows the particles to sufficiently track motion, given PN =25.

The setting for mutation rate m has a similar effect on evolution as in the case of the standard GA. Not enough mutation causes premature convergence to a local optimum, while too much degenerates into a random walk. m should be kept within the range of (0.0 to 0.002).

The parameters, n, d, e, rw, and b affect the rates at which the particles perform the genetic operators. A number of experiments were run, each with one of these parameters changed. A few qualitative observations are given here.

When number of particles n is reduced from 500 to 100, convergence is slow or not at all. When n is increased to 1000, gliders emerge much sooner, although premature convergence is often observed: genetic homogeneity in the CA comes about too fast in the simulation, and little improvement happens after this point. For *resolution* =128, n in the range of (300 to 700) seems to work well.

When rate of genetic exchange e is decreased from 0.1 to 0.0 (eliminating all communication between particles), no gliders emerge, and the CA remains essentially a patchwork of chaotic regimes. When *e* is increased to 0.5, gliders emerge rapidly, although performance is poor, and while local regimes appear to become glider-rich, the CA does not easily converge as a whole. The range (0.02 to 0.4) seems to work well.

Decreasing random weight *rw* from 0.06/*resolution* to 0.0 results in particles being motionless while not riding. Performance is poorer but gliders can still emerge. Increasing *rw* to 0.2/*resolution* causes local regimes to spread faster – gliders still emerge but performance is poorer. This parameter does not have a significant effect on performance, although *rw* should probably stay within the range of (0.02/*resolution* to 0.2/*resolution*).

When rate of depositing genes *d* (reproduction) is decreased from 0.05 to 0.0, there is no effect, as this genetic operator is critical for evolving the CA transition functions. At 0.01, performance is poor and gliders are rare. When increased to 0.2, local regimes emerge with strong homogeneity, convergence is rapid, and gliders still emerge. Staying within the range (0.02 to 0.2) appears to work best.

Reducing birth-rate *b* from 0.001 to 0.0 results in particles never being re-born (re-initialized in new random locations and having *bq* set to 0). This results in stagnation in local regions. Chaotic regimes tend to hoard particles which accumulate and cannot escape. Observe that when *b* is 0.0, the value of *bq* is never refreshed, and so discovery of better ride quality becomes increasingly rare over time.

Increasing *b* to 0.01 on the other hand results in far too rapid re-birth, and consequently poor performance, due to the shorter life-spans of particles, and the inability to stay on the job long enough to discover a good ride and reward local areas accordingly.

A good range for *b* is (0.0005 to 0.005). Consider that the rebirth scheme could instead have been designed around the concept of particle "lifespan" using integer units of time steps (and particles initialized with random "ages" so rebirths are evenly distributed over time). This alternate representation might possibly be less sensitive and more intuitive as a parameter, even though the results would be nearly the same.

These are of course considerations of single parameter changes against the backdrop of the base experiment. Ideally the entire set of swarm-related parameters should be optimized together, potentially using a classical GA.

6.5 Comparison/Contrast with Other Approaches

There appear to be no techniques published thus far that specifically apply particle swarms to CA dynamics as a way to evolve gliders. Many techniques however have been described that use particles to track and identify other kinds of objects, from video sequences. The "Flocks of Features" technique [8] for tracking hands addresses the fact that the view of the human hand can change dramatically in shape over time. Similarly, emerging patterns in an evolving CA can have many possible shapes and sizes that change over time. This technique addresses the fuzzy-objecthood of gliders.

Das et al. [3] have developed a number of techniques that use a GA to evolve CA rules that can solve simple computational tasks. (They refer to the resulting emergent space-time structures as "particles" – and interesting choice of words in this context). Their explorations employ a GA using populations and an objective function to guide evolution towards a computational goal. Techniques cited in section 1, and many others, use explicit generations, individuals, and populations. In contrast, this technique applies the GA operators asynchronously and spatially across a single non-uniform CA lattice, over a single simulation run. It also does not have computation as an explicit objective.

The techniques developed by Wuensche and Lesser [25], for classifying CA rules according to their dynamics, are largely analytical. The technique described here has an analytical component, but there is a deep synthetic component as well: gliders are both *discovered* and *created* at the same time. Because of the tangled, interactive nature of this technique, it serves as a model of co-adaptation, self-organization, and stigmergy (the swarm and the CA serve as each other's adaptive environments) One downside, when compared to other techniques cited, is that it is not good at identifying glider-potential within the entire spectrum of possible CA rules – it is constrained by the selection criterion *S*, the particle physical parameters, the transition function, and the local contingencies of the simulation run.

While a quantitative comparison to both swarm-based systems and evolutionary CA could be made, it is likely to be piecemeal. In general, a conclusion would be that this technique is novel as a model of emergent complexity, and produces a large variety of class 4 CA (*efficiently* – the base experiment runs at 30 Hz on a modern PC as a Java Applet). But it is not an exhaustive, generalized search tool for identifying all possible universal CA.

6.6 Future Work

6.6.1 Other Selection Criteria

In the process of coming up with a good selection criterion, a number of alternatives were explored. One such selection criterion, called "Orthogonal", selects for particle motion that is rapid and orthogonal:

$$q_i = (|vx_i| - |vy_i|)\,/\,C$$

where vx and vy are the two components of $\mathbf{v}$. Another selection criterion, called "Constant", selects for constant velocity without regard to speed:

$$q_i = (|\mathbf{v}_i| - |\mathbf{d}_i|)\,/\,|\mathbf{v}_i|$$

for all $|\mathbf{v}| > 0$, otherwise, $q = 0$.

In all cases, q is clamped $(0 \leq q \leq 1)$.

"Orthogonal" causes rapid convergence to a state of having many (usually small) gliders. But it tends to produce less varieties of gliders. "Constant" was originally expected to have interesting results, but it is poor at selecting for gliders of any kind. The reasons have yet to be explored.

Clearly, more sophisticated selection criteria would help, in some cases including the ability to read the contents of the particle cell neighborhood and compare it to some ideal local environment.

6.6.2 Particle Swarm for Categorizing Glider Interactions

After a glider-rich CA has been evolved, a question remains: is this CA capable of universal computation? While it may not be easy for a particle swarm (or a human) to answer this question, it is possible for a swarm to detect key events that are indicative. As Wuensche points out, "*Once gliders have emerged, CA dynamics could, in principle, be described at a higher level, by glider collision rules as opposed to the underlying CA rules.*" [26]

Figure 6.11 shows a snapshot of a typical evolved glider-rich CA, with cartoon chat-labels overlaid on top, to illustrate the potential of enhanced swarm communication. The particles in this scenario are aware of the general directions of their rides – an easy calculation. Towards the right of the picture is a cluster of particles that have just been left behind after the gliders they were riding had collided, annihilating each other (a typical scenario). One such particle in this cluster is querying another particle (which is in its local neighborhood) about the nature of its most recent ride.

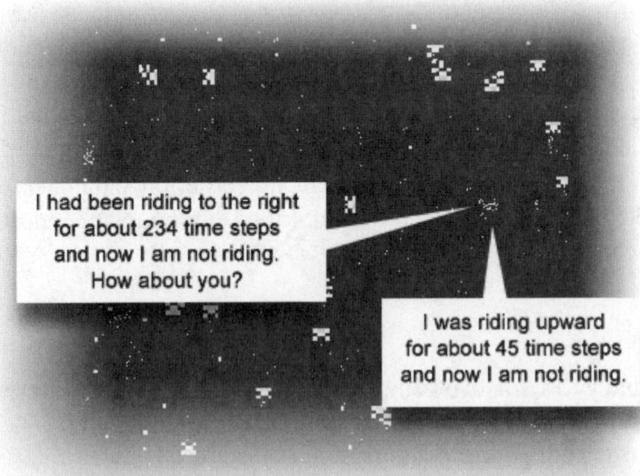

Fig. 6.11. Particles that communicate information about their rides could help with data mining of glider events

While we see illustrated only a few bits of swarm communication, imagine a constant and continual exchange of such information, in which all 500 particles are contributing to an emerging knowledge base of facts about abrupt changes in their rides (indicating glider collisions). This may include the following:

1. What direction was the particle moving before the change?
2. Was there a sudden change in direction? Which direction?
3. Is the particle still riding? (if not: possible glider annihilation)
4. Where did this change occur? (approximation within a range of cells)
5. When did this change occur? (approximation within a range of time steps)
6. How long had the particle been riding "smoothly" before the change?
7. How reliable is this report (how many particles reported a similar event)?
8. etc.

Any resulting events that are above a certain reliability threshold could be plotted over space and time, to gain some knowledge of the journeys of gliders and their interactions. This in itself is still not conclusive of universality, but it may point the way towards some techniques that can be used to better understand universal CA.

6.6.3 Swarm For Measuring Fitness

A variation on this technique which would place it more into a comparative space with other evolutionary CA research would be to apply the swarm over a population of uniform CA, rather than on a single, heterogeneous CA. The individuals in the population would represent different transition functions, and the swarm would be used simply to measure fitness per individual – and not as the agent of evolution. This would be less interesting as an artificial life simulation, but it may yield a more controlled and generalized tool for searching the space of CA rules for gliders.

6.7 Conclusions

We are witnessing many new hybrid forms of evolutionary computation, and biologically-inspired techniques for data mining, image analysis, and other applications, as well as basic research in artificial life. The various combinations of PSO and GA are just one dimension of this hybridization.

This technique particularly addresses the idea of "motion" within CA dynamics, and makes use of a powerful technique modeled on a kind of motion (the movement of agents in a swarm) to identify propagating structures in CA dynamics. But as mentioned before, this technique in fact models a form of co-adaptation, and so the identification of motion is not an end in itself – it is means by which the particle swarm actively changes the rules of the CA so as to generate MORE propagating structures. This technique is thus hybrid in more than one sense: it is a hybridization of GA with PSO, but it is also a hybridization of PSO with Evolutionary CA. It is hoped that this exploration will open another door for further research.

6.8 Symbols Used

Symbol	Description
Q	the quiescent state
t	incrementing time step index
S	the selection criterion
n	number of particles
K	number of cell states
M	maximum neighbor count
R	number of transition function rules
PN	number of cells in the particle cell neighborhood
W	noise wave period
w	noise wave density
ω	inertia weight
aw	attraction weight
rw	random weight
m	mutation rate
b	re-birth rate
c	crossover rate
e	gene exchange rate
d	gene deposit rate
C	speed of light
$\mathbf{a}$	particle attraction force per time step
$\mathbf{r}$	particle random force per time step
$\mathbf{v}$	particle velocity per time step
$\mathbf{d}$	particle change in velocity per time step
$\mathbf{p}$	particle position per time step
q	particle ride quality per time step
bq	particle best ride quality
aq	average q for all *riding particles*

References

1. Berlekamp, E. R., Conway, J. H., and Guy, R. K. (1982) “What Is Life?”. Ch. 25 in ôWinning Ways for Your Mathematical Playsö, Vol. 2. London: Academic Press.
2. Bilotta, E., Lafusa, A., and Pantano, P. (2002) Is self-replication an embedded characteristic of artificial/living matter? Proceedings of the eighth international conference on Artificial life. pages 38û48. MIT Press.
3. Das, R., Mitchell, M. and Crutchfield, J. (1994) “A Genetic Algorithm Discovers Particle-Based Computation in Cellular Automata” In Parallel Problem Solving from Nature-III, Y. Davidor, H.-P. Schwefel, and R. MSnner (eds.), Springer-Verlag 344-353
4. Gardner, M. (1970) Mathematical Games - The fantastic combinations of John Conway’s new solitaire game “life” Scientific American, 223 (October 1970): 120-123.

5. Grassé, P.P. (1959) La reconstruction du nid et les coordinations interindividuelle chez Belli-cositermes natalensis et Cubitermes. La theorie de la stigmergie: Essai d'interpretation des termites constructeurs. Insectes Sociaux, 6:41–83.
6. Kazakov, D., and Sweet, M. (2005) Evolving the Game of Life. Adaptive Agents and Multi-Agent Systems II. eds. D. Kudenko, D. Kazakov, E. Alonso. Lecture Notes in Artificial Intelligence vol. 3394, pp.132-146, Springer - Verlag, Berlin Heidelberg.
7. Kennedy, J. and Eberhart, R. (1995) Particle Swarm Optimization. Proceedings of the 1995 IEEE Conference on Neural Networks.
8. Kolsch, M., Turk, M. (2005) Hand tracking with Flocks of Features. Computer Vision and Pattern Recognition. CVPR 2005. IEEE Computer Society Conference on Computer Vision and Pattern Recognition. Volume 2, 20-25 June 2005 Page(s):1187 vol. 2
9. Langton, C. (1992) Life at the Edge of Chaos, Artificial Life II. Addison-Weskey, pages 41-91
10. Mitchell, M., Hraber, P., and Crutchfield, J. (1993)"Revisiting the Edge of Chaos: Evolving Cellular Automata to Perform Computations", Complex Systems 7 89-130.
11. Mitchell, M. (1998). Computation in Cellular Automata: A Selected Review. In T. Gramss, S. Bornholdt, M. Gross, M. Mitchell, and T. Pellizzari, Nonstandard Computation , pp. 95–140. Weinheim: VCH Verlagsgesellschaft.
12. Packard, N. (1988) Adaptation Towards the Edge of Chaos. In: Kelso, J. A. S., Mandell, A. J., and Shlesinger, M. F. (eds.), Dynamic Patterns in Complex Systems, World Scientific. pp. 293-301, Singapore
13. Poundstone, W. (1985) The Recursive Universe. William Morrow and Company, New York.
14. Ramos, V., and Almeida, F. (2000) Artificial Ant Colonies in Digital Image Habitats - A Mass Behaviour Effect Study on Pattern Recognition, Proceedings of ANTS 2000 û 2nd International Workshop on Ant Algorithms (From Ant Colonies to Artificial Ants), Marco Dorigo, Martin Middendorf and Thomas Stnzle (Eds.), pp. 113-116
15. Ramos, V., and Abraham, A. (2003) Swarms on Continuous Data, in CEC|03 û Congress on Evolutionary Computation, IEEE Press. pp.1370-1375, Canberra, Australia, 8-12 Dec.
16. Rendell, P. (2001) Turing Universality of the Game of Life. Collision-Based Computing, ed. Adamatzky. Springer-Verlag. pages: 513 û 539.
17. Sapin, E., Bailleaux, O., Chabrier, J., and Collet, P. (2004) A New Universal Cellular Automaton Discovered by Evolutionary Algorithms. Lecture Notes in Computer Science. Springer-Verlag. Volume 3102 / 2004. pp. 175 - 187
18. Shi, Y.H., and Eberhart, R.C. (1998) Parameter Selection in Particle Swarm Optimization. Proceedings of 7th Annual conference on Evolutionary Computation, Springer.
19. Smith, A. R. III (1971): Simple computation-universal cellular spaces. Journal of the Association for Computing Machinery, 18, 339û353.
20. Ventrella, J. (2000) Breeding Gliders with Cellular Automata û online at: http://www.ventrella.com/Alife/Cells/cells.html.
21. Ventrella, J. (2005) Gene Pool: Exploring the Interaction Between Natural Selection and Sexual Selection. Artificial Life Models in Software. eds. Adamatzky, A. and Komosinski, M. Springer.
22. Wilson. E. O., (1986) Biophilia, the Human Bond with other Species. Harvard University Press,
23. Wolfram. S. (1984) Universality and Complexity in Cellular Automata. Physica D 10: 1-35.
24. Wolfram. S. (2002) A New Kind of Science. Wolfram Media.

25. Wuensche, A., and Lesser, M. (1992) The Global Dynamics of Cellular Automata. Addison-Wesley
26. Wuensche, A. (2002) Finding Gliders in Cellular Automata. Collision-based Computing, Springer-Verlag, Jan, 2002. pages 381-410

7

Termite: A Swarm Intelligent Routing Algorithm for Mobile Wireless Ad-Hoc Networks

Martin Roth and Stephen Wicker

Wireless Intelligent Systems Laboratory
School of Electrical and Computer Engineering
Cornell University
Ithaca, New York 14850
USA
mhr8@cornell.edu
wicker@ece.cornell.edu

Summary. A biologically inspired algorithm named *Termite* is presented. Termite directly addresses the problem of routing in the presence of a dynamic network topology. In the Termite algorithm, network status information is embedded in the network through the passage of packets. Probabilistic routing decisions are based on this information such that the use of paths of maximum utility is an emergent property. This adaptive approach to routing greatly reduces the amount of control traffic needed to maintain network performance. The stochastic nature of Termite is explored to find a heuristic to maximize routing performance. The analysis focuses on the routing metric estimator, which is known as pheromone in the biological context. The pheromone decay rate is adjusted such that it makes the best possible estimate of the utility of a link to deliver a packet to a destination, taking into account the volatility, or correlation time, of the network. Termite is compared to Ad-hoc On-demand Distance Vector (AODV), showing the former to be superior in many primary performance metrics.

7.1 Introduction

7.1.1 Overview

Routing in mobile wireless ad-hoc networks (MANETs) is complicated by the fact that mobile users cause the network topology to vary over time. This paper shows how a probabilistic routing framework based on the principles of swarm intelligence (SI) directly and successfully addresses this problem. Swarm intelligence is a framework for designing robust and distributed systems composed of many interacting individuals, each following a simple set of rules [1]. The global behavior of the system is an emergent property and is not preprogrammed at any level. Swarm intelligence is strongly related to artificial intelligence algorithms such as reinforcement learning [2] and optimization algorithms such as stochastic gradient descent [3].

M. Roth and S. Wicker: *Termite: A Swarm Intelligent Routing Algorithm for Mobile Wireless Ad-Hoc Networks*, Studies in Computational Intelligence (SCI) **31**, 155–184 (2006)
www.springerlink.com

Traditional approaches to the MANET routing problem use deterministic rules to discover optimal routes. Difficulty arises when the network changes quickly and all routes must be reevaluated. Termite opts for a probabilistic approach in which utility estimates of all routes are maintained simultaneously. Good paths are successively refined as the network environment changes. Control traffic is nearly eliminated by attaching a small amount of route information to each packet, including data. All traffic updates the network as it moves between source and destination.

Termite contains mechanisms for establishing a trade-off between network exploration and exploitation. These mechanisms are considered in the context of finding a heuristic for the optimal pheromone decay rate. Optimizing pheremone decay ensures that received routing information is effectively used to estimate the current routing metric, while maintaining its accuracy over time. Insights leading to performance improvement are achieved in part by modeling the received path utility information from each packet by a non-stationary stochastic process and filtering it to track the mean of the process over time.

The performance of Termite is assessed with a comparison to the standard Ad-hoc On-demand Distance Vector (AODV) routing protocol [4] in a simulated environment.

7.1.2 Previous Work

Traditional MANET Routing

Over the past ten years, an enormous number of MANET routing protocols have been proposed. They are generally based on distance vector or link state routing algorithms suggested nearly fifty years ago [5] [6] [7]. Current protocols are further broken into two categories, proactive and reactive. Examples of the former include DSDV [8] and OLSR [9], and examples of the latter are AODV [4] and DSR [10]. Such algorithms focus on finding and using the best available route at any given time. While able to manage relatively stable environments, this approach can find difficulty in high dynamic environments in which the network topology changes often and a large number of routing updates are forced. A more comprehensive review is available in [11].

Ad-hoc On-demand Distance Vector Routing Protocol (AODV)

Originally based on the proactive DSDV, the Ad-hoc On-demand Distance Vector (AODV) routing protocol is one of the most popular on demand routing protocols for ad-hoc networks. It is in the continuing stages of being standardized by the Internet Engineering Task Force (IETF) MANET working group [4] [12]. AODV was originally introduced in 1999 by Perkins and Royer [13]. It has seen extensive testing and development since its introduction, and is used as a model for many new proposals.

The protocol operates generally as follows. When a data packet arrives at a node with no route to the destination, a route request procedure is initiated. Two variations exist; if a source node has no route to the destination, an expanding ring search is started. A local flood of route request (RREQ) packets is broadcast with a limited time-to-live. If no route reply (RREP) packet is received within a timeout period, another local flood is issued with a larger time-to-live, allowing it to search a larger area. This process is retried a certain number of times until the destination is found or is declared unreachable. Sequence numbers are embedded in the RREQs in order to distribute current information about the source, and to specify a minimum freshness for the information about the destination. Route metrics are also included in RREQ packets in order to create reverse routes to the source. If an intermediate node must search for a destination, this is known as a local repair and is comprised of only a local RREQ flood. If the destination cannot be found, a route error (RERR) is returned to the source to inform it that a new route must be discovered.

Route replies (RREP) are unicast from an intermediate node with an active route to the requested destination or from the destination itself. An active route is guaranteed back to the source since reverse routes are created by the RREQs. An active route to the destination should include a sequence number at least as large as that in the route request. Once a RREP is received, any packets that had been buffered for the destination can be sent to the destination.

A route error (RRER) message is needed to inform the source that an irreparable break has occurred in the route to the destination; a new route should be found. The route error propagates from the node immediately upstream of the break back to the source. The route to the destination is deactivated and each intermediate node will require an equally fresh new route to the destination.

The Social Insect Analogy

Social insect communities have many desirable properties from the MANET perspective. These communities are formed from simple, autonomous, and cooperative organisms who are interdependent for their survival. Despite a lack of centralized planning or any obvious organizational structure, social insect communities are able to effectively coordinate themselves to achieve global objectives. The behaviors which accomplish these tasks are often emergent from much simpler behaviors, or rules, that the individuals are following. The coordination of behaviors is also robust, necessary in an unpredictable world. No individual is critical to any operation and tasks can recover from almost any setback. The complexity of the solutions generated by such simple individual behaviors indicates that the whole is truly greater than the sum of the parts [14].

Such characteristics are desirable in the context of ad-hoc networks. Such systems may be composed of simple nodes working together to deliver messages, while resilient against changes in its environment. The environment of an ad-hoc network might include anything from its own topology to physical layer effects on the communications links, to traffic patterns across the network. A noted difference between biological and engineered networks is that the former have an evolutionary incentive to cooperate, while engineered networks may require alternative solutions to force nodes to cooperate [15] [16].

The ability of social insects to self organize relies on four principles: positive feedback, negative feedback, randomness, and multiple interactions. A fifth principle, stigmergy, arises as a product of the previous four [1]. Such self organization is known generally as swarm intelligence.

How to Build a Termite Hill

A simple example of the hill building behavior of termites provides a strong analogy to the mechanisms of Termite and SI routing in general. This example illustrates the four principles of self organization [17]. A similar analogy is often made with the food foraging behavior of ants.

Consider a flat surface upon which termites and pebbles are distributed. The termites would like to build a hill from the pebbles, ie. all of the pebbles should be collected into one place. Termites act independently of all other termites, and move only on the basis of an observed local pheromone gradient. Pheromone is a chemical excreted by the insect which evaporates and disperses over time.

A termite is bound by these rules:

1. A termite moves randomly, but is biased towards the locally observed pheromone gradient. If no pheromone exists, a termite moves uniformly randomly in any direction.
2. Each termite may carry only one pebble at a time.
3. If a termite *is not* carrying a pebble and it encounters one, the termite will pick it up.
4. If a termite *is* carrying a pebble and it encounters one, the termite will put the pebble down. The pebble will be infused with a certain amount of pheromone.

With these rules, a group of termites can collect dispersed pebbles into one place. The following paragraphs explain how the principles of swarm intelligence interplay in the hill building example.

Positive Feedback

Positive feedback often represents general guidelines for a particular behavior. In this example, a termite's attraction towards the pheromone gradient biases it to adding to large piles. This is positive feedback. The larger the pile, the more pheromone it is likely to have, and thus a termite is more biased to move towards it and potentially add to the pile. The greater the bias to the hill, more termites are also likely to arrive faster, further increasing the pheromone content of the hill.

Negative Feedback

In order for the pheromone to diffuse over the environment, it evaporates. This evaporation consequently weakens the pheromone, lessening the resulting gradient. A diminished gradient will attract fewer termites as they will be less likely to move in its direction. While this may seem detrimental to the task of collecting all pebbles into one pile, it is in fact essential. As the task begins, several small piles will emerge very quickly. Those piles that are able to attract more termites will grow faster. As pheromone decays on lesser piles, termites will be less likely to visit them again, thus preventing them from growing. Negative feedback, in the form of pheromone decay, helps large piles grow by preventing small piles from continuing to attract termites.

In general, negative feedback is used to remove old or poor solutions from the collective memory of the system. It is important that the decay rate of pheromone be well tuned to the problem at hand. If pheromone decays too quickly then good solutions will lose their appeal before they can be exploited. If the pheromone decays too slowly, then bad solutions will remain in the system as viable options.

Randomness

The primary driving factor in this example is randomness. Where piles start and how they end is entirely determined by chance. Small fluctuations in the behavior of termites may have a large influence in future events. Randomness is exploited to allow for new solutions to arise, or to direct current solutions as they evolve to fit the environment.

Multiple Interactions

It is essential that many individuals work together at this task. If not enough termites exist, then the pheromone would decay before any more pebbles could be added to a pile. Termites would continue their random walk, without forming any significant piles.

Stigmergy

Stigmergy refers to indirect communications between individuals, generally through their environment. Termites are directed to the largest hill by the pheromone gradient. There is no need for termites to directly communicate with each other or even to know of each other's existence. For this reason, termites are allowed to act independently of other individual, which greatly simplifies the necessary rules.

Swarm Intelligent MANET Routing

The soft routing protocols proposed to date are essentially probabilistic distance vector protocols. The cost to each destination over each neighbor is estimated by each node based on routing data in received or overheard traffic. The next hop of a packet is chosen based on a distribution proportional to the routing utility of using each neighbor to arrive at the packet's destination. There are thus two key components to any probabilistic routing algorithm, the packet forwarding equation and the metric estimator (pheromone accounting). The former determines the routing distributions based on the metric estimates. The latter is responsible for producing an estimate of the cost (or inversely, the utility) to arrive at each destination through each neighbor. Costs are estimated by probing the network with packets (control or data, depending on the implementation). The network is essentially sampled for changes, with the results tabulated at each node. Because many of the network characteristics are dynamic, the estimates must be continuously updated. Maintaining per neighbor routing information allows multipath routing to be easily implemented.

There are a number of examples showing how SI routing can provide a significant performance gain over traditional deterministic approaches. These include ABC [18], AntNet [19], CAF [20], PERA [21], ARA [22] [23], MABR [24], Termite [25], ANSI [26], and AdHocNet [27] [28] [29]. A wider summary is found in [30].

VWPR and Pheromone Aversion

Virtual-Wavelength Path Routing (VWPR) introduces the concept of source pheromone aversion (SPA) to the SI routing solution [31]. This is an adaptation of the packet forwarding process in order to take advantage of additional routing information already available at each node. Packets usually only follow the pheromone gradient of their destination. SPA forces the packets away from their own source's pheromone, thus simultaneously pulling and pushing the packet towards its destination.

SPA is also used in the Multiple Ant Colony Optimization (MACO) algorithm [30] as a means to find a ranking of good solutions. MACO expands on the ant colony metaphor for routing by placing two competing colonies in a network to find routes; each is repelled by the other, thus forcing them to alternative solutions.

7.1.3 Structure of Paper

Section II provides a detailed explanation of the Termite MANET routing protocol. Some variants are described which allow comparison to Dijkstra's shortest path algorithm. Their performance is then compared to AODV in order to provide a common perspective with an established routing solution. A variety of conditions of interest to ad-hoc network engineers are investigated. Section III develops and tests a heuristic for the pheromone decay rate which maximizes performance. It should help in the selection of parameters to achieve optimal routing performance. A statistical approach is used to explain the performance of the system, and a solution is proposed. Section IV concludes the chapter.

7.2 Termite

The SI MANET routing protocol Termite is presented in detail. It is compared to the Ad-hoc On-demand Distance Vector (AODV) routing algorithm, a state-of-the-art and standards track technology. It is shown that the SI framework can be used to competitively solve the MANET routing problem with a minimal use of control overhead. A small amount of control information is imbedded in every data packet, which is usually sufficient for the network to maintain a current and accurate view of its state. The end result is a routing algorithm requiring only data traffic in the network under many circumstances. The version of Termite presented here takes advantage of a number of enhancements over a generic SI approach, including a generalized packet forwarding equation and source pheromone aversion.

7.2.1 Termite

The routing algorithm presented here is similar hill building proceedure presented earlier. Termite associates a specific pheromone scent with each node in the network. Packets moving through the network are biased to move in the direction of the pheromone gradient of the destination node, as biological termites are biased to move towards a hill. Packets follow the pheromone gradient while laying pheromone for their source on the same links; this is positive feedback. The specific amount of pheromone deposited by a packet on a link, as well as how that pheromone behaves over time, is governed by the pheromone accounting process. Changes in the network environment, such as topological or path quality changes, are accounted for by allowing pheromone to decay over time; this is negative feedback. Paths that are not reinforced are rendered less attractive for future traffic. Each node maintains a table indicating the amount of pheromone for each destination on each of its links. The pheromone table acts as a routing table similar to those found in traditional distance vector routing algorithms. The pheromone table also represents a common area for packets to stigmergetically interact in order to converge on good routes.

Pheromone Table

The pheromone table maintains the amount of pheromone on each neighbor link for each known destination. It serves the same purpose as the routing table of a distance vector routing protocol, with analogous operations on its elements such as neighbor, destination, and routing metric management. Columns represent destinations and rows neighbors. Neighbors also appear in the table as destinations. When a neighbor is gained, an extra row is added to each column with the pheromone initialized to zero. An extra column is added to represent the neighbor, since the new node can also be a packet destination. If a neighbor is lost, the corresponding row is removed from the table. When a destination is gained, the current list of neighbors is replicated for the new destination but with all pheromone values reset to zero. If a destination is lost from the pheromone table, the column is simply removed.

A neighbor row is removed only when the link is explicitly lost through communications failure. There is no HELLO message link maintenance mechanism as is found in other protocols, which retains link information based on the arrival of specially formatted neighbor beacons. Even if the pheromone on that link decays, its row is still retained since the neighbor still exists as a communications option (Presumably. Although in the MANET setting, the neighbor may well have left communications range). A destination is removed if all of the pheromone in its column decays to zero. In this way Termite will know when to issue a route request for that destination. This procedure is initiated if the destination does not exist in the pheromone table.

For all nodes, n, in the network, pheromone values in the pheromone table are referenced with, $P^n_{i,d}$, where $i \in \mathcal{N}^n$, and $d \in \mathcal{D}^n$. These sets represent the current set of neighbors and destinations that node n is aware of. $P^n_{i,d}$ is thus the amount of pheromone at node n for destination node d on the link to neighbor node i.

Packet Forwarding

The forwarding equation with source pheromone aversion is used to determine the next hop neighbor. The forwarding equation models a packet probabilistically following a pheromone gradient to its destination, and repelled by the gradient towards the source. Equation 7.1 maps the destination d pheromone on each outgoing link i at node n, $P^n_{i,d}$, to the "pull" of that link to forward the packet to the destination, $p^n_{i,d}$. The source pheromone distribution, $p^n_{i,s}$, is computed similarly according to Equation 7.2, and represents the "push" of a link away from the source. The specific next hop neighbor is then chosen randomly according to the meta distribution, $\hat{p}^n_{i,d}$, in Equation 7.3, which reflects the total effect of source pheromone aversion. The forwarding equation in each case is a normalization of the resident link pheromone.

Per packet computation of forwarding distributions could become too demanding for small processors handling high packet rates. In such a scenario, it would be possible to implement some optimizations, such as updating the distributions periodically in time of number of received packets.

$$p^n_{i,d} = \frac{\left[P^n_{i,d} e^{-(t-t^n_{i,d})\tau} + K\right]^F}{\sum_{j\in\mathcal{N}^n}\left[P^n_{j,d} e^{-(t-t^n_{j,d})\tau} + K\right]^F} \tag{7.1}$$

$$p^n_{i,s} = \frac{\left[P^n_{i,s} e^{-(t-t^n_{i,s})\tau} + K\right]^F}{\sum_{j\in\mathcal{N}^n}\left[P^n_{j,s} e^{-(t-t^n_{j,s})\tau} + K\right]^F} \tag{7.2}$$

$$\hat{p}^n_{i,d} = \frac{p^n_{i,d}\left(p^n_{i,s}\right)^{-A}}{\sum_{j\in\mathcal{N}^n} p^n_{j,d}\left(p^n_{j,s}\right)^{-A}} \tag{7.3}$$

The exponential term multiplied against each pheromone value is included in order to deemphasize older pheromone values. The current time is t and the last time that a packet arrived from node d (or s) at node n on the link to neighbor i is $t^n_{i,d}$. The pheromone decay rate is τ. There are two primary reasons for such a deemphasis. The first is that according to the original biological inspiration, pheromone always decays; the exponential term updates the current pheromone measurement to model this continuous depreciation. A more technical reason to include such a term is that a method is needed to account for changes in the networking environment since the last received path utility measurement. The exponential term models this diminishing confidence in the accuracy of the metric estimator (the pheromone). This term is based on the network correlation time, as shown in Section III. If no packet is received from a destination for some time, the routing probabilities now reflect the fact that less timely information is available about the path quality to the destination. The routing probabilities tend towards a uniform distribution over all neighbor nodes. Past swarm intelligent routing algorithms have decayed pheromone either on regular intervals or only upon packet arrival [21] [22] [23] [25] [26].

The constants F, K, and A are used to tune the routing behavior of Termite. The *pheromone threshold*, K, determines the sensitivity of the probability calculations to small amounts of pheromone. If $K \geq 0$ is large, large amounts of pheromone will have to be present before an appreciable effect is seen on the routing probability. The nominal value of K is small compared to the expected pheromone levels. The probability of any link is prevented from going to zero, ensuring a minimum level of exploration of the network. Similarly, the *pheromone sensitivity*, $F \geq 0$, modulates the differences between pheromone amounts. $F > 1$ accentuates differences between links, while $F < 1$ will deemphasize them. $F = 1$ yields a simple normalization. The *source aversion sensitivity*, $A \geq 0$, controls the amount by which the packet is repelled by the source pheromone.

Packet Pheromone Accounting

Pheromone carried by a packet is updated immediately upon packet reception to reflect the utility of the previous hop. This is done as shown in Equation 7.4, where γ is the pheromone resident on the packet (equivalent to the packet's total path utiltiy) and $c_{r,n}$ is the cost of the link from previous hop r to current node n. Note that utility is the inverse of cost, only costs are additive (not utilities), and that costs are strictly non-negative (and in any realistic implementation, positive).

$$\begin{aligned} \gamma &\leftarrow \left(\gamma^{-1} + c_{r,n}\right)^{-1} \\ &\leftarrow \frac{\gamma}{1+\gamma c_{r,n}} \end{aligned} \tag{7.4}$$

The link pheromone is then set according to the pheromone update method with the new packet pheromone value.

Pheromone Update Methods

The pheromone table is updated with new pheromone values only upon packet arrival. A pheromone update method describes how the update is managed based on the pheromone carried by a packet which arrives at node n from source node s and previous hop r. If n is not designated as the packet's next hop, the node updates the pheromone table in the way described here and drops the packet. The time at which the packet is received is t, and the last time at which the pheromone for node s on the link to r was updated at node n on is $t^n_{r,s}$.

Three update methods are reviewed, including the γ pheromone filter, the normalized γ pheromone filter, and probabilistic Bellman-Ford. These methods are chosen due to their prevalence in the litterature and their favorable emergent properties.

γ Pheromone Filter (γPF)

γPF is the classic biological model for pheromone update. Current pheromone decays based on the amount of time since the last packet arrived, and the pheromone carried by the new packet, γ, is added to the total. γ is equivalent to the utility of the path that the packet has taken.

$$P^n_{r,s} \leftarrow P^n_{r,s} e^{-\left(t-t^n_{r,s}\right)\tau} + \gamma \tag{7.5}$$

γPF is directly comparable to a biological model of pheromone deposition with exponential decay.

Normalized γ Pheromone Filter ($\bar{\gamma}$PF)

$\bar{\gamma}$PF is a normalized version of γPF. It is a normalized one-tap infinite impulse response averaging filter. Only a fraction of the received pheromone is added based on link observation time. $\bar{\gamma}$PF effectively limits the amount of pheromone on a link with its averaging properties. γPF allows much larger amounts since it is unnormalized.

$$P^n_{r,s} \leftarrow P^n_{r,s} e^{-\left(t-t^n_{r,s}\right)\tau} + \gamma\left[1 - e^{-\left(t-t^n_{r,s}\right)\tau}\right] \tag{7.6}$$

$\bar{\gamma}$PF has no biological equivalent. But by virtue of implementing a low pass filter, it is able to produce an estimate of the average utility of using a particular neighbor to arrive at a particular destination. γPF produces only a relative ranking, and ultimately causes only the best links to be used [2].

Probabilistic Bellman-Ford (pBF)

The probabilistic Bellman-Ford algorithm is designed to turn Termite into an asynchronous version of the Bellman-Ford (link state) routing algorithm. Unlike the filter techniques presented, this one is nonlinear. If a packet is received with information of a better path over the receiving link than is already known (taking into account pheromone decay), then the utility estimate is updated. Otherwise the pheromone table entry is left untouched.

$$if\ P^n_{r,s} e^{-\left(t-t^n_{r,s}\right)\tau} < \gamma,\ P^n_{r,s} \leftarrow \gamma \tag{7.7}$$

Route Discovery

In case there does not exist any destination pheromone at a node for a packet to follow, a route discovery procedure is initiated. Termite uses the traditional flooding approach and does not use any optimizations such as gossiping [32] or the expanding ring search as found in AODV. The reason for this is primarily one of complexity; because Termite requires so little control traffic in the first place, the extra complexity required to manage flooding optimizations is not deemed necessary. The use of flooding optimizations (or some other route discovery scheme alltogether) is entirely possible, and would result in a yet lower control packet overhead. Another tradeoff for route discovery schemes is discovery delay. A route request packet (RREQ) is broadcast and each receiver rebroadcasts it if it cannot answer the query. If the RREQ has been received before, the packet is dropped. Route requests also serve to spread source pheromone into the network. A route reply (RREP) packet is returned if a node is the request destination, or has destination pheromone in its pheromone table. This is done even if other neighbors have already transmitted a route reply to the RREQ source. A RREP is unicast back to the source normally by probabilistically following the source pheromone in the same way that data packets do. A RREP is formatted such that its source is the destination of the RREQ. This creates a destination pheromone trail back to the RREQ source. It is necessary to maintain a list of previously seen RREQ packets according to the source address and a source-unique sequence number in order to prevent the retransmission of previously received RREQs.

The packet that triggers a route request is cached for a route request timeout period before it is dropped. Any additional packets received during this period for the same destination are also cached. If the hold time since the first packet was held is exceeded then all of the held packets for the sought after destination are dropped. If a route reply is received while there are packets cached for the destination, they are processed normally according to the forwarding equation.

Route Repair

Termite has no concept of route repair in the traditional sense. Each next hop is computed online, and every node has an estimate of the utility of each link to deliver a packet to a destination. If a link should fail, the neighbor is simply removed from the routing table, the next-hop probabilities are recomputed for the remaining set, and the packet retransmitted. If all neighbor nodes are found to have disappeared (possibly after many unsuccessful retransmissions), the packet is dropped. There is no such thing as a route error or route error packet (RRER) in Termite.

Loop Prevention

A typical approach to loop prevention is for each node to maintain a list of packets already routed. If the same packet is seen a second time, an error procedure is executed, such as dropping the packet or sending control traffic into the network to fix routing tables [4] [10] [23] [28].

Termite does not make any special effort to prevent loops. All received packets are handled as described, regardless of the number of times they have visited any particular node. SPA helps to mitigate the number of routing loops by making hops towards the source less likely. Nodes closer to the source (measured according to the network metric) will have more source pheromone and less destination pheromone on them; travel towards the source is generally also travel away from the destination.

Termite is also very liberal with other packet reprocessing issues. Consider the case when node n transmits a packet to neighbor h, which then forwards the packet itself. Node n will overhear the retransmission and processes this packet normally.

The effect is not adverse because h will have updated the pheromone on the packet, which will have less an effect on n's table than the original update did since the utility can only go down. Besides, it is important to keep track of underperforming alternative routes in case they improve.

Hello Packets

Some protocols use special hello packets to help nodes advertise their existence to the network. This functionality is also possible in Termite if a node transmits a route request for itself. The time-to-live of such a packet should be set to one. Each receiving node will automatically have pheromone for the requesting node due to the pheromone update procedure, and will be able to return a route reply. The advertising node then also learns about its neighbors (and possibly they about each other). A node broadcasts self requests only when it has no known neighbors (the pheromone table is empty).

The implementation of Termite presented here does not use this HELLO functionality.

Packet Structure

There are three types of packets in Termite, data (DATA), route request (RREQ), and route reply (RREQ). They can be all considered within one generic packet format, with the fields listed below. This list does not assume a pheromone stack, first introduced by AntNet, which maintains the per hop metric in each packet. Certain information such as the previous and next hop IP address can often be obtained from lower layers of the network stack, as seen in the AODV specification [4]. The total size of the header shown here is 24 bytes, excluding any user data. If necessary, an additional four bytes of flags could be added.

- *Packet Type* This field is one byte in size. It's value describes the purpose of the packet, data, route request, or route reply.
- *Source IP Address* This field is four bytes and describes the IP address of the data source.
- *Destination IP Address* This field is four bytes and describes the IP address of the data destination.
- *Previous Hop IP Address* This field is four bytes and describes the IP address of the previous hop.
- *Next Hop IP Address* This field is four bytes and describes the IP address of the next hop.
- *Pheromone* This field is four bytes and describes the amount of pheromone carried by the packet.
- *Time-To-Live (TTL)* This field is one byte and describes the remaining allowed hop-count for the packet. It is initialized to the maximum TTL and decremented at each visited node. The packet is dropped when this counter reaches zero.

- *Data Length* This field if two bytes and describes the length of the data field in this packet; the amount of data carried by this packet.
- *Data* This field contains all of the data carried by the packet.

General Mechanisms

Two general mechanisms are used to enhance the effectiveness of the algorithm.

Piggybacked Routing Information

Packets carrying routing information is not a new technique by any means. However, it should be stressed that this information is piggybacked on *all* packets. Algorithms such as ANSI or AdHocNet use a stack to store more path history, but do so only for control packets. This approach is not currently implemented in Termite in order to reduce the complexity of the algorithm, though it would likely result in a performance increase at the cost of additional packet overhead.

Promiscuous Mode

Nodes are expected to exist in a broadcast (radio) medium. They may eavesdrop on the communications of neighbors and incorporate overheard routing data into their own routing table. This technique is used in many other MANET routing algorithms, and is found to be particularly useful in Termite. In principle it is not obligatory, however it does afford a notable increase in performance.

7.2.2 Simulation

A number of different scenarios are simulated to compare the performance of Termite using the presented pheromone update methods. These results are also compared to the standard MANET Ad-hoc On-demand Distance Vector routing protocol, as described in [4]. Simulations are designed to test the effect of node mobility on the global performance metrics.

Simulation Environment and Parameters

A common test scenario is used in which 100 mobile nodes are distributed uniformly over an area 2200 meters by 600 meters. Each node uses a simulated IEEE 802.11b MAC layer with 2 Mbps data rate and a 250 meter transmission range. The standard MAC layer has been modified to allow promiscuous reception of all in-range transmitted packets, and also to return unsuccessfully transmitted packets back to the routing layer for reassessment. Nodes move according to the random waypoint mobility model with zero pause time and a uniform speed. These parameters are the same as those in [33]. Speeds are varied over 1, 5, 10, and 20 meters per second. All runs are 600 seconds long and all data points are averaged over at least two runs for high speeds and five runs for low speeds. Ten nodes send 512 byte data packets with exponentially distributed interarrival times with a mean of 0.5 seconds to a unique communications partner, which replies to each received packet with an acknowledgement. Both AODV and Termite optimize for path length. AODV parameters are set according to the specifications found in [4]. The maximum time-to-live (TTL) for any packet is 32. Termite parameters include, $K = \frac{1}{32}$, $F = 10.0$, $\tau = 2.0$, and $A = 0.5$. The value of K is determined by the minimum possible received pheromone value. Since the TTL of any packet is 32, the minimum utility of any path is $\frac{1}{32}$. Termite holds packets for as long as AODV's ACTIVE_ROUTE_TIMEOUT parameter, which in this case is 1.28 seconds. The parameters are generally chosen to reflect a generic mobility and communciations scenario using radio characteristics similar to commercially available hardware, and mirror parameter choices made by the ad-hoc routing community. All simulations are completed using Opnet [34].

Evaluation Metrics

A number of metrics are used to determine the utility of the evaluated algorithms. These include data goodput, control packet overhead, control packet distribution, medium load, medium efficiency, medium inefficiency, link failure rate, and end-to-end delay.

Data Goodput

Data goodput is a classic evaluation metric for routing algorithms. It is the fraction of successfully delivered data packets. This metric should remain as high as possible under any circumstances.

Control Packet Overhead

Any routing algorithm should use as little control traffic as possible in order to successfully deliver data packets. Control packet overhead measures the fraction of control packets to the total number of transmitted packets in the system. Successively transmitted packets are counted individually.

Control Packet Distribution

This metric shows how many of each type of control packet were transmitted. This helps to identify the effectiveness of route request and discovery procedures.

Medium Load

This metric characterizes how inefficient the algorithm is in delivering packets. It is the ratio of the total number of packet transmissions, data or control, to the number of data packets successfully delivered. Successively transmitted packets are counted individually. This is not a general metric, but since the algorithms are optimizing for hop count, the fewer the transmissions the better. This metric should be as low as possible, however it will always be larger than the path length.

Medium Efficiency

Medium efficiency is the ratio of the number of transmissions of successfully arriving data packets to the number of total packet transmissions. Multiple transmissions of the same packet are counted individually. Since access to the communications medium often comes at a premium, it is important that it is only accessed in order to move packets that will arrive at their destination. Medium efficiency is a number between zero and one and values close to unity are desired. That would indicate that the only transmissions in the system are those for packets that are delivered.

Medium Inefficiency

This metric is related to the previous two and helps to fill in the full performance picture. Medium inefficiency is the fraction of transmitted packets to the number of data packets offered to the network for delivery. Lower numbers are better. Consideration of this metric should ensure that the routing algorithm is making an effort to deliver all packets, instead of just ones that are easy to deliver.

Link Failure Rate

The link failure rate measures the average number of links that are lost per node per second. It is a relative measure of how fast the network topology is changing, and thus how much time each node has to acquire a sensible local routing pattern before its local topology transforms.

End-To-End Delay

The average end-to-end delay of all successfully delivered data packets is measured. This metric gives another perspective on the overall performance of each algorithm. Delay should be minimized.

Results and Analysis

Data Goodput

As shown in Figure 7.1, the data goodput performance of Termite is higher than that of AODV. The latter is able to deliver at least 90% of its packets, and Termite outperforms it with a moderate 95%. The former sees a more graceful degradation of performance over the latter as node speed increases.

Control Overhead

Termite shows favorable control overhead properties as compared to AODV in Figure 7.2. Not only does Termite have an order of magnitude less control overhead, but it also produces a nearly constant amount over a large range of node mobility. This is true despite the use of flooding for route discovery, and speaks to the effective use of route information caching on the part of Termite. AODV suffers so much overhead because it floods the network with a new route discovery every time a route breaks. This weakness is partially addressed with an expanding ring search [12].

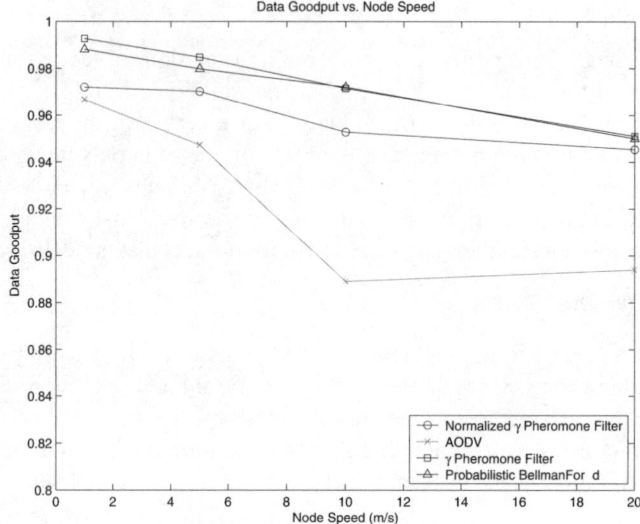

Fig. 7.1. Data Goodput vs. Node Speed

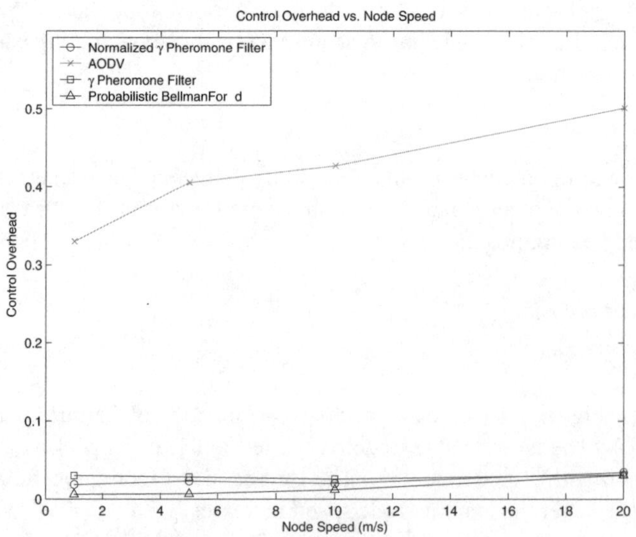

Fig. 7.2. Control Overhead vs. Node Speed

Control Packet Distribution

Termite uses little control traffic compared to AODV. As mentioned above, AODV must issue a route discovery flood whenever there is a route break. Termite is spared this because of its retransmission link repair policy. A full route discovery is almost never needed, which eliminates the majority of the control overhead. For AODV this proportion increases with node speed as links break more often. The most limiting factor is the number of route request packets; there are so many because such packets are flooded which ultimately generates a choking amount of overhead.

Alternatively, Termite produces more route reply packets than route requests. This attests to both the liberal route reply policy (any overhearing node with route information can generate a route reply) and also to the route caching of Termite. Since each node keeps route information about every other node, it is not difficult to find a pheromone trail.

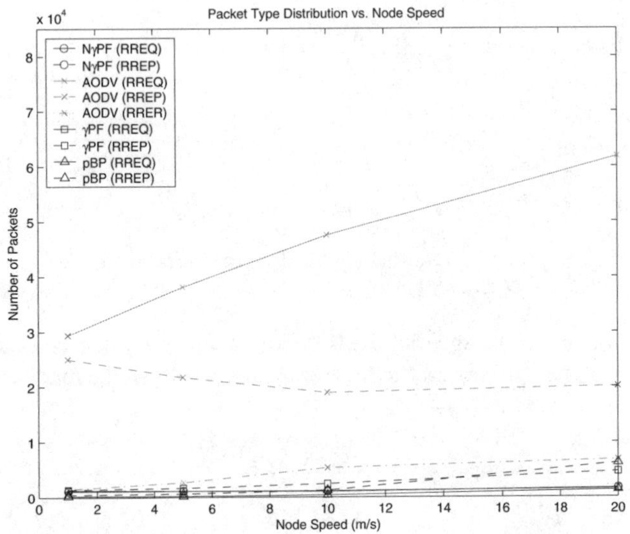

Fig. 7.3. Packet Type Distribution vs. Node Speed

Medium Load

While the control overhead statistics are quite positive, it is ultimately necessary to compare the total amount of access to the medium needed to deliver a packet. Figure 7.4 shows that Termite is able to do better than AODV with regards to the total number of transmissions required to deliver a packet successfully. They perform equally at 20 m/s. Both algorithms show an increasing load on the medium as the network volatility increases. In the case of AODV, this is because of the increasingly large amount of control traffic generated. For Termite, this is because data packets must take longer paths to explore the network as the topology changes faster.

Medium Efficiency

The results of Figure 7.5 show that Termite is able to easily deliver packets at low speeds. Both AODV and Termite expend more effort to deliver a packet successfully,

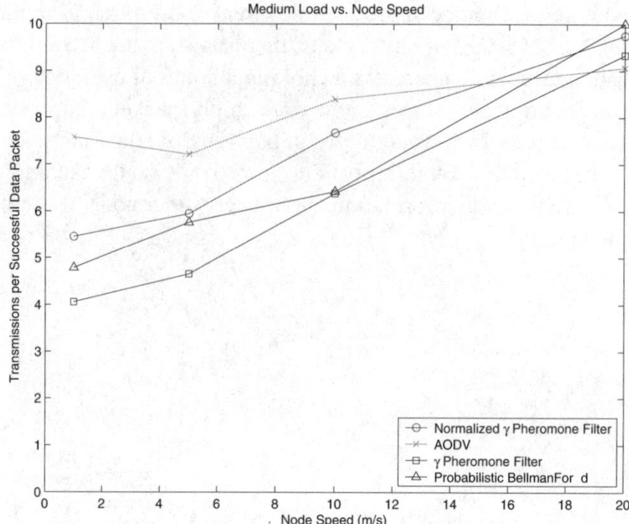

Fig. 7.4. Medium Load vs. Node Speed

again with similar performance at 20 m/s. Medium efficiency decreases linearly with node speed in all cases. This metric reconfirms the results of the medium load.

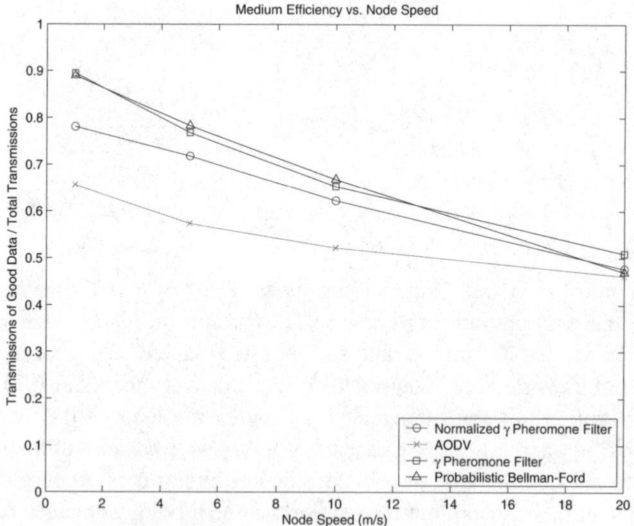

Fig. 7.5. Medium Efficiency vs. Node Speed

Medium Inefficiency

This metric complements the results of the previous two. Figure 7.6 shows that Termite makes a best effort to deliver all packets that are offered. It is able to do so with fewer packet transmissions at low speed but suffers at higher speed by a margin of 13%. The reason for this is due to unsuccessful data packets that wander the network until they exceed their Time-To-Live (TTL). This behavior causes unnecessary transmissions which increases the metric. Naturally this happens more at high speeds when the network topology is more volatile. When the medium inefficiency is readjusted to include only successful data packets (the medium load in Figure 7.4), the outcome is somewhat more even.

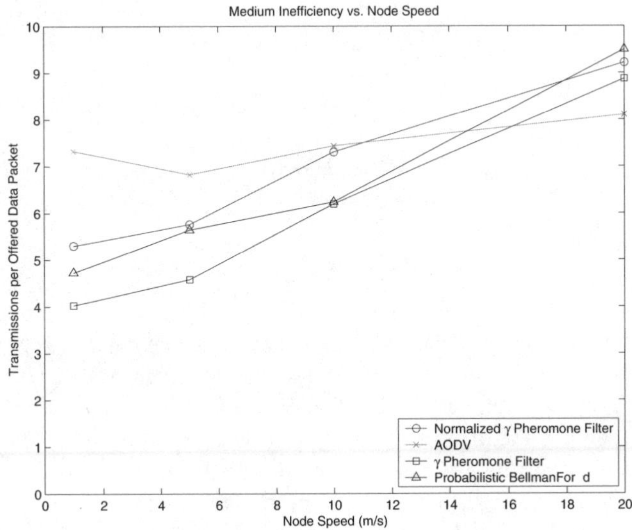

Fig. 7.6. Medium Inefficiency vs. Node Speed

Link Failure Rate

Figure 7.7 shows how the link failure rate changes with node speed. As expected from a mathematical analysis of link lifetime in [35], this trend is linear. The results may be somewhat skewed from standard link failure rates because this data is only measured when a communications attempt fails. That is, when a packet is sent on a link but the receiver is unavailable. Since the packet rate is relatively low, this measured link failure rate may also tend towards the low end. However, this metric does give an idea as to how quickly the network is changing and at what rate the nodes should reconfigure themselves.

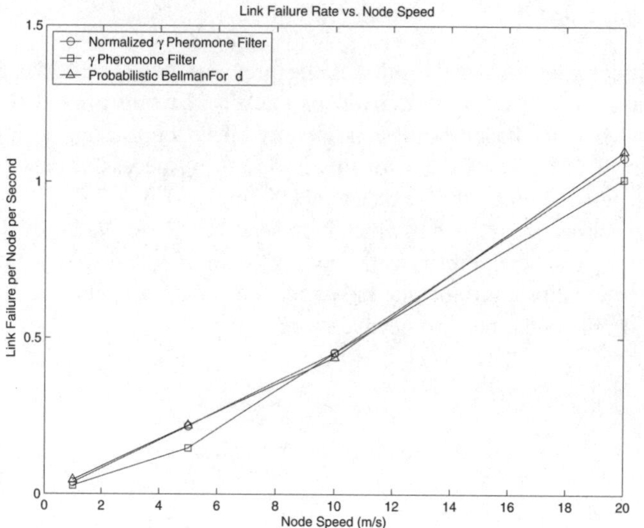

Fig. 7.7. Link Failure Rate vs. Node Speed

End-To-End Delay

Figure 7.8 shows the average end-to-end (ETE) delay for each of the compared routing algorithms. The ETE delay of AODV stays constant regardless of node speed while the delay of the Termite variants grows from less than AODV at low speed to substantially larger at high speed. The AODV results reflect those reported by [33] and is due to the fact that packets are only sent when a full route is known to exist. The Termite delay results are not positive considering that the delay of AODV is nearly an order of magnitude lower at high speeds. However, there are two extraneous issues at work in this situation. The first is that the metric only measures the delay of successful packets. AODV has a lower goodput than Termite, and so packets that might have otherwise timed out in AODV due to link or route discovery failure are delivered by Termite. This extra goodput comes at the expense of some additional delay incurred by perhaps a longer path length or congestion. It can be shown that when the slowest 5% of packets in Termite are not considered (the difference of goodputs between AODV and Termite at 20 m/s), then the delay can be reduced by 66%. This property indicates that Termite's high delay comes from statistical outliers. These are packets that require a great deal of effort to deliver, packets that AODV does not. The second factor is Termite's use of link recovery retransmission. When the network is changing quickly, packets may be retransmitted often due to neighbor loss. If several nodes in the same area are trying to retransmit, this can lead to localized packet storms. This is especially true when using 802.11 which automatically retries up to seven times before reporting a broken link.

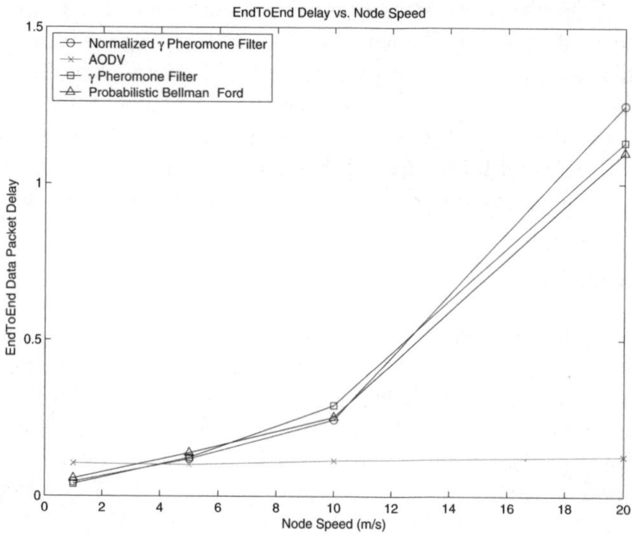

Fig. 7.8. End-To-End Delay vs. Node Speed

Discussion

The results show that Termite is able to outperform AODV primarily due to the lack of control traffic and effectively liberal route caching. AODV often finds itself repairing routes which requires route errors and route request floods. Termite avoids this complexity and effort through the use of piggybacked route information and promiscuous packet reception. However, Termite must find a way to explore the network as well in order to find better routes. It does this by letting data packets do the work. The medium load, efficiency, and inefficiency metrics show what the control overhead does not. The Termite approach works well at low speed but gives fewer performance gains at high speed. The number of packet transmissions per data packet is equal between the two algorithms; they both do the same amount of work to deliver packets. The reason for this is that Termite relies on packet traffic to update routing tables, and this traffic can move erratically over the network it is changing often or little traffic is available. A proactive update procedure such as that found in AdHocNet may be helpful. Termite need the path samples in a timely fashion in order to make good routing decisions. The most noticeable difference between the algorithms at high speed is the end-to-end delay. Termite's packet storms hurt performance tremendously.

7.3 Towards An Optimal Pheromone Decay Rate Heuristic

Termite has several parameters which influence its performance. This section derives a heuristic for the optimal pheromone decay rate, the decay rate which maximizes the performance of Termite, and compares it to simulation data. The heuristic is based only on the correlation time of the network, or how long the network stays relatively the same. This is measured in part by average link lifetime. Additional influences on performance are not considered in the heuristic at this time. If the selected decay rate is larger than the optimal, $\tau > \tau^*$, then the network will forget its state too fast and throw away relevant information. If the decay rate is too low, $\tau < \tau^*$, then the network will retain too much information and also make suboptimal decisions.

7.3.1 The Model

A simple model of the network is first introduced in order to lay an analytical framework for the heuristic. The network is modeled as two communicating nodes with two independent paths available between them [36]. These paths abstract all other connections between the two nodes, including additional nodes, mobility issues, or communications effects. The physical structure is shown in Figure 7.9, and is the same as that used in [2].

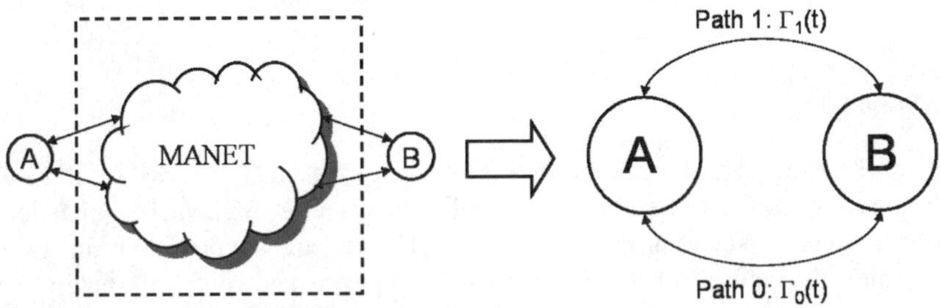

Fig. 7.9. Diagram of the MANET Model

Each node sends packets to the other with independent exponentially distributed interarrival times. The average rate at which node A sends packets to B is λ_A, and λ_B in the opposite direction. Each node decays the pheromone on its links independently. The decay rates at each node are τ_A and τ_B, respectively.

Each path, indexed by v, has a utility characterized by a non-negative random process $\Gamma_v(t)$ with mean $\mu_v(t)$. The pheromone contained in a packet arriving on a link, γ, is a sample of that process. Γ is non-stationary since link utilities change over time due to mobility and other effects, including the fact that the passage of each packet will change local routing probabilities due to pheromone update. Since each packet moving in the same direction passes through the network independently of all other packets, there is no correlation between successive samples of the link utility process. The forwarding equation independently considers each packet.

7.3.2 Optimal Mean Estimation of $\Gamma(t)$

$\Gamma^n_{r,s}(t)$ is a non-stationary stochastic process which models the received pheromone at node n on the link to neighbor r from source s. Because the process is non-stationary, the traditional set of stochastic analysis tools are not helpful. It is assumed that the process may be modeled as stationary over some finite period of time called the correlation time, T seconds. Under such circumstances, and keeping in mind that each received sample of $\Gamma^n_{r,s}(t)$ within T is assumed to be independent and identically distributed, the optimal mean estimator is a simple box filter of a length sufficient to include all and only those samples recieved within the last T seconds [37]. While the accuracy of the resulting estimation is dependant on the number of samples received and thus may vary, there is no better estimator.

7.3.3 Suboptimal Mean Estimation of $\Gamma(t)$

The box filter is an unwieldy approach. The filter length must be continuously updated in order to account for the number of received packets, and all of those pheromone values must be maintained. The mean estimation thus involves possibly a large number of addition operations as well as a division. An estimator requiring less state and computation is desired, though it may be suboptimal.

The biological inspiration for Termite holds the answer. A one-tap infinite impulse response filter is used in place of the optimal box filter, such as γPF or $\bar{\gamma}$PF. The pheromone decay rate can be adjusted in order to account for (changes in) the correlation time of Γ. A heuristic for the optimal pheromone decay rate for the γ filters is developed based on these ideas and then compared to simulation results.

7.3.4 The τ^* Heuristic

The γ pheromone filter and the normalized γ pheromone filter update rules are repeated in Equations 7.8 and 7.9 for reference.

$$P^n_{r,s} \leftarrow P^n_{r,s} e^{-(t-t^n_{r,s})\tau} + \gamma \tag{7.8}$$

$$P^n_{r,s} \leftarrow P^n_{r,s} e^{-(t-t^n_{r,s})\tau} + \gamma\left[1 - e^{-(t-t^n_{r,s})\tau}\right] \tag{7.9}$$

The heuristic is developed by examining the correlation between successive estimates of $E\Gamma(t)$, which is the pheromone resident on a link, $P(t)$, and adjusting the pheromone decay rate in order to minimize any correlation beyond the correlation time of the pheromone process. The current pheromone on a link may be generalized as in Equation 7.10.

$$P(t) = \Gamma(t) * h(t) \tag{7.10}$$

The function $h(t)$ is the impulse response of the estimation filter and the $*$ operator is convolution. The equivalency is shown for both of the pheromone filters in Equations 7.11 and 7.12.

$$h_{\gamma PF}(t) = e^{-\left(t-t_{r,s}^{n}\right)\tau}u(t) \tag{7.11}$$

$$h_{\bar{\gamma} PF}(t) = (1-\beta)\,e^{-\left(t-t_{r,s}^{n}\right)\tau}u(t) \tag{7.12}$$

The constant $(1-\beta)$ in Equation 7.12 is a normalization constant and $u(t)$ is the unit step function. The normalization constant for $h_{\bar{\gamma}PF}(t)$ is derived in [36]. For the purposes of this derivation the constant is unimportant.

The time correlation of the output of the estimation filter, which is the pheromone, can be found based on traditional techniques of statistical signal processing theory [37]. This includes finding the spectral density of $P(t)$. It is important to note in this case that individual samples of $\Gamma(t)$ are independent, since each packet is routed independently. It is also assumed that samples received within a period T of each other are identically distributed. The spectral density of $\Gamma(t)$ is thus flat. The correlation of either of the pheromone filters is shown in Equation 7.13.

$$R_{P_{PF}}(\Delta t) = e^{-|\Delta t|\tau} \tag{7.13}$$

The heuristic for the optimal value of τ, τ^*, can now be determined by finding its required value for the correlation to drop below a threshold at the correlation time. Equation 7.14 formalizes this requirement, where $c \ll 1$ is the threshold.

$$\tau^*(T) : R_{P_{PF}}(T) = e^{-T\tau^*} = c \tag{7.14}$$

The optimal pheromone decay rate is computed and is shown in Equation 7.15.

$$\tau^*(T) = -\frac{\ln c}{T} \tag{7.15}$$

7.3.5 A Return to the Box Filter

It was previously stated that a variable length box filter would be the optimal link utility estimator for this application. The filter should average all of the pheromone samples within the correlation time. It remains to be noted what the correlation structure is of the resulting estimate, such that the forwarding equation can properly account for time uncertainty in the estimate. Following the previous method used to calculate the correlation of the metric estimate, Equation 7.16 shows the result for the box filter.

$$R_{box}(\Delta t) = \begin{cases} 1-\frac{|\Delta t|}{T} & , \quad |\Delta t| \leq T \\ 0 & , \quad |\Delta t| > T \end{cases} \tag{7.16}$$

Since the box filter has only a finite length in time, T, estimates more than T seconds apart have no correlation. In the context of metric estimation, since no node maintains information about the network more than T seconds old, the estimator cannot provide any reliable information about the network at that time; the pheromone is all gone.

7.3.6 Generalization of the Forwarding Equation

When using general estimators, correlation functions must be added to the forwarding equation to account for uncertainty in the estimate introduced by time. Different estimators will have different correlations between successive estimates as seen above. The generalized forwarding equation is shown in Equation 7.17.

$$p_{i,d}^{n} = \frac{\left[P_{i,d}^{n} R\left(t - t_{i,d}^{n}\right) + K\right]^{F}}{\sum_{j \in \mathcal{N}^{n}} \left[P_{j,d}^{n} R\left(t - t_{j,d}^{n}\right) + K\right]^{F}} \tag{7.17}$$

Note that when a pheromone filter is used, the pheromone decay model is recovered exactly. Otherwise the forwarding framework represents a utility estimator, taking into account time uncertainty in the estimator.

7.3.7 τ^* Simulation

Figures 7.10 and 7.11 show how the performance of Termite can vary according to the global pheromone decay rate. The parameter was kept constant in the previous simulations at $\tau = 2.0$ in order to make each scenario directly comparable. The figures test τ over two orders of magnitude at node speeds of 1 m/s and 10 m/s. As is shown, the appropriate selection of this parameter is of critical importance. Using the normalized γ pheromone filter at 1 m/s, $\tau \approx 0.1$ is the best choice. $\tau \approx 1.0$ is best at 10 m/s. This is most easily seen from the achieved goodput and medium efficiency. The performance of Termite was quite good in the first set of simulations, and these results show that the performance could be even better if more appropriate parameters are chosen. The parameters of the original simulations are held constant for the purposes of fair comparison. Otherwise, the parameter space of Termite is so large that it can be difficult to choose those giving the best performance in a given environment.

7.3.8 τ^* Heuristic Analysis

The quality of the heuristic is determined by comparing its predictions to experimental data. The primary difficulty is to determine a good metric for the network correlation time (T) or the network event rate (T^{-1}). This work will use the link failure rate as a lower bound for the event rate (and thus an upper bound on the correlation time). Figure 7.12 takes the results reported in Figures 7.10 and 7.11 and compares them to the τ^* heuristic. A correlation threshold of $c = 0.1$ is used to calculate the heuristic.

The heuristic and the measurements match quite well. The simplicity of the heuristic allows the possibility to dynamically compute the optimal pheromone decay rate locally for each node, or even each link at each node. This would represent a departure from biological possibility, however it also would improve the performance of the system, as seen from a purely mathematical perspective. Unfortunately it

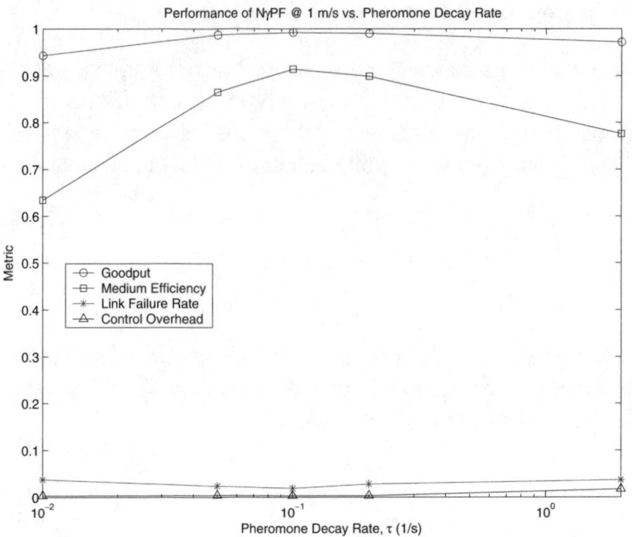

Fig. 7.10. Performance of NγPF @ 1 m/s vs. Pheromone Decay Rate, τ

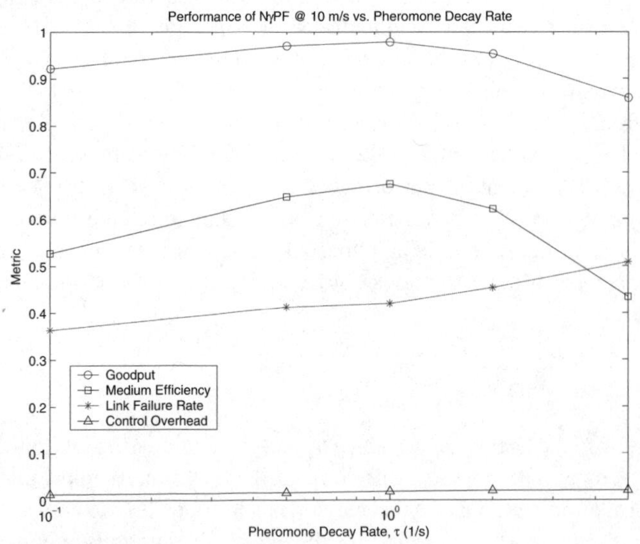

Fig. 7.11. Performance of NγPF @ 10 m/s vs. Pheromone Decay Rate, τ

is not trivial to generate an optimal decay rate in a rigorous way based only on samples of $\Gamma(t)$. A large number of calculations are generally required in addition to having a great deal of data on hand. Perhaps a heuristic could be generated which relates more easily to observed network parameters, such as the link lifetime (or link change rate measured here), to τ^*, as has been done here. There is the caveat

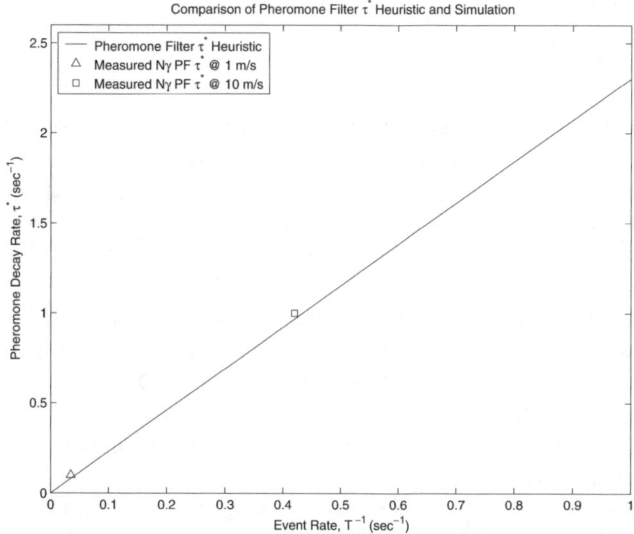

Fig. 7.12. Comparison of Pheromone Filter τ^* Heuristic and Simulation

however, that the link change rate is itself dependant on the decay rate whereas the network correlation time imagined in this work is an independent parameter. The link change rate is dependant because τ affects routing probabilities, which affects what neighbors are used, which affects how often those links are tested for connectivity, which affects the amount of time necessary to detect that a link has broken. The network correlation time is instead dependant on the parameters affecting the metric that the network is optimizing for. In this paper the path length is considered; the parameters controlling the underlying mobility model must be examined. But it is again unfortunate that these parameters, such as node speed, may not be readily available in an implementation.

7.3.9 Full Circle - A Return to Traditional Routing

Termite is a biologically inspired algorithm, but the final algorithm is closely related to well-known distance vector solutions. Traditional implementations update packets with the best known routing metric to its source from a node, instead of letting packets carry their actual path metric as in Termite. This difference is easily reconciled, and is formalized in the context of Termite in Equation 7.18.

$$\gamma \leftarrow \max_{i \in \mathcal{N}^n} P^n_{i,s} R\left(t - t^n_{i,s}\right) \tag{7.18}$$

The correlation term is included in order to properly account for uncertainty introduced by time, at each link. This update of a packet's pheromone occurs immediately before a packet is retransmitted to the next hop; after the packet pheromone is first updated with Equation 7.4, and after the pheromone table update.

7.4 Conclusion

7.4.1 Review

A swarm intelligent routing algorithm named Termite has been presented and its performance evaluated. Termite is an advancement of previous work with its addition of source pheromone aversion, a pheromone decay rate heuristic, and a generalized forwarding equation. A comparison to the standard AODV routing protocol, based on simulation studies, showed the general superiority of Termite. One reason for this superiority lies in the effectiveness of using data to carry routing information, reducing control overhead and maintaining routing information. However, it was seen that the end-to-end delay performance suffers due to the fraction of packets which are difficult to deliver. This may be due to frequent route breaks, route discovery latencies, and other network effects. Ultimately, Termite is shown to deliver more packets with less overhead in more adverse conditions than AODV in a realistic medium access environment.

A heuristic for an optimal pheromone decay rate was developed based on stochastic process theory. The problem was restated to make the question of pheromone update a question of optimal estimation. Optimal parameters were shown to exist through simulation, and the heuristic fits quite nicely with the results. It remains an open question of which independent parameters should be used to determine the optimal estimation parameters.

7.4.2 Future Work

Future work in this area can address a number of open issues. This work has investigated an optimal decay rate heuristic, thought it is still unclear how the other parameters which control the operation of Termite should be set or adjusted. Parameters of interest include the pheromone sensitivity, F, the pheromone threshold, K, and the source aversion sensitivity, A. These parameters could also conceivably be adjusted on a per node per link basis, though it is unclear if such an approach is possible, or even useful.

The results presented here show that Termite can find "acceptable" routing solutions "fast enough." It is still unknown how fast these solutions are reached, and what parameters help shape the solution. The development of a temporal model of SI routing could help to answer such questions, and also help to shed light on their ability to control the behavior of the network.

A discussion of parameter selection and temporal dynamics reveals a more general question regarding the control of networks. If every parameter of the routing algorithm can conceivably be adjusted in real time according the network dynamics, then the entire behavior of the network could perhaps also be controlled, regardless of the demands placed on it. As probabilistic routing is more easily mathematically analyzed than deterministic routing with non-linear rules, using such a framework may allow further insights into the control of large, distributed, dynamic systems.

References

1. E. Bonabeau, M. Dorigo, G. Theraulaz, *Swarm Intelligence: From Natural to Artificial Systems*, Oxford University Press, 1999.
2. D. Subramanian, P. Druschel, J. Chen, *Ants and Reinforcement Learning: A Case Study in Routing in Dynamic Networks*, Proceedings of the International Joint Conference on Artificial Intelligence, 1997.
3. N. Meuleau, M. Dorigo, *Ant Colony Optimization and Stochastic Gradient Descent*, *Artificial Life 8*, 2002.
4. C. Perkins, E. Belding-Royer, S. Das, *Ad-hoc On-demand Distance Vector (AODV) Routing*, http://moment.cs.ucsb.edu/AODV/aodv.html, 2004.
5. R. Bellman, *On a Routing Problem*, Quarterly of Applied Mathematics, 1958.
6. L. Ford Jr., D. Fulkerson, *Maximal Flow Through a Network*, Canadian Journal of Mathematics, 1956.
7. E. Dijkstra, *A note on two problems in connection with graphs*, Numerische Mathematik, Vol. 1, 269-271, 1959.
8. C. Perkins, P. Bhagwat, *Routing over Multihop Wireless Network of Mobile Computers*, SIGCOMM '94, 1994.
9. P. Jacquet, P. Mühlenthaler, T. Clausen, A. Laouiti, A. Qayyum, L. Viennot, *Optimized Link State Routing Protocol for Ad-Hoc Network*, IEEE INMIC, 2001.
10. D. Johnson, D. Maltz, *Dynamic Source Routing in Ad-hoc Wireless Networks*, SIGCOMM '96, 1996.
11. C. Siva Ram Murthy, B. S. Manoj, *Ad Hoc Wireless Networks: Architectures and Protocols*, Prentice Hall PTR, 2004.
12. C. Perkins, E. Belding-Royer, S. Das, *Ad-hoc On-demand Distance Vector (AODV) Routing*, http://www.faqs.org/rfcs/rfc3561.html, 2003.
13. P. Perkins, E. Royer, *Ad-hoc On-demand Distance Vector*, Proceedings of the 2nd IEEE Workshop on Mobile Computing Systems and Applications, 1999.
14. B. Holldobler, E. O. Wilson, *The Ants*, Belknap Press, 1990.
15. A. MacKenzie, S. Wicker, *A Repeated Game Approach to Distributed Control in Wireless Data Networks*, 2003.
16. L. Buttyan, J.-P. Hubaux, *Enforcing Service Availability in Mobile Ad-Hoc WANs*, First Annual Workshop on Mobile and Ad Hoc Networking and Computing, 2000.
17. M. Resnick, *Turtles, Termites, and Traffic Jams: Explorations in Massively Parallel Microworlds*, Bradford Books, 1997.
18. R. Schoonderwoerd, O. Holland, J. Bruten, L. Rothkrantz, *Ant-Based Load Balancing In Telecommunications Networks*, *Adaptive Behavior*, 1996.
19. G. Di Caro, M. Dorigo, *Mobile Agents for Adaptive Routing*,Technical Report, IRIDIA/97-12, Université Libre de Bruxelles, Belgium, 1997.
20. M. Heusse, D. Snyers, S. Guérin, P. Kuntz, *Adaptive Agent-Driven Routing and Local Balancing in Communication Networks*, ENST de Bretagne Technical Report RR-98001-IASC, 1997.
21. J. Baras, H. Mehta, *A Probabilistic Emergent Routing Algorithm for Mobile Ad-hoc Networks*, WiOpt '03: Modeling and Optimization in Mobile, Ad-Hoc, and Wireless Networks, 2003.
22. M. Günes, U. Sorges, I. Bouazizi, *ARA - The Ant-Colony Based Routing Algorithm for MANETs*, Proceedings of the ICPP Workshop on Ad Hoc Networks, 2002.
23. M. Günes, M. Kähmer, I. Bouazizi, *Ant Routing Algorithm (ARA) for Mobile Multi-Hop Ad-Hoc Networks - New Features and Results*, The Second Mediterranean Workshop on Ad-Hoc Networks, 2003.

24. M. Heissenbüttel, T. Braun, *Ants-Based Routing in Large Scale Mobile Ad-Hoc Networks*, Kommunikation in Verteilten Systemen (KiVS), 2003.
25. M. Roth, S. Wicker, *Termite: Ad-hoc Neworking with Stigmergy*, The Second Mediterranean Workshop on Ad-Hoc Networks, 2003.
26. S. Rajagopalan, C. Shen, *A Routing Suite for Mobile Ad-hoc Networks using Swarm Intelligence*, unpublished, 2004.
27. G. Di Caro, F. Ducatelle, L. M. Gambardella, *AdHocNet: An Adaptive Nature-Inspired Algorithm for Routing in Mobile Ad-Hoc Networks*, Technical Report No. IDSIA-27-04-2004, 2004.
28. F. Ducatelle, G. Di Caro, L. M. Gambardella, *Using Ant Agents to Combine Reactive and Proactive Strategies for Routing in Mobile Ad-Hoc Networks*, Technical Report No. IDSIA-28-04-2004, 2004.
29. G. Di Caro, F. Ducatelle, L.M. Gambardella, *Swarm Intelligence for Routing in Mobile Ad-Hoc Networks*, IEEE Swarm Intelligence Symposium, 2005.
30. K. M. Sim, W. H. Sun, *Ant Colony Optimization for Routing and Load-Balancing: Survey and New Directions*, IEEE Transactions on Systems, Man, and Cybernetics - Part A: Systems and Humans, Vol. 33, No. 5, September 2003.
31. G. Varela, M. Sinclair, *Ant Colony Optimization for Virtual-Wavelength-Path Routing and Wavelength Allocation*, Proceedings of the 1999 Congress on Evolutionary Computation, 1999.
32. L. Li, Z. Haas, J. Halpern, *Gossip-Based Ad-hoc Routing*, The 21st Annual Joint Conference of the IEEE Computer and Communications Societies (INFOCOM), 2002.
33. C. Perkins, E. Royer, S. Das, M. Marina, *Performance Comparison of Two On-Demand Routing Protocols for Ad-Hoc Networks*, IEEE Personal Communications, February 2001.
34. http://www.opnet.com/
35. P. Samar, Z. Haas, *On the Behavior of Communication Links of a Node in a Multi-Hop Mobile Environment*, The Fifth ACM International Symposium on Mobile Ad-Hoc Networking and Computing (MobiHoc), 2004.
36. M. Roth, S. Wicker, *Asymptotic Pheromone Behavior in Swarm Intelligent MANETs: An Analytical Analysis of Routing Behavior*, Sixth IFIP IEEE International Conference on Mobile and Wireless Communications Networks (MWCN), 2004.
37. P. J. Brockwell, R. A. Davis, *Introduction to Time Series and Forecasting, Second Edition*, Spring-Verlag, 2002.

8

Stochastic Diffusion Search: Partial Function Evaluation In Swarm Intelligence Dynamic Optimisation

Kris De Meyer[1], Slawomir J. Nasuto[2], and Mark Bishop[3]

[1] King's College London, University of London, UK
kris_demeyer@kcl.ac.uk
[2] Department of Cybernetics, The University of Reading
Whiteknights, Reading, RG6 6AY, UK
s.j.nasuto@reading.ac.uk
[3] Department of Computing, Goldsmiths College
New Cross, London, SE14 6NW, UK
m.bishop@gold.ac.uk

Summary. The concept of partial evaluation of fitness functions, together with mechanisms manipulating the resource allocation of population based search methods, are presented in the context of Stochastic Diffusion Search, a novel swarm intelligence metaheuristic that has many similarities with ant and evolutionary algorithms. It is demonstrated that the stochastic process ensuing from these algorithmic concepts has properties that allow the algorithm to optimise noisy fitness functions, to track moving optima, and to redistribute the population after quantitative changes in the fitness function. Empirical results are used to validate theoretical arguments.

8.1 Introduction

In recent years there has been growing interest in a distributed mode of computation utilising interaction between simple agents, (e.g., evolutionary algorithms; particle swarm optimisation; ant algorithms etc.). Certain of these "swarm intelligence" systems have been directly inspired by observing interactions between social insects, such as ants and bees. For example, algorithms inspired by the behaviour of ants – ant algorithms – typically use the principle of communication via pheromone trails to successfully tackle hard search and optimisation problems (see [19] for a recent review). This indirect form of communication, based on modification of the physical properties of the environment, has been termed stigmergetic communication. The problem solving ability of these algorithms emerges from the positive feedback mechanism and spatial and temporal characteristics of the pheromone mass recruitment system they employ. Other swarm intelligence methods

K.D. Meyer et al.: *Stochastic Diffusion Search: Partial Function Evaluation In Swarm Intelligence Dynamic Optimisation*, Studies in Computational Intelligence (SCI) **31**, 185–207 (2006)
www.springerlink.com

explore mechanisms based on biological evolution, flocking behaviour, brood sorting and co-operative transport, [28].

Independently of the above mechanisms, Stochastic Diffusion Search (SDS) was proposed in 1989 as a population-based pattern-matching algorithm [3] [4]. Unlike stigmergetic communication employed in ant algorithms, SDS uses a form of direct communication between agents (similar to the tandem calling mechanism employed by one species of ants, Leptothorax Acervorum, [33]).

SDS uses a population of agents where each agent poses a hypothesis about the possible solution and evaluates it partially. Successful agents repeatedly test their hypothesis while recruiting unsuccessful agents by direct communication. This creates a positive feedback mechanism ensuring rapid convergence of agents onto promising solutions in the space of all solutions. Regions of the solution space labelled by the presence of agent clusters can be interpreted as good candidate solutions. A global solution is thus constructed from the interaction of many simple, locally operating agents forming the largest cluster. Such a cluster is dynamic in nature, yet stable, analogous to "a forest whose contours do not change but whose individual trees do" [1].

Optimisation problems with stochastic and dynamically changing objectives pose an interesting challenge to many swarm intelligence algorithms which require repeated (re)evaluations of the fitness function. For certain applications the computational cost of these evaluations can prove prohibitive: e.g., for online tracking of rapidly changing objectives, or for computationally expensive fitness functions. In addition, in the case of genetic optimisation of dynamically changing objectives, an additional complication comes from the tendency of selection mechanisms to reduce diversity in the population (population homogeneity), potentially resulting in inadequate responses to subsequent changes in the fitness function. The first issue has previously been addressed by methods that attempt to reduce the amount of evaluation work performed, e.g., by estimating fitness values or by evaluating cheap, approximative fitness functions instead. The second issue has typically been addressed by methods introducing or preserving diversity in the population.

SDS handles these two problems in a related, but slightly different manner: firstly, it utilises the radically different concept of *partial evaluation* of fitness functions to save on the computational cost of repeated evaluations, reminiscent of the partial information available to individual social insects as they engage in recruitment behaviour. Secondly, the variation and selection mechanisms employed by SDS offer a new solution to the population homogeneity problem providing an alternative mechanism to balance the tradeoff between a wide *exploration* of all feasible solutions and a detailed *exploitation* of a small number of them.

This chapter introduces SDS in the context of swarm intelligence algorithms and demonstrates its applications in the field of stochastic and dynamic optimisation. The chapter is structured as follows: Sect. 8.2 discusses interaction mechanisms in social insects. Section 8.3 introduces partial evaluation of the fitness function and the balance between exploration and exploitation in search (the allocation of resources). In Sect. 8.4, an in-depth account of the standard SDS algorithm is provided. Section

8.5 examines the similarities and differences between SDS and Swarm Intelligence algorithms. Alternative mechanisms for the manipulation of resource allocation in SDS are discussed in Sect. 8.6. Section 8.7 illustrates the use of SDS in a few simple stochastic and dynamic optimisation problems. Finally, discussion and conclusions are presented in Sect. 8.8 and Sect. 8.9 respectively.

8.2 Mechanisms of Interaction in Social Insects

Swarm intelligence views the behaviour of social insects – ants, bees, termites and wasps – as offering a powerful problem solving metaheuristic with sophisticated collective intelligence. Composed of simple interacting agents, this intelligence lies in a network of interactions among the individuals and between the individuals and the environment [6].

Social interaction in ants [24] and honey bees [21] [43] has evolved an abundance of different recruitment strategies with the purpose of assembling agents at some point in space for foraging or emigration to a new nest site.

Such recruitment forms can be local or global, one to one or one to many, deterministic or stochastic. The informational content of the interaction ranges from very simple to complex and can be partial or complete. However, all such recruitment mechanisms propagate useful information through the colony as a whole.

Often, the recruitment is based on exchange of very simple stimulative information to trigger a certain action. Although the stimulative effect of a recruitment signal is typically mixed with the directional function of the signal, they actually constitute different functions: the stimulative function is merely used to induce following behaviour in other individuals, whereas the directional function conveys the information of where exactly to go.

In ants, chemical communication through the use of pheromones constitutes the primary form of recruitment. From an evolutionary viewpoint, the most primitive strategy of recruitment seems to be tandem running: a successful foraging ant will, upon its return to the nest, attract a single ant (different strategies exist - chemical, tactile or through motor display) and physically lead this ant to the food source.

In so-called group recruitment, an ant summons several ants at a time, then leads them to the target area. In more advanced group recruitment strategies, successful scouts lay a pheromone trail from the food source to the nest. Although this trail in itself does not have a stimulative effect, ants that are stimulated by a motor display in the nest can follow the trail to the food source without additional cues from the recruiter.

Finally, the most developed form of recruitment strategy is mass recruitment. Stimulation occurs indirectly: the pheromone trail from nest to food source has both a stimulative and directional effect. Worker ants encountering the trail will follow it without the need for additional stimulation. Individual ants deposit an amount of pheromones along the trail, dependent on the perceived quality or type of the food source. The outflow of foragers is dependent on the total amount of pheromone discharged. Recruitment strategies during emigration to new nest sites show a similar wide variety of physiology and behaviour.

In honeybees, both stimulation and orientation occur primarily via motor display. Bees that have successfully located a source of nectar or pollen will engage in so called *waggle dances*. The direction of the dance indicates the direction of the food source, whereas the velocity of the dance depends on the distance to the find. The perceived quality and accessibility of the food source influence the probabilities that a particular forager becomes a dancer, continues exploiting the food source without recruiting or abandons the food source and becomes a follower. A follower bee follows the dance of one randomly chosen dancing bee, then tries to find the food source indicated by that bees dance.

When compared to the stimulative function of recruitment strategies in ants, bees can be said to practice group recruitment: each bee directly recruits several other bees during its time on the dance floor. However, the directional function is very different. Whereas ants either have to lead the follower to the food source - which is time consuming - or leave signposts along the way; bees do neither. They have evolved a form of symbolic communication, more adapted to their specific conditions.

Different foraging and recruitment strategies induce different quantitative performances. For ants, it was demonstrated that tandem recruitment is slower than group recruitment, which in turn is slower than mass recruitment [12]. Also, the degree of accuracy - how many ants reach the food source for which they have been recruited - is dependent on the type of communication used and differs greatly from species to species [17].

Whatever the exact details of the recruitment behaviour, it leads to a dynamical balance between exploration of the environment and exploitation of the discovered food sources. Abstracting the social interaction and recruitment mechanisms observed in insect societies has inspired the design of many of the artificial swarm intelligence methods. The next section will concentrate on one such heuristic abstracted from natural systems – that of partial information exchange – and discuss its implications for search efficiency.

8.3 The Concept of Partial Evaluation

Many functions that have commonly been used as benchmark problems for swarm intelligence algorithms (e.g., evolutionary algorithms, particle swarm optimisation, etc.) typically have relatively small evaluation costs [18, 44]. This stands in stark contrast with real-world applications, which are not necessarily so well-behaved – for several possible reasons: the evaluation cost of a single candidate solution may be a rapidly-increasing function of the number of parameters, as e.g., for some problems in seismic data interpretation [44]; or, even an evaluation cost that is linear in the number of function parameters can be excessively high: for example, the selection of sites for the transmission infrastructure of wireless communication networks can be regarded as a *set-cover* problem [20] with an evaluation cost of candidate solutions that is linear in the number of sites; however, the evaluation of a single site involves costly radio wave propagation calculations [25]. Hence for swarm intelligence algorithms which explicitly evaluate costly fitness functions, it is not only important to limit the total number of fitness evaluations, but also the amount of computational work that is performed during the evaluation of a single candidate solution. This is exceedingly true for stochastic and dynamically changing problems, which may require multiple and continuing function evaluations.

The problem of costly function evaluations has been addressed many times independently for static and dynamic, noisy and noise-free problem settings (see [26] for a recent review). Two somewhat different approaches exist: firstly, the fitness of part of the individuals can be *estimated* – rather than calculated – from the fitness of other individuals or individuals from previous generations using tools from statistics [9, 27]. In the second line of approach, the costly fitness function is replaced with a cheaper, approximate fitness function, which is evaluated instead; when the search has started to converge, the computational process can switch to evaluating the original fitness function to ensure correct convergence [26].

In contrast, by analogy to the partial information about the environment available to individuals in insect societies, the approach advocated here capitalises on the fact that many fitness functions are *decomposable* into components that can be evaluated independently. An evaluation of only one or a few of these components – a *partial evaluation* of the fitness function – may still hold enough information for optimisation purposes. The next section will introduce a metaheuristic based on partial evaluation of fitness function.

8.4 Stochastic Diffusion Search

Stochastic Diffusion Search (SDS) is an efficient generic search method, originally developed as a population-based solution to the problem of best-fit pattern matching. SDS uses a one-to-one recruitment system akin to the tandem-running behaviour found in certain species of ants. In this section we will introduce the SDS algorithm and subsequently demonstrate that efficient global decision making can emerge from interaction and communication in a population of individuals each forming hypotheses on the basis of partial evidence.

We start by providing a simple metaphor, the restaurant game, that encapsulates the principles of SDS behaviour.

8.4.1 The restaurant game

A group of delegates attends a long conference in an unfamiliar town. Each night they have to find somewhere to dine. There is a large choice of restaurants, each of which offers a large variety of meals. The problem the group faces is to find the best restaurant, that is the restaurant where the maximum number of delegates would enjoy dining. Even a parallel exhaustive search through the restaurant and meal combinations would take too long to accomplish. To solve the problem delegates decide to employ a Stochastic Diffusion Search.

Each delegate acts as an agent maintaining a hypothesis identifying the best restaurant in town. Each night each delegate tests his hypothesis by dining there and randomly selecting one of the meals on offer. The next morning at breakfast every delegate who did not enjoy his meal the previous night, asks one randomly selected colleague to share his dinner impressions. If the experience was good, he also adopts this restaurant as his choice. Otherwise he simply selects another restaurant at random from those listed in 'Yellow Pages'.

Using this strategy it is found that very rapidly significant number of delegates congregate around the best restaurant in town. Abstracting from this algorithmic process:

```
Initialisation phase
  whereby all agents (delegates) generate
  an initial hypothesis (restaurant)
loop
  Test phase
    Each agent evaluates evidence for its hypothesis
    (meal degustation). Agents divide into active
    (happy diners) and inactive (disgruntled diners).
  Diffusion phase
    Inactive agents adopt a new hypothesis by either
    communication with another agent (delegate) or, if the
    selected agent is also inactive, there is no information
    flow between the agents; instead the selecting agent
    must adopt a new hypothesis (restaurant) at random.
endloop
```

By iterating through test and diffusion phases agents stochastically explore the whole solution space. However, since tests succeed more often on good candidate solutions than in regions with irrelevant information, an individual agent will spend more time examining good regions, at the same time recruiting other agents, which in turn recruit even more agents. Candidate solutions are thus identified by concentrations of a substantial population of agents.

Central to the power of SDS is its ability to escape local minima. This is achieved by the probabilistic outcome of the partial hypothesis evaluation in combination with reallocation of resources (agents) via stochastic recruitment mechanisms. Partial hypothesis evaluation allows an agent to quickly form its opinion on the quality of the investigated solution without exhaustive testing (e.g. it can find the best restaurant in town without having to try all the meals available in each).

Terminology

In the original formulation of SDS a population of *agents* searches for the best solution to a given optimisation problem. The set of all *feasible* solutions to the problem forms the *solution space* $\mathcal{S}$. Each point in $\mathcal{S}$ has an associated *objective* value. The objective values taken over the entire solution space form an *objective function* f. For simplicity reasons, it is assumed that the objective is to minimise the sum of n $\{0,1\}$-valued *component functions* f_i:[4]

$$\min_{\forall s \in \mathcal{S}} f(s) = \min_{\forall s \in \mathcal{S}} \sum_{i=1}^{n} f_i(s) \qquad f_i : \mathcal{S} \rightarrow \{0,1\} \ . \tag{8.1}$$

Although this may seem as a serious restriction, many optimisation problems can actually be transformed into (8.1) – as explained in [31]. Section 8.7 will also give an example of such a transformation. During operation, each agent maintains a *hypothesis* about the best solution to the problem; a hypothesis is thus a candidate solution, or designates a point in the solution space. No a-priori assumptions are made about the representation of hypotheses: they can be binary strings, symbolic strings, integer numbers, or even (at least in theory) real numbers.

Algorithm

Agents in the original SDS algorithm operate synchronously. They undergo various stages of operation, which are summarised in the algorithm below

```
Initialise(Agents);
repeat
  Test(Agents);
  Diffuse(Agents);
until (Halting Criterion);
```

Initialise

As a first step, agents' hypothesis parameters need to be initialised. Different initialisation methods exist, but their specification is not needed for the basic understanding of the algorithm; a discussion can be found in [31].

Test

Each agent randomly selects a single component function f_i, $i \in \{1,\ldots,n\}$, and evaluates it for its particular hypothesis $s_h \in \mathcal{S}$. Based on the outcome of the evaluation, agents are divided into two groups: *active* and *inactive*. For active agents, $f_i(s_h) = 0$; for inactive agents, $f_i(s_h) = 1$. Please note that, by allowing f_i to be probabilistic, it is possible that different evaluations of $f_i(s_h)$ have a different outcome. The test phase is described in pseudo-code below.

```
for agent = 1 to (All Agents)
  cf = Pick-Random-Component-Function();
  if (cf(agent.hypothesis) == 0)
    agent.activity = TRUE;
```

[4] Component functions f_i can be deterministic or probabilistic.

```
  else
    agent.activity = FALSE;
  end
end
```

Diffuse

During the *diffusion phase*, each inactive agent chooses at random another agent for communication. If the selected agent is active, then the selecting agent copies its hypothesis: *diffusion* of information. If the selected agent is also inactive, then there is no flow of information between agents; instead, the selecting agent adopts a new random hypothesis. Active agents, from their side, do not start a communication session in standard SDS. The diffusion phase is summarised below.

```
for agent = 1 to (All Agents)
  if (agent.activity == FALSE)
    agent2 = Pick-Random-Agent(Agents);
    if (agent2.activity == TRUE)
      agent.hypothesis = agent2.hypothesis;
    else
      agent.hypothesis = Pick-Random-Hypothesis();
    end
  end
end
```

Halt

Several different types of halting criteria exist [31]; their specification is not needed for the understanding of the algorithm. The most simple halting criterion could be based on reaching a prescribed threshold of a total number of active agents.

From agent operation to population behaviour

The algorithmic description of agent operation is insufficient to understand how SDS solves optimisation problems. Therefore, it is necessary to consider what happens with the population as a whole: by iterating through test and diffusion phases individual agents continually explore the entire solution space. Since tests succeed more often in points in the solution space with good objective values, agents spend on average more time examining high-quality solutions, at the same time attracting other agents, which in turn attract even more agents – a mechanism that causes dynamic yet stable clusters of agents to form in certain points in the solution space. However, the limitedness of resources (the finite population size) ensures that only the *best* solution discovered so far is able to maintain a stable cluster of agents. It is this disproportionate allocation of resources that eventually allows the optimal solution to be identified from the largest cluster of agents, without any single agent ever evaluating the full objective function explicitly.

The stochastic process underlying the resource allocation in standard SDS – an ergodic Markov chain – has been thoroughly analysed [36]. The behaviour of the process is determined by probabilities of producing active agents during the test phase. For each candidate solution, these probabilities, averaged over all component functions, form the *test score* of the optimisation problem. The v test score does not only depend on the values of the objective function, but also on the particular test procedure used. Convergence times and average cluster size are functions of population size and the test score [36].

8.4.2 Previous Work on SDS

SDS was introduced in [3] [4] and subsequently applied to a variety of real-world problems: locating eyes in images of human faces [5]; lip tracking in video films [23]; self-localisation of an autonomous wheelchair [2] and site selection for wireless networks [25]. Furthermore, a neural network model of SDS using spiking neurons has been proposed [37]; [38]. Emergent synchronisation across a large population of neurons in this network can be interpreted as a mechanism of attentional amplification [16]. The analysis of SDS includes the characterisation of its steady state resource allocation [36], the proven convergence to the globally optimal solution [39] and linear time complexity [40].

8.5 Similarities and Differences between SDS and Social Insects Algorithms

8.5.1 Comparison with social insects

Contrary to the stigmergetic communication used in most ant algorithms, SDS uses a one-to-one recruitment system akin to the tandem-running behaviour found in certain species of ants. With reference to SDS it is claimed that efficient global decision making can emerge from interaction and communication in a population of individuals each forming hypotheses on the basis of partial evidence.

The recruitment process in real insects is much more complex than that used in SDS where the process of communicating a hypothesis has been completely abstracted. An agent does not have to go through a lengthy and possibly erroneous process of tandem running or waggle dancing to communicate its hypothesis parameters to another agent.

Although no ant or bee species matches exactly the recruitment behaviour of inactive or active agents in SDS, Pratt et al [42] describe the collective decision making strategy of a species of ants that use a similar tandem running recruitment strategy during nest migration. They come to the conclusion that these ants need higher individual cognitive abilities - such as the ability to compare the quality of two nest sites - to come to an optimal solution, as opposed to ants using stigmergetic forms of communication.

Nevertheless, the fundamental similarity between SDS and social insects suggests that global and robust decision making in both types of systems emerges quickly from the co-operation of constituent agents, each of which individually would not be able to solve the problem within the same time frame.

8.5.2 Comparison with Ant Algorithms

Both SDS and ant algorithms are population-based approaches to search and optimisation that use a form of communication reminiscent of communication in real ants. However, most ant algorithms, and especially the ones described by the ant colony optimisation metaheuristic [19], rely on the idea of stigmergetic communication. Good solutions emerge from temporal and spatial characteristics of the recruitment strategy: short routes receive more pheromones because it takes less time to travel them. In SDS, communication is direct, one-to-one and immediate; solutions do not emerge from temporal aspects of the recruitment system, but merely from the end result of recruitment - the spatial clustering of agents.

Non-stigmergetic ant algorithms have also been proposed. It was shown in [29] that a tandem running recruitment mechanism improves the foraging efficiency of a colony of robots. Further, an optimisation algorithm based on the foraging strategy of a primitive ant species has also been proposed, [34]. This algorithm - called API - alternates between evaluation phases and nest replacement phases. During evaluation, ants explore random points in a certain area around the nest site and remember the best sites. The evaluation phases allow for recruitment between ants: an ant with a better solution can summon an ant with a poorer solution to help it explore its area. However, recruitment on this level did not seem to improve significantly the results obtained. Nest replacement in API can also be considered as a form of recruitment: all the ants are summoned to the optimal point found so far, then start exploring anew. Although on a much slower time scale, the alternation between evaluation and nest replacement in API has similarities with the test and diffusion phases in SDS.

8.6 Variations on a Theme

Many variations of the standard SDS algorithm are possible: agent updates can occur synchronously for the whole population or asynchronously; the choice of another agent during diffusion can be restricted to agents in a certain neighbourhood or to the whole population; the activity of agents can be binary, integer or even real values, possibly reflecting the history of the agent; during testing, agents can vary the amount of evidence needed for a positive test of a hypothesis. During diffusion, agents can have different reactions to information from other agents, e.g. active agents could choose to communicate and modify their hypothesis according to the state of the contacted agent etc. Some of these modifications have been previously documented [2], [36], [14], [16]. Each of them has a distinct effect on the convergence and steady-state behaviour of the algorithm. However, it can be said that in all cases a dynamical balance between exploration of the solution space and exploitation of discovered solutions naturally emerges.

8.6.1 Manipulating the Resource Allocation Process

The resource allocation process of SDS can be manipulated in a number of ways by altering properties of the test and diffusion phase [31]. This section focusses on two modifications that are useful for application towards dynamic problems.

Shifting the balance towards local exploration

Standard SDS has no mechanism to exploit *self-similarity* in the objective function – a regularity exhibited by many real-world problems: namely the fact that nearby solutions in the solution space often have similar objective function values [13]. However, a mechanism introducing small variations on the diversity of hypotheses already present in the population can be easily incorporated into the algorithm. One possibility is to perturb the copying of hypotheses parameters by adding a small random offset during replication of a hypothesis in the diffusion phase, much like mutation in evolutionary algorithms. The effect thereof is to smear out large clusters of agents over neighbouring locations in the solution space. It allows the SDS process to implicitly perform hill-climbing – resulting in improved convergence times in solution spaces with self-similarity [31] – as well as tracking of moving peaks. An example in Sect. 8.7 will demonstrate the latter point.

Shifting the balance towards global exploration

The conflicting demands of a continued wide exploration of the solution space (especially in dynamic environments), versus the need for a stable cluster exploiting the best solution discovered so far, are not necessarily satisfied in the most optimal way by standard SDS. Its allocation process is very greedy: once a good solution is detected, a large proportion of the population is allocated towards its exploitation, making these agents unavailable for further exploration. A mechanism that frees up part of these resources without severely disrupting the stability of clusters would increase the efficiency of SDS for many classes of problems, including dynamic ones. One such mechanism is *context-sensitive* SDS [36]. The sole difference with standard SDS resides in the diffusion phase for *active* agents: each active agent selects one agent at random; if the selected agent is active and supports the same hypothesis, then the selecting agent becomes inactive and picks a new random hypothesis. This self-regulatory mechanism counteracts the formation of large clusters: the probability that two active agents with the same hypothesis communicate during the diffusion phase increases with relative cluster size. This introduces a mechanism of negative selection or negative feedback to the original algorithm. For certain test scores, it also allows the formation of clusters on multiple similar, near-optimal solutions.

8.6.2 Standard SDS and stochastic objective functions

Certain types of noise in the objective function may be completely absorbed in the probabilistic nature of the partial evaluation process, and do not influence the search performance of SDS: i.e., they have no effect on convergence times and stability of clusters. More formally, noise that introduces or increases variance in the evaluation of component functions f_i – without altering the averaged probabilities of the *test score* – has no effect on the resource allocation process.

Only when noise changes the values of the test score can the resource allocation process be affected, with a potential for positive as well as negative consequences: a bias which pushes the best test score values more up than poor test score values is likely to accelerate convergence and increase the stability of clusters; conversely, a bias that increases lower test scores more than the test score of the optimal solution will hamper search performance. In a worst case scenario, the bias could disturb the order of the test score and make SDS converge to a false optimum. However, without any knowledge about the probability distribution generating the noise, no optimisation method would be able to correct such noise. Section 8.7 presents an example demonstrating the robustness of SDS search performance to moderate amounts of noise.

8.6.3 Standard SDS and dynamic objective functions

In principle, standard SDS is immediately applicable to dynamically changing objective functions. The probabilistic outcome of the partial evaluation process, in combination with a continued random re-sampling of the solution space, means that the search process can reallocate its resources from a global optimum that has become sub-optimal to the new global optimum. Allocation of resources in standard SDS is *dynamic* and *self-regulatory*; however, it need not be *optimal*: for instance, no variational mechanism for the tracking of slowly moving peaks is present in the original formulation of SDS. However, as section 8.6.1 demonstrates, such a mechanism is easily included.

8.7 Examples

In general, synthetic dynamic benchmarks (as introduced in [7, 35]) make no assumptions about the computational costs of function evaluations. In other cases, objective functions that allow cheap function evaluations and that have often been used to benchmark optimisation algorithms in static, noise-free conditions – such as the DeJong test suite – have been adapted to reflect noisy [30] or dynamic [41] conditions. These two approaches do not allow to demonstrate the potential gain in algorithmic efficiency of partial function evaluation. It is therefore necessary to construct an alternative objective function that allows partial evaluation. Such a function can be constructed from the elementary image alignment problem depicted

Fig. 8.1. Image alignment problem. The task is to align a small image, taken from another image which was photographed from a slightly different angle, with this large image. The best alignment of the two images is indicated by the black rectangle

in Fig. 8.1. Please note that this example is meant as *proof of principle*, rather than an attempt to construct one optimised solution to a specific problem.

The problem consists of locating a small image within a larger image by finding the (x,y) transformation coordinates that produce the best match between the small image and a similar-sized part of the larger image. The small image is taken from another large image which was photographed from a slightly different angle. Sub-pixel sampling is not performed, meaning that the search space is discrete. The size of the solution space – all admissible combinations of x and y values – corresponds to the size of the large photograph, (300 by 860 pixels). The size of the small image is 30 by 40 pixels. The images are RGB colour images, meaning that 3 colour intensity values are available per pixel.

The measure to determine the degree of matching between the two images is the Manhattan distance over all colour intensity values R, G and B:

$$f(x,y) = \sum_{k,l}(|r_{kl} - R_{kl}(x,y)| + |g_{kl} - G_{kl}(x,y)| + |b_{kl} - B_{kl}(x,y)|) \tag{8.2}$$

Here r_{kl} stands for the red colour intensity value of pixel (k,l) in the small image, and $R_{kl}(x,y)$ for the red colour intensity value of pixel $(x+k, y+l)$ in the large image. The image alignment problem then consists of finding a solution to the problem:

$$\min_{x,y} f(x,y) \tag{8.3}$$

The motivation for choosing this particular problem is threefold: firstly, the solution space is small enough in size and number of dimensions so that it can be visualised (Fig. 8.2); secondly, the shape of the resulting landscape is more complex than is easily attainable with artificially constructed benchmark problems; thirdly,

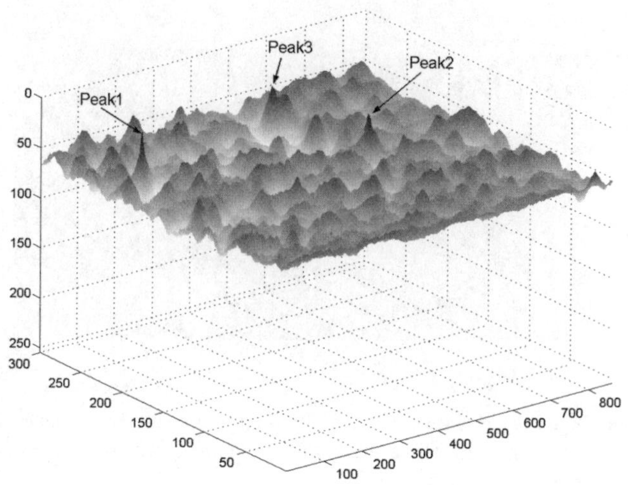

Fig. 8.2. Objective function generated from the image matching problem in Fig. 8.1. Peak 1 is the optimal solution. Peak 2 and 3 are of slightly lower quality. Peak 2 and 3 have been manually increased to make the problem more challenging

the objective function can be *decomposed* into *component functions* f_i (a single term of the summation in (8.2)) that can be evaluated independently.

The solution space $\mathcal{S}$ is 2-dimensional and discrete, with $x \in \{1,\ldots,860\}$ and $y \in \{1,\ldots,300\}$. The size of the solution space is $860 * 300 = 258000$. The number of component functions f_i is determined by the number of terms in the summation of (8.2) and hence by the size of the small image and the different colour intensity values: $30 * 40 * 3 = 3600$. Component functions f_i are discrete each with an integer range $[0, 255]$.

Minimisation problem (8.3) is easily transformed into problem (8.1); for component i and solution hypothesis (x, y), the test procedure should calculate the quantity:

$$t_i(x,y) = \frac{f_i(x,y)}{255} \tag{8.4}$$

The test procedure then needs to output 0 with probability $t_i(x,y)$, and 1 with probability $1 - t_i(x,y)$. This procedure ensures that the transformation of objective function values to test score values is strictly order-preserving, a sufficient condition for a correct optimisation of the objective function f [31].

Characterisation of the search problem

Unlike well-known benchmark problems such as the DeJong test suite or Schaffer's F6 function the structure of this specific problem is not well characterised in terms of its search difficulty. This section provides an empirical assessment of search difficulty by comparing the behaviour of SDS with several common optimisation algorithms on a noise free problem: random search, a multi-start best-improving hill climber, and a standard implementation of the particle swarm optimisation (PSO) algorithm[5]. The performance of SDS for noisy and dynamic perturbations of the objective function will be discussed in subsequent sections.

Random Search proceeds by choosing a solution at random and evaluating it until the optimal solution of Peak 1 in Fig. 8.2 is found.

Hill Climber A solution is chosen at random and evaluated. In subsequent iterations, the eight surrounding solutions are evaluated, and the search moves to the solution offering the greatest improvement in objective value. If no such improvement is possible (the search has arrived at a local optimum), then it is restarted in a new, randomly chosen location. These steps are performed until the optimal solution of Peak 1 is discovered.

PSO This algorithm follows the *local constriction* variant of PSO [28]. The algorithm runs until the optimal solution of Peak 1 has been discovered by at least 1 particle. Following parameters have been used: constriction coefficient $\chi = 0.729$; cognitive and social parameters $c1 = c2 = 2.05$; 200 particles with a neighbourhood radius of 1; and $\max V_x = 100$ and $\max V_y = 200$. These parameters have been chosen to give optimal search performance of the PSO algorithm. The reader is referred to [28] for details of the implementation.

SDS A standard SDS algorithm with a population size of 1000 agents has been used. A small mutational mechanism, as described in Sect. 8.6.1, has also been employed: during copying, the hypothesis is perturbed by adding a randomly generated offset to the (x,y) parameters of the active agent. The offset o_j is generated independently for x and y by:

$$o_j = \left[\frac{r}{s} \right] \tag{8.5}$$

where r is a normally-distributed random variable with zero mean and standard deviation 1, s is a parameter controlling the standard deviation of o_j, and $\left[\cdot \right]$ denotes rounding to the nearest integer. For this particular experiment, $s = 4$, resulting in an average copying accuracy of 92%, or in a mutation rate of 8%. The search is said to be converged when the optimal solution of Peak 1 has attracted a cluster of 1/3 of all agents. No other parameters need to be defined.

[5]Because noisy and dynamic conditions have led to several alternative PSO formulations that outperform the standard PSO algorithm under these specific conditions, this comparison is performed for noise-free and static conditions only. This will ensure that the characterisation of the search problem difficulty is not biased by the relatively poor performance of the standard PSO under these conditions.

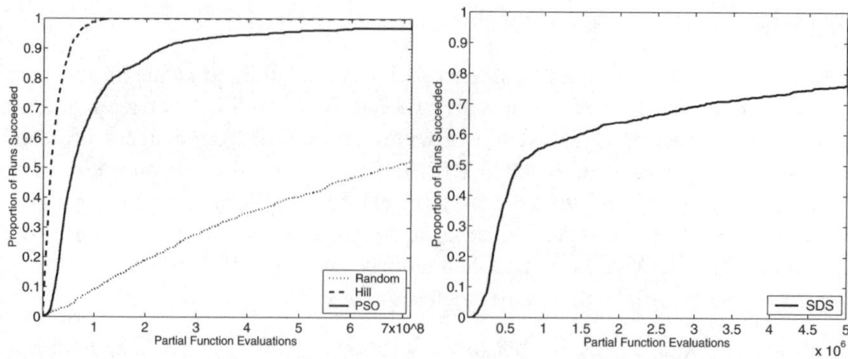

Fig. 8.3. Comparison of random search, hill climbing, PSO and SDS on the search problem of Fig. 8.2. Results are averaged over 1000 runs for each algorithm

Figure 8.3 compares the search behaviour of these four algorithms: Figure 8.3a shows the cumulative distribution of total number of *partial* function evaluations for random search, hill climbing and PSO. Fig. 8.3b shows the cumulative distribution of total number of *partial* function evaluations for SDS. The efficiency of partial evaluation can be illustrated by comparing the evaluation cost of SDS with that of the three other algorithms. For example, random search needs around 191000 complete evaluations of (8.2) to attain a 50% success rate of locating the global optimum, this corresponds to $191000 * 3600 = 687600000$ evaluations of component functions f_i. In contrast, SDS has a median of around 683000 component evaluations, a difference of three orders of magnitude. For comparison, PSO needs 16000 full function evaluations and hill climbing 4000.

However, rather than just comparing these numbers, it is interesting to see from how many component functions f_i onwards SDS starts to outperform the other algorithms. The probabilistic, partial evaluation mechanisms in SDS transforms the search into a stochastic dynamical process that is independent of the number of component functions in the objective function, and only depends on the exact shape of the landscape. In other words, whether the landscape of Fig. 8.2 is generated by a function consisting of 100, 1000 or 10000 component functions, the averaged search behaviour of SDS is always the same. For this particular landscape, SDS would outperform random search for objective functions consisting of $683000/191000 \approx 4$ or more component functions f_i. For PSO this number becomes $683000/16000 \approx 43$, and for hill climbing $683000/4000 \approx 171$. The relatively poor performance of PSO compared to the hill climber can be explained by the fact that the swarm consisted of 200 particles each performing full function evaluations. It is likely that the performance of PSO relative to hill climber would improve if, for example, the dimensionality of the problem were increased.

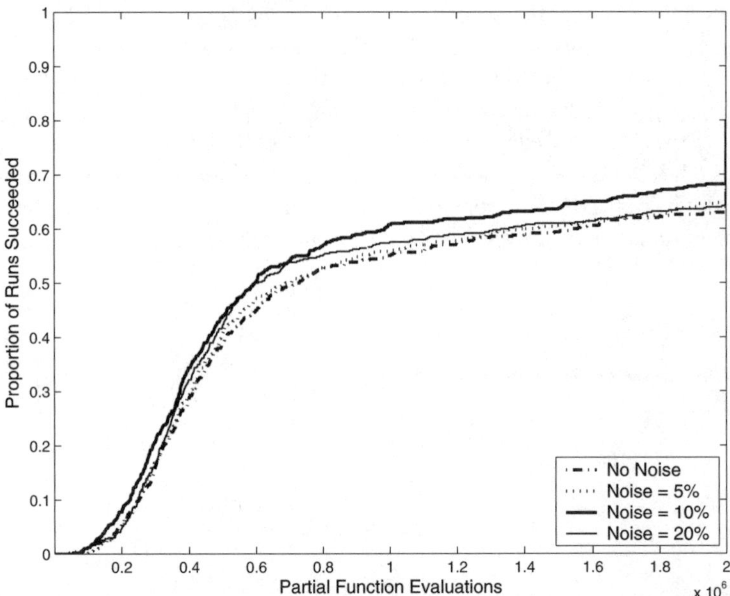

Fig. 8.4. Influence of noise on the cumulative distribution of convergence times. Results are averaged over 1000 runs for each of the noise levels

The effect of noisy perturbations

To illustrate that SDS is relatively immune to certain types of noise, the following experiment was conducted: during every evaluation of a particular f_i, the outcome of the evaluation is perturbed with normally distributed noise with zero mean value and different levels of standard deviation: 5%, 10% and 20% of the actual function value of $f_i(x,y)$. The parameter settings for the standard SDS algorithm were the same as for the previous experiment. The cumulative distribution of convergence times is reported in Fig. 8.4. It can be seen that there is hardly any effect of the noise levels on the cluster formation process. Increasing the noise levels even further, beyond 20%, introduces a negative bias into the test score. Because of the non-linear properties of the SDS process, the effect on this particular landscape is to accelerate the search. However, this is not necessarily so for all objective functions. A detailed discussion of the ramifications of using such mechanisms to improve search performance is beyond the scope of this paper.

Moving peaks

To illustrate that a cluster of agents is able to track a moving peak, the following experiment was performed: the entire objective function of Fig. 8.2 was shifted one location to the left and one location to the front every 50 iterations of an SDS simulation. A population of 1000 context-sensitive agents with mutation parameter

$s = 2$ (resulting in a copying accuracy of 47%, or a mutation rate of 53%) was run for 10000 iterations. Figure 8.5 summarises the results: the left graph depicts the total number of active agents (higher curve) and the size of the cluster at the location of the moving Peak 1 (lower curve). The right graph depicts the location of the *largest cluster* of agents every 50 iterations, just before a new shift of the objective function. The results show that the largest cluster of agents follows the movement of Peak 1 almost perfectly. 50 iterations of 1000 agents constitute 50000 evaluations of component functions, equivalent to only 14 evaluations of (8.2).

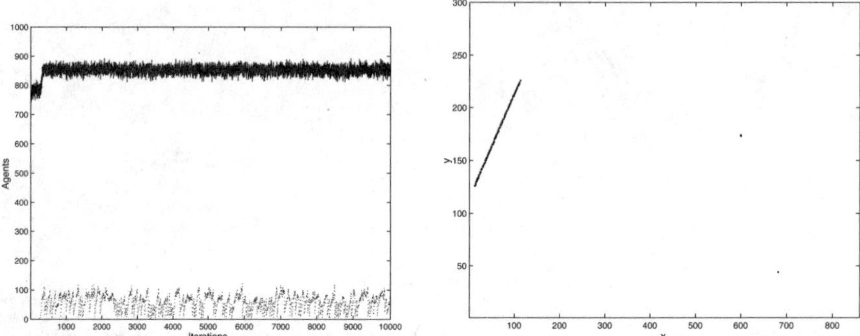

Fig. 8.5. Tracking of a moving peak. The left graph depicts overall activity and the size of the cluster at the moving location of Peak 1. The right graph depicts the (x, y) location of the largest agent cluster every 50 iterations

Changing peaks

To illustrate that a cluster of SDS agents can reallocate itself successfully when optimal solutions become sub-optimal, the following experiment was performed: 1000 context-sensitive SDS agents[6] were simulated for 5000 iterations, while peaks in the landscape were slowly decreased or increased. The results of this experiment can be seen in Fig. 8.6. After 1000 iterations, Peak 1 starts to deteriorate, with as consequence a gradually decreasing cluster size at that location. Peak 2 remains constant, while Peak 3 grows gradually in height and width. Shortly after Peak 1 becomes lower than Peak 2, there is a sudden shift of the dominant cluster towards the location of Peak 2. When Peak 3 grows larger than Peak 2, a similar shift occurs to the location of Peak 3.

8.8 Discussion

Pratt [42] suggests that Leptothorax Alpipennis require extra cognitive abilities in order to efficiently compare different nest sites. Although it could be that these

[6] With the same perturbation of hypothesis-copying as in the previous experiment.

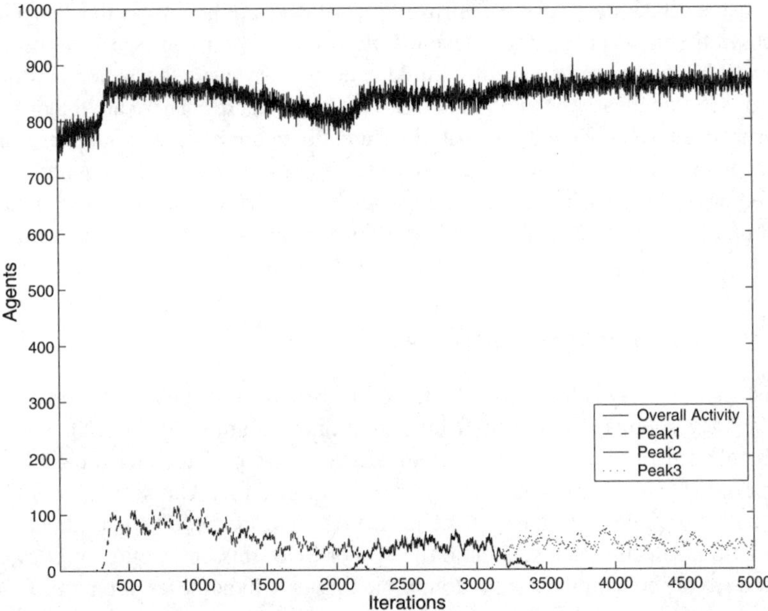

Fig. 8.6. Changing peaks. Depicted are the overall activity in the population and the cluster sizes at Peak 1, 2 and 3, for a population of 1000 agents run for 5000 iterations. After 1000 iterations, Peak 1 starts slowly decreasing in height and width, while Peak 3 starts slowly increasing. Peak 2 remains the same throughout the experiment. At iteration 1700, Peak 1 becomes lower than Peak 2. At iteration 2300, Peak 3 becomes higher than Peak 2, and keeps growing until iteration 4000. The height of the peaks changes very slowly: e.g. for Peak 1 only 0.3% of the function range every 100 iterations. However, even these subtle changes are reflected in the resource allocation of the agent population

ants need higher cognitive abilities because the exact dynamics of their recruitment process do not allow convergence on the best site in a fast enough time span, experience with SDS shows that these abilities are *in principle* not required. As long as one of the two nest sites has a higher probability of inducing recruitment, ants can come to a global decision about the best site without the ability of comparing the two sites directly.

Differences in the operation of SDS and the bulk of ant algorithms has resulted in their application in different types of search and optimisation problems. In Mitchell [32], a taxonomy of search problems has been proposed:

- Pattern matching problems, in which the goal is to locate a predefined target in a larger solution space.
- Optimisation problems, in which the goal is to select a solution from a set of candidates such that a given cost function is optimised.
- Path planning problems, in which the goal is to construct a path to reach a specified target.

Whereas SDS in its present form seems mostly applicable to the first type of search problems, ant algorithms have mostly been used for solving the second type. As such, both approaches seem complementary. However, the general principles behind SDS can clearly be applied to other problem classes. These are the principles of partial evaluation of candidate solutions and direct communication of information between agents. Using these principles, SDS can be defined as a new generic search method or *metaheuristic*, applicable to other types of problems outside the pattern-matching domain, such as model fitting; robust parameter estimation; and Inductive Logic Programming. Research in these areas is ongoing.

8.8.1 SDS and evolutionary algorithms

At first sight, the SDS algorithm, described in a language of agents, test and diffusion phases, may seem far removed from evolutionary algorithms. Indeed, it did not originate from metaphors about biological evolution, but from the field of neural networks [4]. However, SDS and algorithms inspired by Darwinian evolution fit both within a general framework of processes that are governed by mechanisms of variation, selection and replication [11]. For SDS, this description applies to the perspective of the hypotheses: randomly picking new hypotheses and perturbing the copying process constitute mechanisms of variation, similar to random immigrants and mutation mechanisms in evolutionary algorithms; the rejection of hypotheses in the diffusion phase is a form of "death" for the hypotheses; hypothesis copying is a form of reproduction. Good hypotheses are more likely to survive test phases for longer, and are able to spread more to other agents. Finally, resources are limited, in that there is only a finite number of agents which hypotheses can occupy.

There are, of course, differences. Firstly, there is no explicit fitness-based selection: selection is the consequence of agent interaction, resulting in the most radical form of tournament selection. Secondly, because of the indirect and continual evaluation of individual hypotheses, SDS can be thought to simulate evolutionary processes *on a different timescale* than other types of evolutionary algorithms. Thirdly, because single agents lack the capacities to judge the quality of solutions on their own, good solutions need to be identified by clusters of agents. This means that SDS explicitly needs at least some level of convergence, whereas this is not necessarily true for other evolutionary algorithms.

8.9 Conclusions

It has been shown that SDS is in principle applicable to stochastic and dynamic optimisation problems. The algorithmic concepts of partial evaluation and mechanisms for altering the balance between exploration and exploitation – together with a well-developed understanding of how these influence the behaviour of the stochastic process underlying SDS – can be of potential interest to the swarm intelligence community at large. Although SDS has been applied to different types of optimisation problems, e.g., [2, 25], it has never before been applied explicitly to stochastic or dynamic optimisation problems. To this end, the present work draws on the expanded understanding of SDS developed in [31].

Future work should include a more precise characterisation of the influence of external noise on the search performance, in the context of the mathematical models of SDS developed in [36], as well as methods to estimate for specific types of problems how much – if anything – can be gained in computational efficiency from partial evaluation. Hybridisation with explicit hill-climbing strategies – already employed in [22] – and multi-population implementations – as described for evolutionary algorithms in [8] – may prove to be invaluable extensions to the simple methods presented here. Finally, a better understanding of the more complex, *focussed* SDS mechanisms, as employed in [2, 25], can render SDS useful for stochastic and dynamic problems of much larger scale than the ones described here.

References

1. Arthur, W B,(1994) Inductive Reasoning and Bounded Rationality (The El Farol Problem). Amer. Econ. Rev. Papers and Proceedings 84: 406
2. Beattie, P, Bishop, J (1998) Self-localisation in the SENARIO autonomous wheelchair. Journal of Intelligent and Robotic Systems 22: 255–267
3. Bishop, J M (1989) Anarchic Techniques for Pattern Classification. Chapter 5. PhD Thesis, University of Reading
4. Bishop, J (1989) Stochastic searching networks. In: 1st IEE Conf. ANNs, 329331 London
5. Bishop, J M, Torr, P (1992) The Stochastic Search Network. In: Lingard, R, Myers, D J, Nightingale, C Neural Networks for Images, Speech and Natural Language. Chapman and Hall, New York, 370387
6. Bonabeau, E, Dorigo, M, Theraulaz, G (2000) Inspiration for Optimization from Social Insect Behaviour. Nature 406: 3942
7. Branke, J (1999) Memory-enhanced evolutionary algorithms for dynamic optimization problems. In: Congress on Evolutionary Computation. Volume 3., IEEE 1875–1882
8. Branke, J, Kaußler, T, Schmidt, C, Schmeck, H (2000) A multi-population approach to dynamic optimization problems. In Parmee, I., ed.: Adaptive Computing in Design and Manufacture, Springer 299–308
9. Branke, J, Schmidt, C, Schmeck, H (2001) Efficient fitness estimation in noisy environments. In Spector, L., ed.: Genetic and Evolutionary Computation Conference, Morgan Kaufmann 243–250
10. Branke, J (2003) Evolutionary approaches to dynamic optimization problems – introduction and recent trends. In: Branke, J, ed. Proceedings of EvoDOP
11. Campbell, D (1974) Evolutionary epistemology. In Schilpp, P, ed. The Philosophy of Karl Popper. Open Court 413–463
12. Chadab, R, Rettenmeyer, C (1975) Mass Recruitment by Army Ants. Science 188:11241125
13. Christensen, S, Oppacher, F (2001) What can we learn from no free lunch? a first attempt to characterize the concept of a searchable function. In: Spector et al., L, ed. Genetic and Evolutionary Computation Conference, San Fransisco, Morgan Kaufmann 1219–1226
14. De Meyer, K (2000) Explorations in Stochastic Diffusion Search: Soft- and Hardware Implementations of Biologically Inspired Spiking Neuron Stochastic Diffusion Networks, *Technical Report KDM/JMB/2000/1*, University of Reading

15. De Meyer, K, Bishop, J M, Nasuto, S J (2002) Small-World Effects in Lattice Stochastic Diffusion Search, Proc ICANN2002 Madrid, Spain
16. De Meyer, K, Bishop, J M, Nasuto S J (2000) Attention through Self-Synchronisation in the Spiking Neuron Stochastic Diffusion Network. Consciousness and Cognition 9(2)
17. Deneuborg, J L, Pasteels, J M, Verhaeghe, J C (1983) Probabilistic Behaviour in Ants: a Strategy of Errors? Journal of Theoretical Biology 105:259271
18. Digalakis, J, Margaritis, K (2002) An experimental study of benchmarking functions for evolutionary algorithms. International Journal of Computer Mathemathics 79:403–416
19. Dorigo, M, Di Caro, G, Gambardella, L M (1999) Ant Algorithms for Discrete Optimization. Artificial Life 5(2):137172
20. Garey, M R, Johnson, D S (1979) Computers and Intractability: a guide to the theory of NP-completeness. W. H. Freeman
21. Goodman, L J, Fisher, R C (1991) The Behaviour and Physiology of Bees, CAB International, Oxon, UK
22. Grech-Cini, E, McKee, G (1993) Locating the mouth region in images of human faces. In: Schenker, P, ed. SPIE - The International Society for Optical Engineering, Sensor Fusion VI 2059, Massachusetts
23. Grech-Cini, E (1995) Locating Facial Features. PhD Thesis, University of Reading
24. Holldobler, B, Wilson, E O (1990) The Ants. Springer-Verlag
25. Hurley, S, Whitaker, R (2002) An agent based approach to site selection for wireless networks. In: ACM symposium on Applied Computing, Madrid, ACM Press
26. Jin, Y (2005) A comprehensive survey of fitness approximation in evolutionary computation. In: Soft Computing, 9:3–12.
27. El-Beltagy, M A, Keane, A J (2001) Evolutionary optimization for computationally expensive problems using Gaussian processes. In: Arabnia, H, ed. Proc. Int. Conf. on Artificial Intelligence'01, CSREA Press 708–714
28. Kennedy, J, Eberhart, R C (2001) Swarm Intelligence. Morgan Kaufmann
29. Krieger, M J B , Billeter, J-B, Keller, L (2000) Ant-like Task Allocation and Recruitment in Cooperative Robots. Nature 406:992995
30. Krink, T, Filipic, B, Fogel, G B, Thomsen, R (2004) Noisy Optimization Problems – A Particular Challenge for Differential Evolution? In: Proc. of 2004 Congress on Evolutionary Computation, IEEE Press 332–339
31. De Meyer, K (2003) Foundations of Stochastic Diffusion Search. PhD thesis, University of Reading
32. Mitchell, M (1998) An Introduction to Genetic Algorithms. The MIT Press
33. Moglich M, Maschwitz U, Holldobler B (1974) Tandem calling: a new kind of signal in ant communication. Science 186(4168):1046-7
34. Monmarch, N, Venturini, G, Slimane, M (2000) On How Pachycondyla Apicalis Ants Suggest a New Search Algorithm. Future Generation Computer Systems 16:937-946
35. Morrison, R W, DeJong, K A (1999) A test problem generator for non-stationary environments. In: Congress on Evolutionary Computation. Volume 3., IEEE 2047–2053
36. Nasuto, S J (1999) Resource Allocation Analysis of the Stochastic Diffusion Search. PhD Thesis, University of Reading
37. Nasuto, S J, Bishop, J M (1998) Neural Stochastic Diffusion Search Network - a Theoretical Solution to the Binding Problem. Proc. ASSC2, Bremen
38. Nasuto, S J, Dautenhahn, K, Bishop, J M (1999) Communication as an Emergent Methaphor for Neuronal Operation. Lect. Notes Art. Int. 1562:365380
39. Nasuto, S J, Bishop, J M (1999) Convergence Analysis of Stochastic Diffusion Search. Parallel Algorithms and Applications 14(2):89107

40. Nasuto, S J, Bishop, J M, Lauria, S (1998) Time Complexity of Stochastic Diffusion Search. Neural Computation (NC98), Vienna, Austria
41. Parsopoulos, K E, Vrahatis, M N, (2005) Unified Particle Swarm Optimization in Dynamic Environments, Lect. Notes Comp. Sci. 3449:590-599
42. Pratt, S C, Mallon, E B, Sumpter, D J T, Franks, N R (2000) Collective Decision- Making in a Small Society: How the Ant Leptothorax Alpipennis Chooses a Nest Site. Proc. of ANTS2000, Brussels, Belgium
43. Seeley, T D (1995) The Wisdom of the Hive. Harvard University Press
44. Whitley, D, Rana, S B, Dzubera, J, Mathias, K E (1996) Evaluating evolutionary algorithms. Artificial Intelligence 85:245–276

9

Linear Multi-Objective Particle Swarm Optimization

Sanaz Mostaghim[1,2] and Werner Halter[1] and Anja Wille[2]

[1] Institute of Isotope Geochemistry and Mineral Resources
[2] Computational Laboratory-CoLab
Swiss Federal Institute of Technology (ETH), CH-8092 Zürich, Switzerland

Summary Linear Multi-Objective Particle Swarm Optimization (LMOPSO) has been proposed in this chapter as a methodology to solve linearly constrained multi-objective optimization problems. In the presence of the linear (equality and inequality) constraints, the feasible region can be specified by a polyhedron in the search space. LMOPSO is designed to explore the feasible region by using a linear formulation of particle swarm optimization. This method guarantees the feasibility of solutions by using feasibility preserving methods. This is different from the existing constraint handling methods in multi-objective optimization in the way that it never produces infeasible solutions. LMOPSO has been studied on several test problems with different degrees of difficulties. As a real-world application, LMOPSO has been utilized to identify the parameters of a class of multi-component chemical systems in mineralogy. This has been tested on the system $Na_2O - SiO_2$.

9.1 Introduction

Due to their efficient search strategy, Particle Swarm Optimization (PSO) methods have become popular during the last few years and have been used to solve many optimization problems [8]. PSO can be categorized as a subfamily of Evolutionary Algorithms, as it inherits several properties such as the population based iterative search technique from the EAs. PSO has also been used to solve Multi-Objective Problems [1, 15] which appear in many real world applications. These methods are called Multi-Objective Particle Swarm Optimization (MOPSO) methods. The first step in solving real world optimization problems with most evolutionary algorithms is to define the search space. Despite of having a very large search space, many real-world applications have constraints which divide the space into a small feasible and a large infeasible region. Therefore, an efficient approach to find optimal solutions would be to only search the feasible part of the search space. Unfortunately in the presence of the nonlinear constraints, it is very difficult to specify the feasible regions

S. Mostaghim et al.: *Linear Multi-Objective Particle Swarm Optimization*, Studies in Computational Intelligence (SCI) **31**, 209–238 (2006)
www.springerlink.com

which has led to many different strategies to handle such constraints[3]. The well-known techniques include the Penalty approach [3, 9], stochastic ranking [20], and selection in terms of feasibility [3]. In the presence of linear constraints, the feasible region is often a very small portion of the search space and if we do not apply an efficient exploration technique, most of our candidate solutions will be located in the infeasible part of the search space. Therefore, a high computational time will be required to achieve a few good solutions.

Here, LMOPSO method is proposed to guarantee the feasibility of solutions during the optimization with linear constraints. This is achieved by finding and exploring the feasible region and preserving the feasibility of solutions. After finding the feasible region and defining a set of feasible population members in this region, a good exploring method must be utilized to preserve the feasibility of those solutions. Linear PSO (LPSO) can be used to explore the feasible region. It can also guarantee the feasibility of solutions in terms of the linear equality constraints. LPSO has been studied by Pauqet and Engelbrecht [17] for single objective problems with equality constraints. There is a critical point in using the LPSO in [17], as the feasibility of solutions cannot be preserved. Even in the mere presence of equality constraints, it is possible that some solutions move out of the boundaries of the search space. Indeed, the boundaries of the search space can be considered as inequality constraints.

In this chapter, three feasibility preserving methods have been studied to guarantee the feasibility of solutions. The main difference between LMOPSO and the other existing feasibility preserving methods, e.g., [1], is that the feasible region in the presence of constraints may be different from the search space. If a solution violates any of the constraints it must be moved back into the feasible region and not only in the search space or on the boundaries of the search space.

Furthermore, a real-world application which contains linear equality and inequality constraints has been studied. These kinds of constraints may appear in chemical reactions. We study multi-component chemical systems and try to find their parameters. For some of these systems, there are reference data in the literature. We compare our solutions with the reference data and apply the method to the other systems for which there is no reference data available.

The idea of LMOPSO is a new idea in exploring the feasible region and preserving the feasibility of solutions while improving the convergence of solutions during the generations. This is due to the new formulation of MOPSO method. A similar idea in a single objective Genetic Algorithm has been proposed by Michalewicz and Janikow [12] where they change the genetic operators to linear operators in order to produce feasible solutions. However, producing such solutions does not imply improvement in terms of the convergence. Also, the feasibility of the solutions has not been preserved.

This chapter is organized as follows. In the following of this section, linear constrained multi-objective problems are described. Section 3 is a short description

---

[3]One of the most complete listings of publications about the constraint handling in MOEA field is maintained in a database at the web address http://www.cs.cinvestav.mx/~constraint/ by Dr. Carlos Coello Coello. Currently, this database contains over 470 citations.

about Multi-Objective Particle Swarm Optimization and a background on constraint handling. This section explains the basics of LMOPSO. In Section 4, LMOPSO is explained in three parts and Section 5 is dedicated to experiments on several test functions. In Section 6, a real-world application to multi-component chemical systems has been studied. Finally, conclusion is provided in Section 7.

9.1.1 Linear Constrained Multi-Objective Problems

Typically a linearly constrained Multi-Objective Problem (MOP) can be written as follows:

$$\begin{aligned} \text{minimize/maximize} \quad & \mathbf{f}(\mathbf{x}) \\ \text{subject to} \quad & \mathbf{A}\mathbf{x} \leq \mathbf{b} \\ & \mathbf{A}_{eq}\mathbf{x} = \mathbf{b}_{eq} \end{aligned} \tag{9.1}$$

A solution $\mathbf{x}$ is a vector of n decision variables which are bounded by a set of **boundary conditions**: $\forall i \in \{0, \cdots, n\}$: $x_i^{(L)} \leq x_i \leq x_i^{(H)}$. These bounds constitute a decision variable space S (also called parameter or search space). We denote the image of S by $Z \subset \Re^m$ and call it the objective space. The elements of Z are called objective vectors: $\mathbf{f}: \Re^n \rightarrow \Re^m$ which are optimized at the same time. Without loss of generality we assume that all of the objective are to be minimized. $\mathbf{b} \in \Re^k$, $\mathbf{b}_{eq} \in \Re^{k_{eq}}$, and $\mathbf{A}$ and $\mathbf{A}_{eq}$ are $k \times n$ and $k_{eq} \times n$ matrices.

Since we are dealing with MOPs, there is generally not one global optimum but a set of so-called **Pareto optimal solutions**. A decision vector $\mathbf{x}_1 \in S$ is called **Pareto-optimal** if there is no other decision vector $\mathbf{x}_2 \in S$ that **dominates** it. An objective vector is called Pareto-optimal if the corresponding decision vector is Pareto-optimal. In the following some definitions in multi-objective optimization are listed:

Definition 1. $\mathbf{x}_1$ *is said to* **dominate** $\mathbf{x}_2$ *iff* $\mathbf{x}_1$ *is not worse than* $\mathbf{x}_2$ *in all objectives and* $\mathbf{x}_1$ *is strictly better than* $\mathbf{x}_2$ *in at least one objective.*

Definition 2. $\mathbf{x}_1$ **weakly dominates** $\mathbf{x}_2$ *iff* $\mathbf{x}_1$ *is not worse than* $\mathbf{x}_2$ *in all objectives.*

Definition 3. $\mathbf{x}_1$ *is said to* **ε-dominate** $\mathbf{x}_2$ *for some* $\varepsilon > 0$ *iff* $f_i(\mathbf{x}_1)/(1+\varepsilon) \leq f_i(\mathbf{x}_2)$ $\forall i = 1, \ldots, m$ *and* $f_i(\mathbf{x}_1)/(1+\varepsilon) < f_i(\mathbf{x}_2)$ *for at least one* $i = 1, \ldots, m$.

Constraints divide the search space into two divisions - **feasible** and **infeasible** regions. In the case of linear constraints, the feasible region can be defined by a polyhedron as a set described by a finite number of linear equality and inequality constraints. In particular, the feasible region is convex, i.e., if $\mathbf{x}_1$ and $\mathbf{x}_2$ are two feasible solutions, then all vectors described by $(1-\lambda)\mathbf{x}_1 + \lambda\mathbf{x}_2$ for $\lambda \in [0, 1]$ are also feasible solutions. This property must be specifically explored while searching the search space for the optimal solutions.

Definition 4. *Consider a polyhedron defined by linear equality and inequality constraints, and let* $\mathbf{x}^*$ *be an element of* $\Re^n$. *The vector* $\mathbf{x}^*$ *is a* **basic solution** *if*

(a) All equality constraints are active[4]*;*
(b) Out of the constraints that are active at $\mathbf{x}^*$*, there are* n *of them that are linearly independent.*

If a basic solution satisfies all of the constraints, it is said that it is a **basic feasible solution.**

It can be proved that for a given number of linear inequality constraints, there are a finite number of basic feasible solutions [2].

Example 1. Let's consider a multi-objective optimization problem of three parameters and the following constraints.

$$x_1 + x_2 + x_3 = 1 \tag{9.2}$$

$$2x_1 + 3x_2 - x_3 \leq 2 \tag{9.3}$$

$$0 \leq x_1 \leq 1 \tag{9.4}$$

$$0 \leq x_2 \leq 1 \tag{9.5}$$

$$0 \leq x_3 \leq 1 \tag{9.6}$$

These constraints build the feasible region which is bounded by three basic feasible solutions. The basic feasible solutions which also indicate the extreme points are $(1,0,0)$, $(0,\frac{3}{4},\frac{1}{4})$, and $(0,0,1)$. Indeed, these basic feasible solutions satisfy all of the constraints. The search space and the basic feasible solutions are illustrated in Figure 9.1. The shaded surface in Figure 9.1 shows the feasible region.

If we changed the constraint (9.3) to an equality constraint, then the non-negative part of the dashed line would be the feasible region and the basic feasible solutions would be $(1,0,0)$ and $(0,\frac{3}{4},\frac{1}{4})$.

□

In order to compute the basic feasible solutions, all of the inequality constraints must be changed to equality forms by introducing **slack**[5] variables. Thus, we only have constraints of the form $\mathbf{A}'\mathbf{x} = \mathbf{b}$ where $\mathbf{A}'$ is the $\mathbf{A}$ matrix together with the slack variables and thereby, $\mathbf{A}'$ is a $k \times (n+k)$ matrix. Let us assume that $\mathbf{A}'$ has k independent rows. A vector $\mathbf{x} \in \Re^n$ is said to be a basic solution if and only if we have $\mathbf{A}'\mathbf{x} = \mathbf{b}$ and there exist indices $\mathbf{B}(1), \cdots, \mathbf{B}(m)$ such that:

(a) the columns $\mathbf{A}'_{B(1)}, \cdots, \mathbf{A}'_{B(k)}$ are linearly independent;
(b) if $i \neq \mathbf{B}(1), \cdots, \mathbf{B}(m)$, then $x_i = 0$.

Now we can describe a procedure to construct basic solutions as follows:

1. Choose k linearly independent columns $\mathbf{B} = [\mathbf{A}'_{\mathbf{B}(1)}, \cdots, \mathbf{A}'_{\mathbf{B}(k)}]$ ($\mathbf{B}$ is a $k \times k$ matrix).
2. Let $x_i = 0$, for all $i \neq \mathbf{B}(1), \cdots, \mathbf{B}(m)$ and $\mathbf{x}_B = (x_{\mathbf{B}(1)}, \cdots, x_{\mathbf{B}(k)})$.
3. Solve the system of k equations $\mathbf{x}_B = \mathbf{B}^{-1}\mathbf{b}$.

In Example 1, only the non-negative constructed basic solutions construct the basic feasible solutions due to the constraints in Equations (4), (5), and (6).

[4] The constraint i, defined by $a_i\mathbf{x} \leq b_i$ is active at $\mathbf{x}^*$, if $a_i\mathbf{x}^* = b_i$.
[5] For more details refer to the linear programming references or [2].

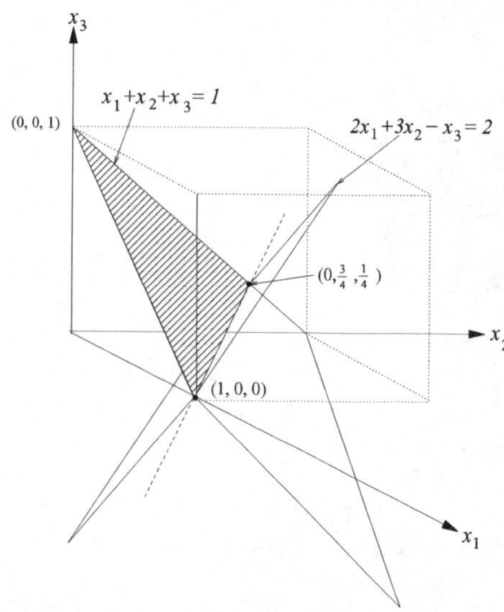

Fig. 9.1. Search space of Example 1. Two surfaces show the constraints 9.2 and 9.3 and the shaded surface is the feasible region. This example contains three basic feasible solutions: $(1,0,0)$, $(0,\frac{3}{4},\frac{1}{4})$, and $(0,0,1)$.

9.1.2 Boundaries of the Feasible Region

Basic feasible solutions are the extreme points of the feasible region. Also, the feasible region is bounded by the active constraints at the basic feasible solutions. All the active constraints i must satisfy $a_i\mathbf{x}_f = b_i$ for at least one of the basic feasible solutions denoted by $\mathbf{x}_f$. Therefore, in order to find the feasible region, the basic feasible solutions must be found.

Example 2. Figure 9.2 (a) shows the bounded region of Example 1 by the dashed lines. We observe that the constraint $x_2 \leq 1$ is inactive at all of the basic feasible solutions. This can be observed easily, by computing the maximum of the basic feasible solutions in terms of the x_2 parameter: $\frac{3}{4}$. If we changed the inequality constraint in the Equation (9.3) to the equality form, then we would have two inactive constraints at the basic feasible solutions, namely $x_2 \leq 1$ and $x_3 \leq 1$. The new boundaries for the basic feasible region would be $0 \leq x_1 \leq 1$, $0 \leq x_2 \leq \frac{3}{4}$, and $0 \leq x_3 \leq \frac{1}{4}$ which are made from the basic feasible solutions (Figure 9.2 (b)).

□

9.2 Multi-objective Particle Swarm Optimization

Algorithm 9.1 shows one typical structure of a Multi-Objective Particle Swarm Optimization (MOPSO). The Algorithm starts with a set of uniformly distributed random initial individuals (also called particles) defined in the search space S.

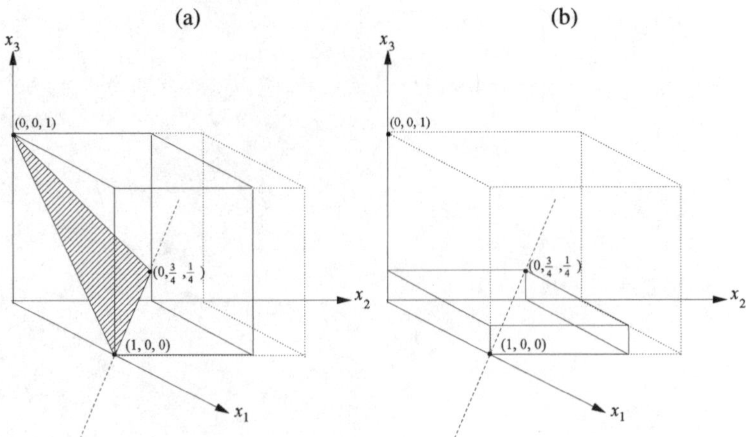

Fig. 9.2. Feasible region in the presence of equality and inequality constraints (a), and only equality constraints (b)

A set of N particles are considered as a population P_t at the generation t. Each particle i has a position defined by $\mathbf{x}^i = (x_1^i, x_2^i, \cdots, x_n^i)$ and a velocity defined by $\mathbf{v}^i = (v_1^i, v_2^i, \cdots, v_n^i)$ in the variable space S.

Beside the population, another set (called Archive) A_t can be defined in order to store the obtained non-dominated solutions. Due to the presence of an archive, good solutions are preserved during generations and therefore, convergence might be guaranteed [19]. In Step 2 of the Algorithm, the individuals are evaluated and the non-dominated solutions are added to the archive. Thereby, the archive is kept domination-free. Obviously, during the execution of the function *Update*, dominated solutions must be removed. This is done in Step 3 of the Algorithm 9.1.

In Step 4, the particles are moved to the new positions in the space. The velocity and position of each particle i is updated as below:

$$\begin{aligned} v_{j,t+1}^i &= wv_{j,t}^i + c_1R_1(p_{j,t}^i - x_{j,t}^i) + c_2R_2(p_{j,t}^{i,g} - x_{j,t}^i) \\ x_{j,t+1}^i &= x_{j,t}^i + v_{j,t+1}^i \end{aligned} \tag{9.7}$$

where $j = 1, \ldots, n$, $i = 1, \ldots, N$, c_1 and c_2 are two positive constants, R_1 and R_2 are random values in the range $[0, 1]$ and

- w is the so called **inertia weight** of the particle. This is employed to control the impact of the previous history of velocities on the current velocity, thus to influence the trade-off between global and local exploration abilities of the particles [21, 8]. A larger inertia weight w facilitates global exploration while a smaller inertia weight tends to facilitate local exploration to fine-tune the current search area. Suitable selection of the inertia weight w can provide a balance between global and local exploration abilities requiring fewer iterations for finding the optimum on average [21, 8]. A nonzero inertia weight introduces

Algorithm 9.1 : MOPSO Algorithm

Require: N
Ensure: A
1. Initialization: Initialize population P_t, $t = 0$:
for $i = 1$ to N **do**
  Initialize $\mathbf{x}_t{}^i$, $\mathbf{v}_t{}^i = \mathbf{0}$ and $\mathbf{p}_t{}^i = \mathbf{x}_t{}^i$
end for
Initialize the archive $A_t := \{\}$
2. Evaluate: $Evaluate(P_t)$
3. Update: $A_{t+1} := Update(P_t, A_t)$
4. Move: $P_{t+1} := Move(P_t, A_t)$
for $i = 1$ to N **do**
  $\mathbf{p}_t{}^{i,g} := FindBestGlobal(A_{t+1}, \mathbf{x}_t{}^i)$
  for $j = 1$ to n **do**
    $v^i_{j,t+1} = wv^i_{j,t} + c_1 R_1 (p^i_{j,t} - x^i_{j,t}) + c_2 R_2 (p^{i,g}_{j,t} - x^i_{j,t})$
    $x^i_{j,t+1} = x^i_{j,t} + v^i_{j,t+1}$
  end for
  if $\mathbf{x}^i_{t+1} \prec \mathbf{p}_t{}^i$ **then**
    $\mathbf{p}^i_{t+1} = \mathbf{x}^i_{t+1}$
  else
    $\mathbf{p}^i_{t+1} = \mathbf{p}_t{}^i$
  end if
end for
5. Termination: Unless a *termination criterion* is met $t = t + 1$ and *goto* Step 2

the preference for the particle to continue moving in the same direction as in the previous iteration.

- $c_1 R_1$ and $c_2 R_2$ are called **control parameters** [8]. These two control parameters determine the type of trajectory the particle travels. If R_1 and R_2 are 0.0, it is obvious that $v = v + 0$ and $x = x + v$ (for $w = 1$). It means the particles move linearly. If they are set to very small values, the trajectory of x rises and falls slowly over time.
- $\mathbf{p}_t{}^{i,g}$ is the position of the **global best particle** in the population, which guides the particles to move towards the optimum. The important part in MOPSO is to determine the best global particle $\mathbf{p}_t{}^{i,g}$ for each particle i of the population. In single-objective PSO, the global best particle is determined easily by selecting the particle that has the best position. But in MOPSO, $\mathbf{p}_t{}^{i,g}$ must be selected from the updated set of non-dominated solutions stored in the archive A_{t+1}. Selecting the best local guide is achieved in the function $FindBestGlobal(A_{t+1}, \mathbf{x}_t{}^i)$ for each particle i [15].
- $\mathbf{p}^i_t$ is the best position that particle i could find so far. This is like a memory for the particle i and keeps the non-dominated (best) position of the particle by comparing the new position $\mathbf{x}^i_{t+1}$ in the objective space with $\mathbf{p}_t{}^i$ ($\mathbf{p}_t{}^i$ is the last non-dominated (best) position of the particle i).

The steps of a MOPSO are iteratively repeated until a termination criterion is met, such as a maximum number of generations, or when there has been no change in the set of non-dominated solutions for a given number of generations. The output of an elitist MOPSO method is the set of non-dominated solutions stored in the final archive.

In MOPSO, we also define a parameter called **turbulence factor** which is basically designed to avoid the local optima. With a probability value equal to the turbulence factor, a particle is moved to a random position in the search space. It is obvious that if we increase the turbulence factor, the number of random solutions increases.

This algorithm illustrates a simple method to solve problems without constraints. In the following, some methods which can be used to solve constrained problems are studied. These are known as **constraint handling** methods:

9.2.1 Constraint Handling

Constraint handling in PSO have been studied, e.g., by G. Toscano Pulido, C. Coello Coello [23] and U. Paquet and A. P. Engelbrecht [17], and A. E. Munoz Zavala et al. [16]. Paquet and Engelbrecht [17] suggest a methodology called Converging Linear PSO (CLPSO) to handle linear equality constraints for single objective problems. CLPSO is used to change the position of each particle linearly. Also, it is designed to efficiently explore the feasible space. But the main criticism to this method is that they only study the equality constraints. Even with only equality constraints, the particles should be kept inside the search space. Depending on the boundaries of the search space or the feasible region, it may happen quite often that a particle moves out of the boundaries and this has not been studied in [17]. Toscano Pulido and Coello Coello [23] have studied a general form for nonlinear constraint handling. This mechanism is applied when selecting a leader or the global best guide and uses the simple criterion based on closeness of a particle to the feasible region in order to select a leader.

Also in the context of Evolutionary Algorithms, there is a large number of constraint handling methods, e.g. [11, 12].

All of these methods have been studied for single objective problems. This is different from constraint handling in Multi-Objective Optimization. It must be emphasized that in MOPs, we must preserve a good diversity among the optimal solutions and the constraint handling method has a great impact on this diversity. In the following, we briefly explain the ideas of some existing methods used in Multi-Objective Evolutionary Algorithms (MOEA).

Ignoring infeasible solutions: A simple way to handle constraints is to ignore any solution that violates any of the constraints [3]. This approach is very simple to implement, but finding a feasible solution which satisfies all the constraints is a major problem, when the feasible region is small in comparison with the search space.

Penalty function approach: In this approach, the constraint violation will be calculated for each candidate solution and will be added to the fitness value. The constraint violation value is a measurement of the distance of each solution to each of the constraints boundaries. The infeasible solutions which are very far from the constraints boundaries will have a higher constraint violation value than the infeasible solutions close to the boundaries of the feasible region. The constraint violation is then the sum of all constraints violations times a penalty factor. The quality of the solutions and the amount of infeasible solutions in each iteration depends highly on the Penalty factor.

There are also other methods like Constrained Tournament Method [3], Ray-Tai-Soew's [18], and Jimenez-Verdegay-Gomez-Skarmeta's methods [7, 24] which are based on a comparison between feasible and infeasible solutions.

Altogether, the existing methods in MOEA are designed for a general form of having no constraints and the constraints are then handled separately. This is a common method to deal with nonlinear constraints. However, in some applications we may encounter linear constraints in addition to nonlinear objectives.

9.3 Linear Multi-Objective Particle Swarm Optimization

Linear Multi-Objective Particle Swarm Optimization (LMOPSO) is developed as a new method in handling linear constraints. The basic idea is not to generate infeasible solutions at all and only explore the feasible regions. This idea differs from the existing methods in the way that infeasible solutions are not allowed to enter the population. For realization of this idea, the feasible region must be found and the exploration of the particles has to be controlled so that they never violate the constraints. In this chapter, this is achieved in three parts: (a) finding and exploring the feasible region, (b) updating the solutions in the feasible region (Linear PSO), and (c) applying feasibility preserving methods. Thereby, we are able to guarantee the feasibility of the solutions.

9.3.1 Exploring the Feasible Region

The first step in optimizing a linearly constrained optimization problem is to find the feasible region. Due to its convexity, the feasible region of a linear constrained problem can be explored by defining linear combinations of the basic feasible solutions. A feasible solution can be described as:

$$\mathbf{x}_{new} = a_1 \mathbf{x}_{f1} + \cdots + a_J \mathbf{x}_{fJ} \tag{9.8}$$

where $\mathbf{x}_{fj}$ $(j = 1, \cdots, J)$ are the basic feasible solution, a_j are scalar values, and J is the number of feasible solutions. If we set $a_j \geq 0$ for all j and $\sum_{j=1}^{J} a_j = 1$, the exploration will only take place in the feasible region between the basic feasible solutions.

Example 3. Let's consider Example 1. In this example we have three basic feasible solutions, i.e., $J = 3$. Therefore, the vectors $\mathbf{a} = (\frac{1}{3}, \frac{1}{3}, \frac{1}{3})$ and $\mathbf{a} = (0.8, 0.1, 0.1)$ would make new feasible solutions $(\frac{1}{3}, \frac{1}{4}, \frac{5}{12})$ and $(\frac{8}{10}, \frac{3}{40}, \frac{5}{40})$ respectively.

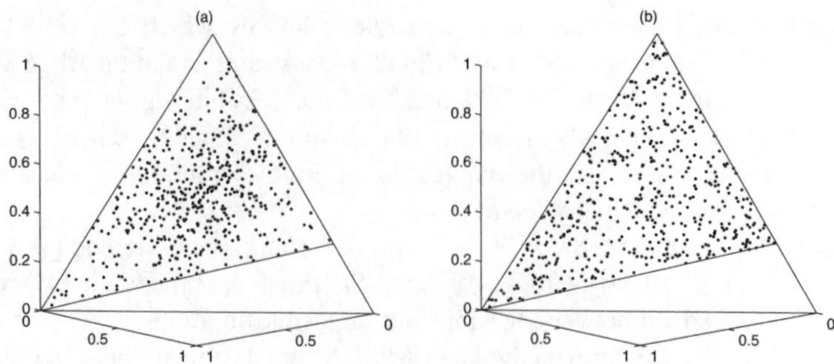

Fig. 9.3. Defining random initial population of feasible solutions (500 solutions are shown).

□

After finding the feasible region, the next step is to initialize an initial population which is a random linear combination of the basic feasible solutions. In the presence of linear constraints, the way the initial population is initialized has a great impact on the distribution of solutions. For instance, Figure 9.3 shows two populations of the size 500, which are both made by linear combinations of basic feasible solutions from Example 1. We observe that the solutions in Figure 9.3(a) are not as well distributed as in Figure 9.3(b). In the following, the two methods which produced these solutions are being described.

As it has been explained, we can explore a feasible region by defining a vector **a**, where $\sum_{j=1}^{J} a_j = 1$. Now, if we want to define N random feasible solutions, N random vectors **a** must be defined:

Method (a):

Select J random values of $b_j \in [0,1]$, $j = \{1,2,\cdots J\}$ and scale them to have sum 1, i.e., set $a_j = b_j/B$ with $B = \sum_{j=1}^{J} b_j$. 500 random feasible solutions obtained from this method are shown in Figure 9.3(a). As the values a_j are independent, the sampled population is unevenly dispersed over the search space.

Method (b):

Select $J-1$ random values in $[0,1]$ and order them from from the smallest to the largest value. One obtains $J-1$ values $b_j \in [0,1]$, $j = \{1,2,\cdots J-1\}$ with $b_j \leq b_{j+1}$. Set the elements of vector **a** to $a_1 = b_1$, $a_j = b_j - b_{j-1}, \forall j = \{2,3,\cdots,J-1\}$ and $a_J = 1 - b_{J-1}$.

By this approach, the interval $[0,1]$ is randomly split into J subintervals and the correct correlation structure between the elements a_j is established. The results of this method is illustrated in Figure 9.3(b) for 500 random feasible solutions.
Therefore, for sampling random vectors **a**, it is important to reflect the pairwise negative correlation between the elements a_j. This is very important in handling

linear constraints, as in most of the MOPs the feasible region is a small portion of the search space and defining a well-distributed initial population results in a better exploration.

9.3.2 Linear Particle Swarm Optimization

Linear PSO [17] can be used to guarantee the feasibility of solutions in terms of the equality constraints. Let us consider only linear equality constraints $\mathbf{A}_{eq}\mathbf{x} = \mathbf{b}_{eq}$ with no boundaries and the initial velocity vector of each particle to be zero. Therefore, in the first generation ($t = 0$) the new velocity would be as follows:

$$\mathbf{v}^i_{t+1} = w\mathbf{v}^i_t + c_1R_1(\mathbf{p}^i_t - \mathbf{x}^i_t) + c_2R_2(\mathbf{p}^{i,g}_t - \mathbf{x}^i_t) \tag{9.9}$$

Note that in this equation, the velocity has been considered as a vector [17] and $\mathbf{v}^i_{t=0} = \mathbf{0}$ (In Equation 9.8 each element is updated separately). Here, because $\mathbf{p}^i_t$, $\mathbf{x}^i_t$, and $\mathbf{p}^{i,g}_t$ are feasible, $\mathbf{d}_1 = (\mathbf{p}^i_t - \mathbf{x}^i_t)$ and $\mathbf{d}_2 = (\mathbf{p}^{i,g}_t - \mathbf{x}^i_t)$ constitute two vectors with $\mathbf{A}_{eq}\mathbf{d}_1 = \mathbf{0}$ and $\mathbf{A}_{eq}\mathbf{d}_2 = \mathbf{0}$.

Therefore, we have $\mathbf{A}_{eq}\mathbf{v}^i_t = \mathbf{0}$. This is valid for all velocity vectors in the next generations, if and only if $\mathbf{x}^i_t$ vectors be feasible. Thereby, the updated $\mathbf{x}^i_{t+1}$ is always feasible if we update $\mathbf{x}^i_t$ as a vector:

$$\mathbf{x}^i_{t+1} = \mathbf{x}^i_t + \mathbf{v}^i_{t+1} \tag{9.10}$$

The updated $\mathbf{x}^i$ is always feasible, since

$$\mathbf{A}_{eq}\mathbf{x}^i_{t+1} = \mathbf{A}_{eq}\mathbf{x}^i_t + \mathbf{A}_{eq}\mathbf{v}^i_{t+1} = \mathbf{b}_{eq} \tag{9.11}$$

9.3.3 Feasibility Preserving Method

LPSO guarantees the feasibility of solutions in terms of the equality constraints. But even if we only consider the equality constraints, the search space is generally bounded and a linear combination of basic feasible solutions may lead some particles to the infeasible part of the search space. This can be formulated in a general form of considering both the inequality and equality linear constraints. For example in Example 1, the solutions are only allowed to move on the surface made by the three feasible solutions and inside the positive part of the coordinate axis.

Example 4. Consider Example 1. We might obtain a velocity $\mathbf{v}$ from two of the basic feasible solutions: $\mathbf{v} = (1,0,0) - (0,\frac{3}{4},\frac{1}{4}) = (1,\frac{-3}{4},\frac{-1}{4})$. For a feasible solution $\mathbf{x}_1 = (\frac{8}{10},\frac{3}{20},\frac{1}{20})$, we obtain a new position of $\mathbf{x}_{1,new} = \mathbf{x}_1 + \mathbf{v} = (\frac{9}{5},\frac{-3}{5},\frac{-1}{5})$ which is infeasible.

□

Here, some feasibility preserving methods must be applied to the particles who move to the infeasible part of the search space. Therefore, if the new position of a particle is infeasible, the following methods are proposed:

Pbest method

Consider the best position that particle i could find so far ($\mathbf{p}_t^i$) as the new position and set the new velocity to zero. This may cause a very low improvement of convergence, when the Pareto optimal front or a local Pareto optimal front is close to one of the boundaries.

Random method

Ignore the infeasible position, find a new random feasible position and set the new velocity to zero. A random feasible position can be found by a random linear combination of the basic feasible solutions. This is like a random search method and if there are many solutions which violate the constraints this has a great impact on the convergence of solutions and degrades the convergence and diversity of the solutions.

Boundary method

Move the infeasible solution back to the closest feasible solution which lies on the boundary and reverse its velocity. This is to let the solutions move so that they never violate the constraints. Indeed, if the updated solution $\mathbf{x}+\mathbf{v}$ is infeasible, we add a rescaled velocity $\lambda\mathbf{v}$ to the position $\mathbf{x}$ so that the new position $\mathbf{x}_{final}$ is feasible. For the next generation, the velocity $\lambda\mathbf{v}$ will be reversed:

$$\begin{aligned} \mathbf{x}_{final} &= \mathbf{x}+\lambda\mathbf{v} \\ \mathbf{v}_{final} &= -\lambda\mathbf{v} \end{aligned} \tag{9.12}$$

Here, we should choose an appropriate $\lambda \in [0,1]$. If $\lambda = 0$, $\mathbf{x}_{final} = \mathbf{x}$. If the new position $\mathbf{x}+\mathbf{v}$ violates at least one of the constraints, the appropriate $\lambda \in [0,1]$ can be found as follows (Figure 9.4):

(1) Compute two vectors $\mathbf{c}$ and $\mathbf{d}$: $\mathbf{Ax} = \mathbf{c}$ and $\mathbf{A}(\mathbf{x}+\mathbf{v}) = \mathbf{d}$
(2) $\lambda = \min_i \frac{b_i - c_i}{d_i - c_i}$
(3) The final position is $\mathbf{x}_{final} = \mathbf{x}+\lambda\mathbf{v}$ and $\mathbf{v}_{final} = -\lambda\mathbf{v}$.

It must be emphasized that the Boundary method is different from the other Boundary methods, e.g. in [1]. Here, the boundary of the feasible region is not necessarily the same as the search space and because we want to preserve the feasibility of the solution, the **feasible boundary** must be found and not the boundary of the search space.

9.3.4 Discussion

Every particle that violates the constraints contains a memory which includes the best position for that particle and the velocity. Depending on the value of the inertia weight, the velocity has a great impact on the constraint violation in the next iteration. Consider Figure 9.4. This shows the Boundary method. In the case that we move the

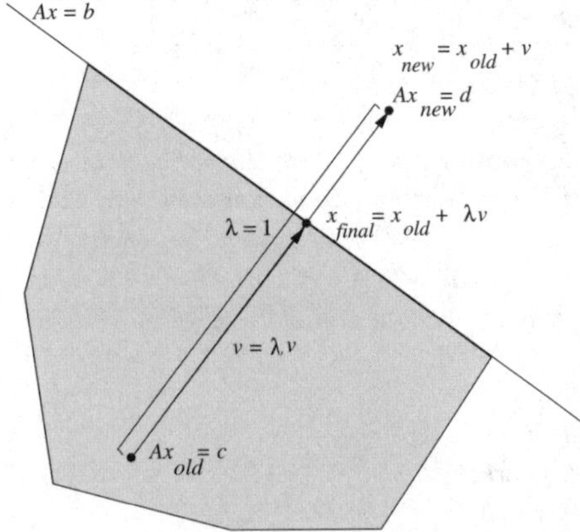

Fig. 9.4. This figure shows the boundary method in which the infeasible particle $\mathbf{x}_{new}$ will be moved to the boundary $\mathbf{x}_{final}$. dark surface indicates the feasible region.

particle to the boundary, the velocity vector will change to $\lambda\mathbf{v}$. In the next generation this velocity times the inertia weight will be again used in computing the new velocity vector and this may lead the particle to move again to the infeasible part of the search space. In order to deal with this problem the velocity must be reversed. This is proposed for the Boundary method, because the solution is kept on the boundary and by the reverse velocity in the next iteration it tends to go towards the feasible region and not the infeasible region.

9.3.5 Linear Constraint Handling

Algorithm 9.2 shows the summary of LMOPSO. This structure is very similar to the Algorithm 9.1 with the difference that here the feasibility of solutions is guaranteed. In Step 1 of this algorithm, J basic feasible solutions $\mathbf{x}_{fj}$ $j = 1, \cdots, J$ are found in the function $FindBasicFeasible(\mathbf{A}, \mathbf{b})$. The inputs to this functions are the matrix $\mathbf{A}$ and vector $\mathbf{b}$ from the constraints. In Step 2, N random feasible particles are created. These can be produced by the methods explained in Section 9.3.1. Steps 3 and 4 are the same as in Algorithm 9.1. In Step 5, the particles are updated vector-wise as proposed in Section 9.3.2. If a solution violates one of the constraints, i.e., $\mathbf{A}\mathbf{x}^i_{t+1} \nleq \mathbf{b}$, which can only happen to the inequality constraints, one of the feasibility preserving method proposed in Section 9.3.3 is applied to this particle. This is done in the function $PreserveFeasibility(\mathbf{x}^i_{t+1})$.

9.4 Experiments

In this section, LMOPSO has been studied on several multi-objective test problems with different degrees of difficulties.

Algorithm 9.2 : LMOPSO Algorithm

Require: $N, \mathbf{A}, \mathbf{b}$
Ensure: A
1. Find Basic Feasible Solutions: $\mathbf{x}_{f,j} = FindBasicFeasible(\mathbf{A}, \mathbf{b})$
2. Initialization: Initialize a feasible population P_t, $t = 0$:
for $i = 1$ to N **do**
  Find a random vector **a**: $\sum_{j=0}^{J} a_j = 1$
  Initialize $\mathbf{x}_t{}^i = \sum_{j=0}^{J} a_j x_{fj}$, $\mathbf{v}_t{}^i = \mathbf{0}$ and $\mathbf{p}_t{}^i = \mathbf{x}_t{}^i$
end for
Initialize the archive $A_t := \{\}$
3. Evaluate: $Evaluate(P_t)$
4. Update: $A_{t+1} := Update(P_t, A_t)$
5. Move: $P_{t+1} := Move(P_t, A_t)$
for $i = 1$ to N **do**
  $\mathbf{p}_t{}^{i,g} := FindBestGlobal(A_{t+1}, \mathbf{x}_t{}^i)$
  $\mathbf{v}_{t+1}^i = w\mathbf{v}_t^i + c_1 R_1(\mathbf{p}_t^i - \mathbf{x}_t^i) + c_2 R_2(\mathbf{p}_t^{i,g} - \mathbf{x}_t^i)$
  $\mathbf{x}_{t+1}^i = \mathbf{x}_t^i + \mathbf{v}_{t+1}^i$
  if $\mathbf{x}_{t+1}^{\ i} \prec \mathbf{p}_t{}^i$ **then**
    $\mathbf{p}_{t+1}^{\ i} = \mathbf{x}_{t+1}^{\ i}$
  else
    $\mathbf{p}_{t+1}^{\ i} = \mathbf{p}_t{}^i$
  end if
  if $\mathbf{A}\mathbf{x}_{t+1}^{\ i} \not\leq \mathbf{b}$ **then**
    $(\mathbf{x}_{t+1}^{\ i}, \mathbf{v}_{t+1}^{\ i}) = PreserveFeasibility(\mathbf{x}_{t+1}^{\ i})$
  end if
end for
6. Termination: Unless a *termination criterion* is met $t = t + 1$ and *goto* Step 3

9.4.1 Test functions

The test functions are listed in Table 9.1. The constraints of these test functions are based on the constraints of the Example 1. The approximated optimal solutions of these test functions in the search and the objective space are illustrated in Figures 9.5 - 9.7. Each of these test functions contains difficulties as follows:
Test (1)-(2): One part of the optimal front of the Test (1) is located on the boundary (Figure 9.5). This causes a difficulty for solving this problem, because many of particles will violate the constraints. Test (2) is another variation of the Test (1), where the optimal front is completely located on the boundary. It will be shown that because the optimal front is on the boundary many particles tend to explore the infeasible part of the search space and therefore we obtain a high constraint violation rate[6].
Test (3)-(4): The optimal front of the Test (3) is a continuous surface located in the middle of the parameter space. This is designed to test the ability of

[6]Here, the constraint violation rate indicates the total number of solutions which violate the constraints.

Table 9.1. Test functions

Test Objectives	Constraints
(1) $f_1(\mathbf{x}) = x_1^2 + x_2^2 + x_3^2$ $f_2(\mathbf{x}) = (x_1-1)^2 + (x_2-1)^2 + (x_3+1)^2$	$0 \le x_i \le 1, i = 1,2,3$ $x_1 + x_2 + x_3 = 1$ $2x_1 + 3x_2 - x_3 \le 2$
(2) $f_1(\mathbf{x}) = x_1^2 + x_2^2 + x_3^2$ $f_2(\mathbf{x}) = (x_1-1)^2 + (x_2-1)^2 + (x_3+1)^2$	$0 \le x_i \le 1, i = 1,2,3$ $x_1 + x_2 + x_3 = 1$ $2x_1 + 5x_2 - x_3 \le 2$
(3) $f_1(\mathbf{x}) = (x_1-0.05)^4 + (x_2-0.05)^2 + (x_3-0.05)^2$ $f_2(\mathbf{x}) = (x_1+0.05)^2 + (x_2+0.05)^4 + (x_3+0.05)^2$ $f_3(\mathbf{x}) = (x_1-0.05)^2 + (x_2+0.05)^2 + (x_3-0.05)^4$	$0 \le x_i \le 5, i = 1,2,3$ $x_1 + x_2 + x_3 = 1$ $2x_1 + 3x_2 - x_3 \le 2$
(4) $f_1(\mathbf{x}) = (x_1-0.05)^4 + (x_2-0.05)^2 + ((x_3-0.7)-0.05)^2$ $f_2(\mathbf{x}) = (x_1+0.05)^2 + (x_2+0.05)^4 + ((x_3-0.7)+0.05)^2$ $f_3(\mathbf{x}) = (x_1-0.05)^2 + (x_2+0.05)^2 + ((x_3-0.7)-0.05)^4$	$0 \le x_i \le 5, i = 1,2,3$ $x_1 + x_2 + x_3 = 1$ $2x_1 + 3x_2 - x_3 \le 2$
(5) $f_1(\mathbf{x}) = x_1$ $f_2(\mathbf{x}) = x_2$ $f_3(\mathbf{x}) = 3.5 - \sum_{i=1}^{n-1}(2x_i)\sin(3\pi x_i)(1+x_3^2)$	$0 \le x_i \le 1, i = 1,2,3$ $x_1 + x_2 + x_3 = 1$ $2x_1 + 3x_2 - x_3 \le 2$
(6) $f_1(\mathbf{x}) = (1+x_3)(x_1^3x_2^2 - 10x_1 - 4x_2)$ $f_2(\mathbf{x}) = (1+x_3)(x_1^3x_2^2 - 10x_1 + 4x_2)$ $f_3(\mathbf{x}) = 3(1+x_3)x_1^2$	$0 \le x_1 \le 3.5$ $0 \le x_2 \le 2$ $0 \le x_3 \le 0.5$ $x_1 + x_2 + x_3 \le 1$

different techniques in constraint handling. Test (4) has the same constraints with the difference in the objective functions. The optimal front of this test function is shifted to a small region very close to two boundaries (Figure 9.6).
Test (5): The optimal front is made of three different sized disconnected surfaces which also contain boundaries of the feasible region (Figure 9.7 top).
Test (6): This test function does not contain any equality constraint, therefore the feasible region is larger than the other test functions. One part of the optimal front is located on the boundary (Figure 9.7 bottom).

Parameter Setting

LMOPSO is implemented into the Multi-objective Particle Swarm Optimization method from [15]. In the experiments, three different methods in preserving the feasibility of solutions has been studied; namely the Pbest, the Random, and the Boundary methods from Section 9.3.3. The evaluations are based on 100 runs with different initial seeds and the influence of the inertia weight w on the diversity, convergence and constraint violation has been investigated. The population size and number of generations are selected equal to 50 (for both) with a turbulence factor of 0.1. c_1R_1 and c_2R_2 are random values between 0 and 1.

During the optimization the size of the archive is fixed to an upper-bound. Usually, in MOEAs if the size of the archive exceeds a maximum amount, a clustering, niching or crowding method must be used [3, 24, 13]. Here, ε-dominance

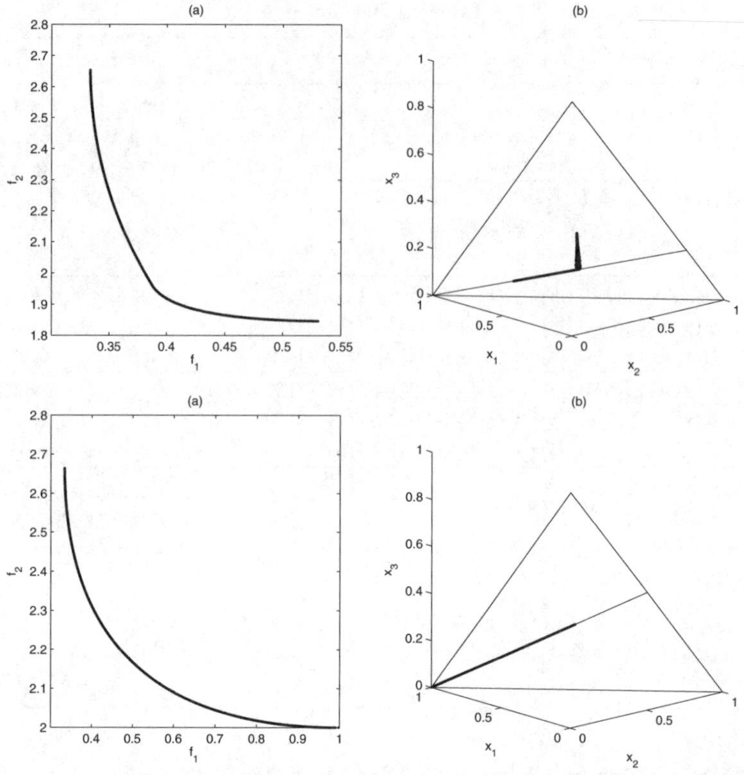

Fig. 9.5. First and second rows show two approximated optimal fronts of the Test (1) and Test (2), respectively. (a) and (b) indicate the objective and the search space.

technique which is based on ε-domination defined in Section 9.1.1 has been used. The ε-domination makes an upper bound for the archive size [14, 10]. This method tries to keep a good diversity of solutions. Keeping the size of the archive constant or rather small, is very important in multi-objective particle swarm optimization. This helps to maintain a good diversity among the archive members which are the global best guides for the particles in the population. The ε values are set to 0.001, 0.0005, 0.02, 0.03, 0.03, and 0.03 for the Tests (1) to (6), respectively. By these values of ε, the archive size is bounded to a maximum of about 100.

9.4.2 Comparisons

The results of different methods have been evaluated by two measurements. The first measure is the ***Cmetric*** [24], which compares the convergence rate of two non-dominated sets A and B:

$$C(A,B) = \frac{|\{\mathbf{b} \in B | \exists \mathbf{a} \in A : \mathbf{a} \preceq \mathbf{b}\}|}{|B|} \tag{9.13}$$

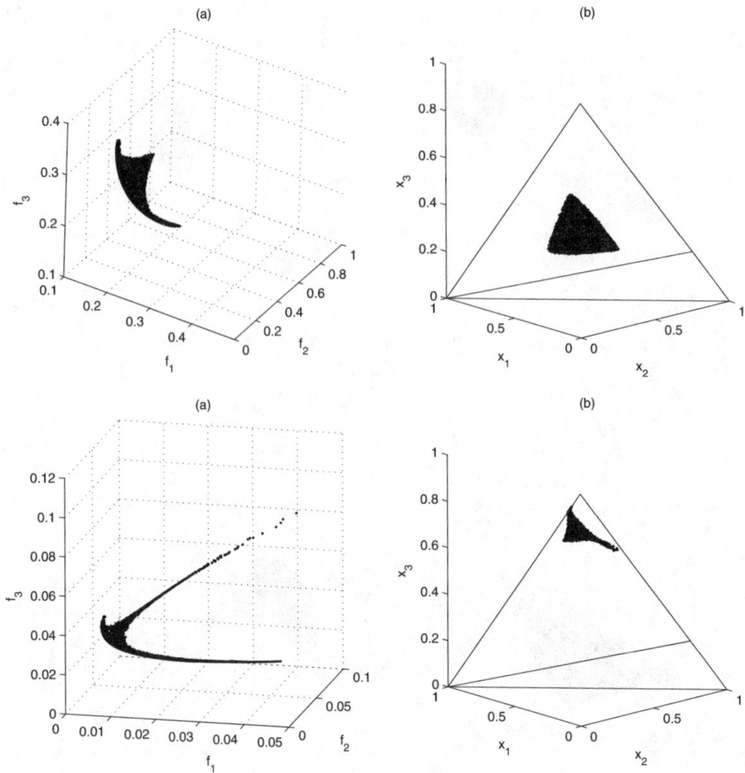

Fig. 9.6. First and second rows show two approximated optimal fronts of the Test (3) and Test (4), respectively. (a) and (b) indicate the objective and the search space.

where $\mathbf{a} \preceq \mathbf{b}$ denotes $\mathbf{a}$ weakly dominates $\mathbf{b}$. The value of $C(A,B) = 1$ means that all the members of B are weakly dominated by the members of A. We can also conclude that $C(A,B) = 0$ means that none of the members of B is weakly dominated by the members of A. $C(A,B)$ is not equal to $1 - C(B,A)$, and both $C(A,B)$ and $C(B,A)$ must be considered for comparisons.

ε-*S*metric

The *S*metric [24] measures the hyper-volume of a region made by a non-dominated set A and a *reference point* in the objective space (Figure 9.8 (a)). If the non-dominated set A has a better diversity and convergence than a non-dominated set B, then the hyper-volume computed for the set A is bigger than the hyper-volume of the set B. Here, the hyper-volume is not computed in the way is done in the original *S*metric [24]. Instead, it is suggested to approximate the hyper-volume by building a grid in the objective space as follows:

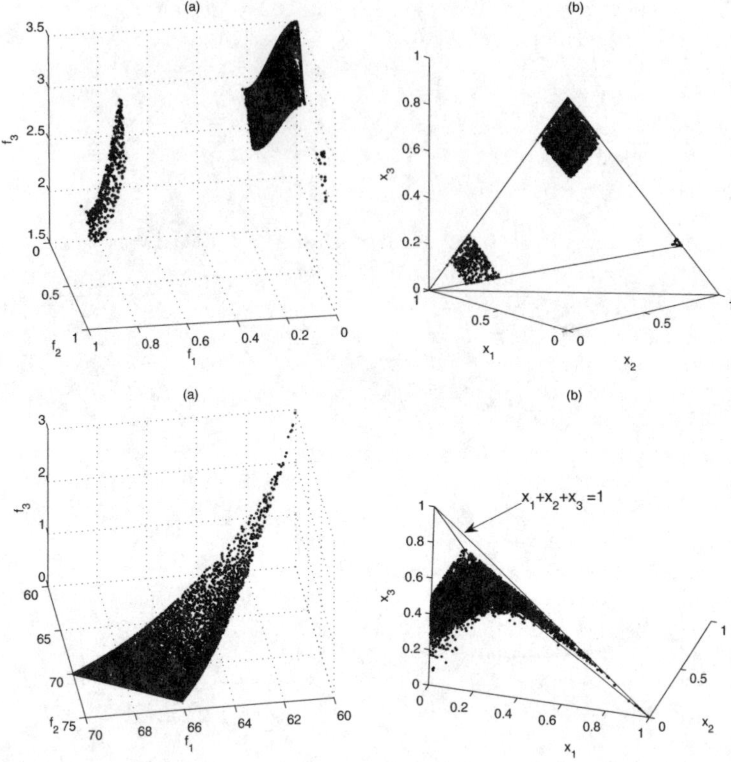

Fig. 9.7. First and second rows show two approximated optimal fronts of the Test (5) and Test (6), respectively. (a) and (b) indicate the objective and the search space.

(a) Define a reference point and then build a grid between the origin and the reference point in the objective space.
(b) Count the grid points which are at least dominated by one of the non-dominated solutions.

A non-dominated set with a good diversity and convergence dominates more solutions than another set with worse diversity and convergence. Actually, this method is very similar to the hyper-volume method with the difference that here we approximate the volume and therefore the computational time is less than the original method. This method is also very easy to implement.

The grid resolution and the reference point will have a great impact on this measurement. Figure 9.8 (a) shows an example for a two-objective non-dominated front. □, o, and x show the reference point, the non-dominated solutions, and the grid points which are dominated by the non-dominated solutions, respectively. The hyper-volume is shown by the dark region.

Here, because the ε dominance method has been used to keep a fixed sized archive, ε-dominated grid points are computed. An example of this idea is shown in

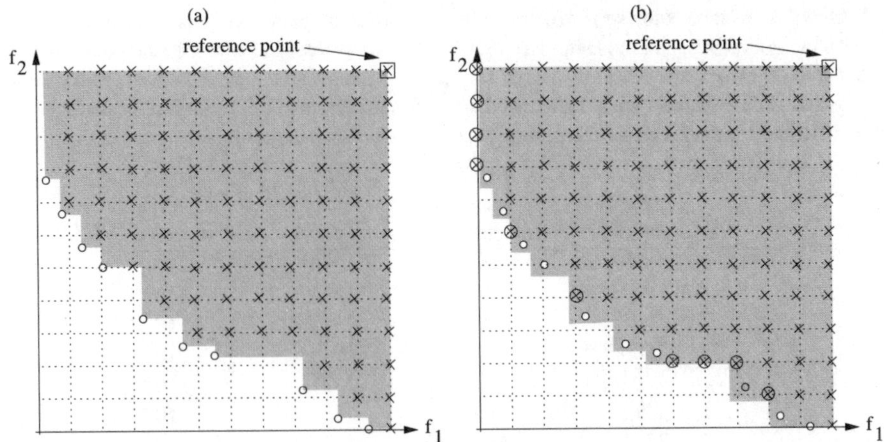

Fig. 9.8. Computing the diversity and convergence of the solutions shown by o. A grid is built between a reference point (□) and the origin. Those grid points which are dominated by the solutions (x) are counted. ⊗ indicates the ε-dominated grid points.

Figure 9.8 (b). In comparison with the Figure 9.8 (a), we observe that the number of ε-dominated grid points are more than the dominated ones. This is called ε-*S*metric.

In the experiments, a 200×200 grid for two-objective and a $100 \times 100 \times 100$ grid for the three-objective test functions are selected. For simplicity, the non-dominated solutions and the grid are normalized to unity.

9.4.3 Evaluations

In order to test the Random, the Pbest, and the Boundary methods and compare them with each other, 100 runs are carried out on each test function. Each run provided *S*metric and *C*metric values. The summary results are shown in Tables 9.2 and 9.3, and in Figure 9.9.

Table 9.2 shows the mean and standard deviations for the *S*metric values for the Random (*SR*), the Pbest (*SP*), and the Boundary (*SB*) method. The sampling distributions of Smetric values are given in Figure 9.9. In general, the Random or the Boundary method score better than the Pbest method. Therefore, we focus on the comparison between *SR* and *SB* values here. It can be seen that for Test (1), (2), (4), and (6), the *S*metric values are higher on average for the Boundary method than for the Random method. For Test (5), the Boundary method achieves higher *S*metric values for an inertia weight $w = 1$.

In order to test whether the *S*metric values for the Boundary method are in fact larger than for the Random method, a (two-sided) paired Wilcoxon signed rank test has been used. In this statistical testing procedure, it is assumed that Boundary and Random method score equally well, i.e. *SB* and *SR* values have the same distribution. It is then evaluated with which probability it can happen by chance that the *SB* values exceed the *SR* values by the same amount and frequency as in the 100 test runs.

Table 9.2. This table shows the *S*metric values for different inertia weights *w* for the Random (SR), the Pbest (SP), and the Boundary (SB) methods. p-values indicate the statistical support that the *SB* values are generally larger (or smaller) than *SR* values. p-values < 0.05 show strong statistical support for a difference between methods, p-values ≥ 0.05 state that there is no difference in the *S*metric values between both methods.

Test	w	SR mean (stdev)	SP mean (stdev)	SB mean (stdev)	SB≠SR p-value
(1)	0	18998(61)	18963(71)	19101(27)	$< 10^{-4}$
	0.4	19061(25)	19033(43)	19117(21)	$< 10^{-4}$
	0.6	19074(27)	19043(39)	19128(15)	$< 10^{-4}$
	1	19065(30)	19055(43)	19133(11)	$< 10^{-4}$
(2)	0	13058(76)	13031(89)	13216(50)	$< 10^{-4}$
	0.4	13124(29)	13146(42)	13258(12)	$< 10^{-4}$
	0.6	13140(24)	13151(45)	13261(13)	$< 10^{-4}$
	1	13143(21)	13157(40)	13270(6)	$< 10^{-4}$
(3)	0	446230(2760)	446010(2676)	445852(2960)	0.46
	0.4	447489(2704)	447774(2435)	447577(2897)	0.99
	0.6	448597(2227)	447589(2588)	448083(2366)	0.10
	1	449588(1987)	448015(2408)	449280(1904)	0.14
(4)	0	777159(2034)	776786(2294)	779583(2044)	$< 10^{-4}$
	0.4	779406(1907)	778384(1725)	780321(1718)	$< 10^{-4}$
	0.6	779349(1781)	778713(1834)	780557(1723)	$< 10^{-4}$
	1	779312(1877)	779212(1840)	780542(1700)	$< 10^{-4}$
(5)	0	652905(5973)	652427(5159)	653679(4436)	0.43
	0.4	655276(4590)	652894(4698)	654633(4737)	0.36
	0.6	656332(4707)	652544(5237)	655637(5218)	0.58
	1	657551(4247)	653682(4691)	662068(4592)	$< 10^{-4}$
(6)	0	492721(17354)	486757(19344)	495554(18600)	0.18
	0.4	508233(15340)	501355(15179)	517541(17089)	$< 10^{-4}$
	0.6	516909(10202)	510470(12852)	530234(10200)	$< 10^{-4}$
	1	520212(9423)	516370(11827)	543106(7214)	$< 10^{-4}$

This probability is called p-value. The p-value ranges from 0 to 1. For test functions and inertia weights for which the p-value of the paired Wilcoxon signed rank test is smaller than 0.05, there is a strong statistical support that *SB* values are generally larger than *SR* values. Table 9.2 also lists the p-values for the tests between *SB* and *SR* values. For the Tests (1), (2), (4), (6), and (5) when $w = 1$, we have a strong statistical support that *SB* values are generally larger than *SR* values (p-values $< 10^{-4}$). For the Test (3), there is no statistically significant difference in the *S*metric values between the Boundary method and the Random method. As the optimal front of the Test (3) is inside the feasible solution space, it is not surprising that the Random method, which moves particles away from the boundary, has a good performance in comparison with the Boundary method. However, statistically, the Random method does not provide uniformly better results than the Boundary method. Therefore, we can come to the conclusion that the Boundary and Random method have similar performance for

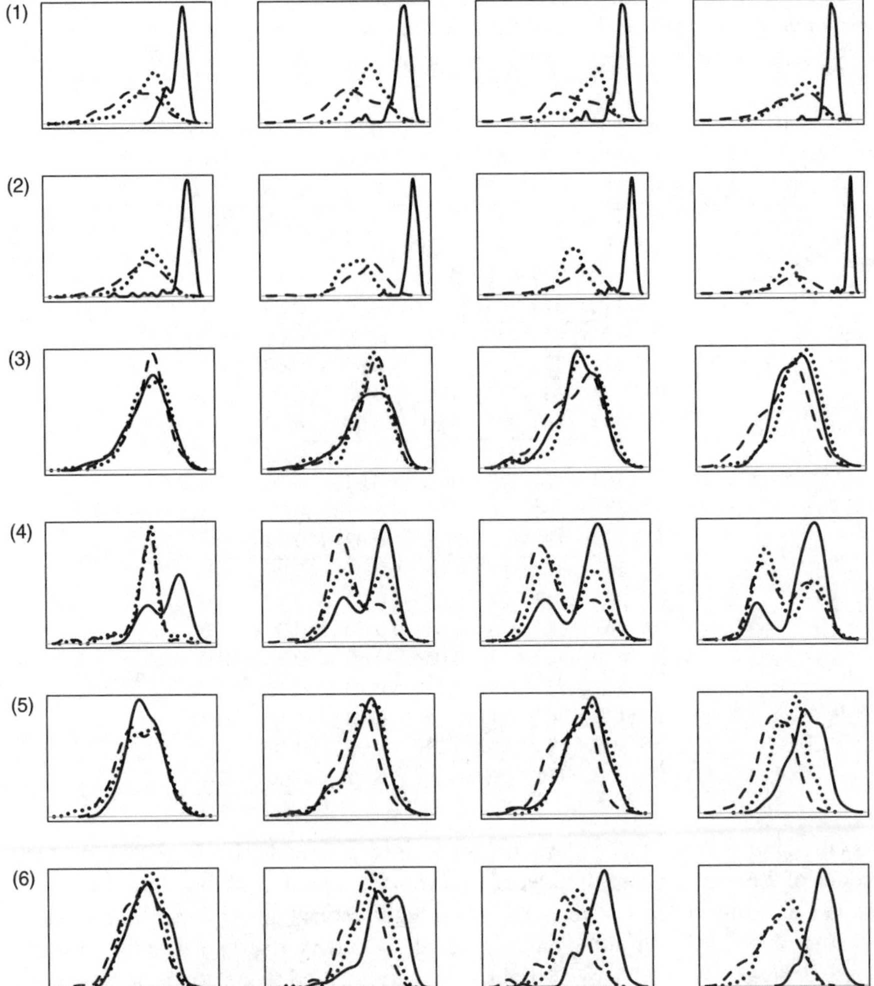

Fig. 9.9. This figure illustrates the sampling distributions of the *S*metric values given in Table 9.2. Solid, dashed, and dotted lines indicate the results of the Boundary, Pbest and Random methods, respectively. Rows indicate the results of the Tests (1) to (6) from top to bottom. In each row the results of different inertia weights 0, 0.4, 0.6 and 1 are illustrated from left to right.

optimal fronts within the feasible solution space but that the Boundary method can be expected to outperform the Random method when the optimal front includes parts of the boundary.

The same conclusion can be drawn when looking at the *C*metric values and the constraint violation rates in Table 9.3. Constraint violation rates indicate the total number of solutions which violate the constraints. We observe that the constraint

Table 9.3. *C*metric values and the constraint violation rates (*N*) for different inertia weights *w* and for the Random (R), the Pbest (P), and the Boundary (B) methods

Test	*w*	*N*R	*N*P	*N*B	*C*(B,R)	*C*(R,B)	*C*(B,P)	*C*(P,B)
(1)	0	128	77	84	0.33	0.13	0.36	0.12
	0.4	509	221	272	0.34	0.11	0.36	0.11
	0.6	714	323	389	0.32	0.10	0.38	0.08
	1	1011	585	860	0.28	0.11	0.38	0.10
(2)	0	114	74	85	0.78	0.04	0.79	0.03
	0.4	524	226	292	0.82	0.01	0.81	0.01
	0.6	762	339	421	0.80	0.01	0.80	0.01
	1	1103	647	935	0.81	0.00	0.83	0.00
(3)	0	14	13	12	0.00	0.00	0.00	0.00
	0.4	41	35	38	0.00	0.00	0.00	0.00
	0.6	124	90	119	0.00	0.00	0.00	0.00
	1	503	283	903	0.00	0.00	0.00	0.00
(4)	0	92	70	77	0.02	0.02	0.02	0.02
	0.4	564	225	277	0.02	0.01	0.03	0.01
	0.6	783	294	390	0.03	0.01	0.03	0.01
	1	1088	480	852	0.03	0.02	0.03	0.02
(5)	0	61	51	53	0.04	0.01	0.04	0.01
	0.4	272	142	164	0.03	0.00	0.03	0.00
	0.6	462	204	273	0.03	0.00	0.03	0.00
	1	848	362	863	0.03	0.00	0.04	0.00
(6)	0	37	33	39	0.12	0.10	0.12	0.10
	0.4	140	96	131	0.14	0.10	0.15	0.09
	0.6	300	195	315	0.17	0.09	0.19	0.08
	1	656	416	1160	0.26	0.05	0.30	0.04

violation rate increases when we increase the inertia weight. This is because the high values of the inertia weight produce large velocity vectors and therefore they cause more constraint violations. Interestingly, a higher constraint violation rate leads to a better diversity for all three methods as the boundary region of feasible solution space is better explored. However, the Boundary method seems to profit more from the constraint violation of a single solution since information on the position and velocity is not completely lost.

The *C*metric values recorded in Table 9.3 show that for 3-objective test functions, *C*metric cannot be the only comparison metric. This is because the solution sets of the size less than 100 which are basically approximating a surface in a three dimensional space are being compared. It is obvious that these solutions do not have similar diversities and therefore the *C*metric value is very low. Our comparisons for the 3-objective test functions rely on *S*metric values.

9.5 Multi-objective Optimization of Multi-component Chemical Systems

In this section, a real-world application in identifying the parameters of a Multi-component chemical system has been studied. This application contains a linearly constrained multi-objective optimization problem. In the following, a short description about multi-component systems, the problem, and the results are being studied. More details about these systems are provided in [5].

9.5.1 Background

Dynamic processes occurring in volcanic or other magmatic systems depend significantly on the viscosity of the silicate liquid. Thus, knowledge of the viscosity is an essential piece of information to evaluate hazards associated with volcanic eruptions. The viscosity depends primarily on the relative abundance x_i of **species**, i.e., the **speciation** and this case study calculates the speciation of a liquid in the system $Na_2O - SiO_2$ at various temperatures T. We distinguish five types of species, each formed of a tetrahedra with a central silica atom surrounded by four atoms of oxygen. These oxygen atoms can either serve as bridges between the tetrahedra (**bridging** oxygen) or be connected to a sodium cation. In the latter case, the bridges between tetrahedra are broken and the oxygen is **non-bridging**. The species are named Q_i, where i is the number on bridging oxygen atoms. Thus, in the system $Na_2O - SiO_2$ the species are:

$$
\begin{aligned}
Q_4 &: SiO_2 \\
Q_3 &: Na_2Si_2O_5 \\
Q_2 &: Na_4Si_2O_6 \\
Q_1 &: Na_6Si_2O_7 \\
Q_0 &: Na_8Si_2O_8
\end{aligned}
$$

The species are schematically represented in Figure 9.10. The relative abundance of the species is controlled by the following **speciation reactions**:

$$
\begin{aligned}
2Q_1 &= Q_0 + Q_2 \\
2Q_2 &= Q_1 + Q_3 \\
2Q_3 &= Q_2 + Q_4
\end{aligned}
\tag{9.14}
$$

The free energy of a species Q_i is given by

$$G_i = H^o_{i,T} - TS^o_{i,T} + RTln(x_i) = G^o_{i,T} + RTln(x_i) \tag{9.15}$$

where $H^o_{i,T}$ is the standard state enthalpy of the species i at the temperature T, $S^o_{i,T}$ is the standard state entropy of i at T, R is the gas constant, and $G^o_{i,T}$ is the standard state free energy of i at T. At equilibrium the changes in the free energy ΔG of each speciation reaction is zero:

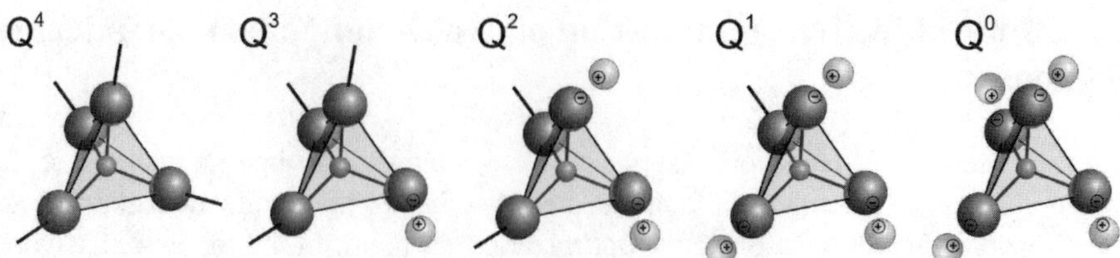

Fig. 9.10. species Q_4 to Q_0. Lines represent covalent bonds between bridging oxygens and Si, and ⊕ ⊖ indicates the ionic bonds between non-bridging oxygens and Na.

$$\Delta G_T = \Delta G_T^o + RT lnK = 0 \tag{9.16}$$

where ΔG_T^o is the change in the standard state free energy at the temperature T and K is the equilibrium constant of the speciation reaction. K and ΔG_T^o for the speciation reactions are given by:

$$\Delta G_T^o = G_{Q0,T}^o + G_{Q2,T}^o - 2G_{Q1,T}^o, \; K = \frac{x_0 x_2}{x_1^2} \tag{9.17}$$

$$\Delta G_T^o = G_{Q1,T}^o + G_{Q3,T}^o - 2G_{Q2,T}^o, \; K = \frac{x_1 x_3}{x_2^2} \tag{9.18}$$

$$\Delta G_T^o = G_{Q2,T}^o + G_{Q4,T}^o - 2G_{Q3,T}^o, \; K = \frac{x_2 x_4}{x_3^2} \tag{9.19}$$

Gurmann [4] suggested that the ΔG^o for the three speciation reactions are equal. Thus, at equilibrium, three objective functions which should minimized at the same time:

$$\begin{aligned}
\text{minimize} \quad & f_1(\mathbf{x}) = \frac{x_0 x_2}{x_1^2} - C \\
& f_2(\mathbf{x}) = \frac{x_1 x_3}{x_2^2} - C \\
& f_3(\mathbf{x}) = \frac{x_2 x_4}{x_3^2} - C \\
& C = \exp -2\frac{(\Delta H^o - T\Delta S^o)}{RT}) \\
\text{subject to} \quad & x_0 + x_1 + x_2 + x_3 + x_4 = XSiO_2 \\
& 4x_0 + 3x_1 + 2x_2 + x_3 = 2(1 - XSiO_2) \\
& 0 \le x_i \le 1 \;\; \forall i = 0, 1, \cdots, 4
\end{aligned} \tag{9.20}$$

where ΔH^o and ΔS^o are change in enthalpy and entropy, respectively, of the speciation reactions. These values are constant in the system at 7722 and -7.1422. $XSiO_2$ is the mole fraction of SiO_2:

$$XSiO_2 = \frac{MSiO_2}{MSiO_2 + MNaO_{0.5}} \tag{9.21}$$

where M represents the number of moles.

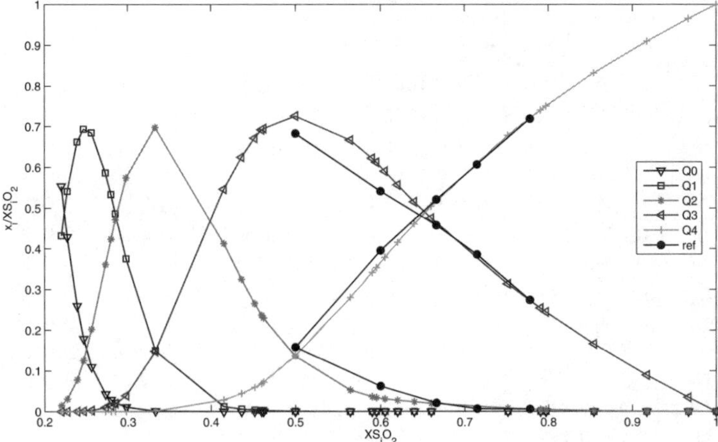

Fig. 9.11. This figure shows the optimized parameters x_i obtained for all the $XSiO_2$ values. • shows the reference data available for the system Na_2O-SiO_2.

9.5.2 Experiments

The experiments are done to solve the multi-objective optimization problem in Equation (9.20). This problem contains 3 objectives, 5 parameters[7], and two linear equality constraints. T and $XSiO_2$ are two constants from the database recorded in [5]. This database contains 34 different values and the experiments are done for all of these values by using the Boundary method with 100 population size, 1000 generations, turbulence factor of 0.1, and the inertia weight equal to 0.6.

The first step in solving this problem is to find the basic feasible solutions (extreme points) of the feasible region. The initial boundary conditions are $0 \leq x_i \leq 1 \ \forall i = 0, 1, \cdots, 4$. After computing the basic feasible solutions, we obtain new boundaries as in Table 9.4 for different values of $XSiO_2$ and T. The new boundaries are in some cases much less than the initial boundaries indicating that the feasible region is smaller than the search space. The stopping criteria for this problem is when at least one optimal solution obtains all of the objective values less than $1e-10$. If several solutions obtain this rate of convergence one of them is selected at random. This must be obtained for all of the $XSiO_2$ and T values. Figure 9.11 illustrates the obtained results together with the reference data from [5] for the system Na_2O-SiO_2.

[7]It is possible to reduce one of the parameters by using the two equality constraints. But this brings more nonlinearity inside the objective functions and because the number of parameters is not high, 5 parameters are kept.

Table 9.4. New boundaries of parameters for different $XSiO_2$ and T (qtz: quartz, tri: tridimite, and crist: cristobalite).

Composition	$XSiO_2$	T	x_0	x_1	x_2	x_3	x_4
$Na_6Si_2O_7$	0.2203	1298	[0.118,0.194]	[0,0.101]	[0,0.050]	[0,0.033]	[0,0.025]
$Na_6Si_2O_7$	0.2274	1338	[0.090,0.193]	[0,0.137]	[0,0.068]	[0,0.045]	[0,0.034]
$Na_6Si_2O_7$	0.2392	1365	[0.043,0.190]	[0,0.196]	[0,0.098]	[0,0.065]	[0,0.049]
$Na_6Si_2O_7$	0.2469	1388	[0.012,0.188]	[0,0.234]	[0,0.117]	[0,0.078]	[0,0.058]
$Na_6Si_2O_7$	0.2563	1399	[0,0.185]	[0,0.247]	[0,0.140]	[0,0.093]	[0,0.070]
$Na_6Si_2O_7$	0.2731	1338	[0,0.181]	[0,0.242]	[0,0.182]	[0,0.121]	[0,0.091]
$Na_6Si_2O_7$	0.2796	1294	[0,0.180]	[0,0.240]	[0,0.199]	[0,0.132]	[0,0.099]
$Na_4Si_2O_6$	0.2796	1294	[0,0.180]	[0,0.240]	[0,0.199]	[0,0.132]	[0,0.099]
$Na_4Si_2O_6$	0.2849	1308	[0,0.170]	[0,0.238]	[0,0.212]	[0,0.141]	[0,0.106]
$Na_4Si_2O_6$	0.2978	1338	[0,0.175]	[0,0.234]	[0,0.244]	[0,0.163]	[0,0.122]
$Na_4Si_2O_6$	0.3331	1362	[0,0.166]	[0,0.222]	[0,0.332]	[0,0.221]	[0,0.166]
$Na_4Si_2O_6$	0.4152	1228	[0,0.146]	[0,0.194]	[0,0.292]	[0,0.358]	[0,0.269]
$Na_4Si_2O_6$	0.4362	1177	[0,0.140]	[0,0.187]	[0,0.281]	[0,0.393]	[0,0.295]
$Na_4Si_2O_6$	0.4518	1135	[0,0.137]	[0,0.182]	[0,0.274]	[0,0.419]	[0,0.314]
$Na_4Si_2O_6$	0.4600	1113	[0,0.130]	[0,0.180]	[0,0.270]	[0,0.433]	[0,0.325]
$Na_2Si_2O_5$	0.4600	1113	[0,0.135]	[0,0.180]	[0,0.270]	[0,0.433]	[0,0.325]
$Na_2Si_2O_5$	0.4623	1124	[0,0.134]	[0,0.179]	[0,0.268]	[0,0.437]	[0,0.327]
$Na_2Si_2O_5$	0.4998	1147	[0,0.125]	[0,0.166]	[0,0.250]	[0,0.499]	[0,0.374]
$Na_2Si_2O_5$	0.5641	1102	[0,0.108]	[0,0.145]	[0,0.217]	[0,0.435]	[0.128,0.455]
SiO_2qtz	0.5900	1076	[0,0.102]	[0,0.136]	[0,0.205]	[0,0.410]	[0.180,0.487]
SiO_2qtz	0.5950	1078	[0,0.101]	[0,0.135]	[0,0.202]	[0,0.405]	[0.190,0.493]
SiO_2qtz	0.6051	1111	[0,0.098]	[0,0.131]	[0,0.197]	[0,0.394]	[0.210,0.506]
SiO_2tri	0.6202	1178	[0,0.094]	[0,0.126]	[0,0.189]	[0,0.379]	[0.240,0.520]
SiO_2tri	0.6397	1268	[0,0.090]	[0,0.120]	[0,0.180]	[0,0.360]	[0.279,0.549]
SiO_2tri	0.6598	1348	[0,0.085]	[0,0.113]	[0,0.170]	[0,0.340]	[0.319,0.574]
SiO_2tri	0.7518	1638	[0,0.062]	[0,0.082]	[0,0.124]	[0,0.248]	[0.503,0.689]
SiO_2tri	0.7905	1723	[0,0.052]	[0,0.069]	[0,0.104]	[0,0.209]	[0.581,0.738]
SiO_2tri	0.7969	1743	[0,0.050]	[0,0.067]	[0,0.101]	[0,0.203]	[0.593,0.746]
SiO_2crist	0.8544	1823	[0,0.036]	[0,0.048]	[0,0.072]	[0,0.145]	[0.708,0.818]
SiO_2crist	0.9170	1866	[0,0.020]	[0,0.027]	[0,0.041]	[0,0.083]	[0.834,0.896]
SiO_2crist	0.9664	1919	[0,0.008]	[0,0.011]	[0,0.016]	[0,0.033]	[0.932,0.958]
SiO_2crist	1.0000	1996	[0,0]	[0,0]	[0,0]	[0,0]	[1,1]

9.5.3 Evaluations

The available reference data includes the optimal values for the $XSiO_2$ equal to 0.5, 0.6, 0.666, 0.715, and 0.778. It can be observed in Figure 9.11 that the obtained values are very *close* to the reference data for the same $XSiO_2$. From mineralogical point of view this closeness is acceptable. However, there is another source of information to evaluate the results as follows. The reference values have been obtained by W. Halter and B. Mysen [5] by using spectroscopic techniques. They observed that the maximum abundance of the Q_3 species is approximately 0.7 at the glass transition temperature. Here, the obtained results for Q_3 have the maximum

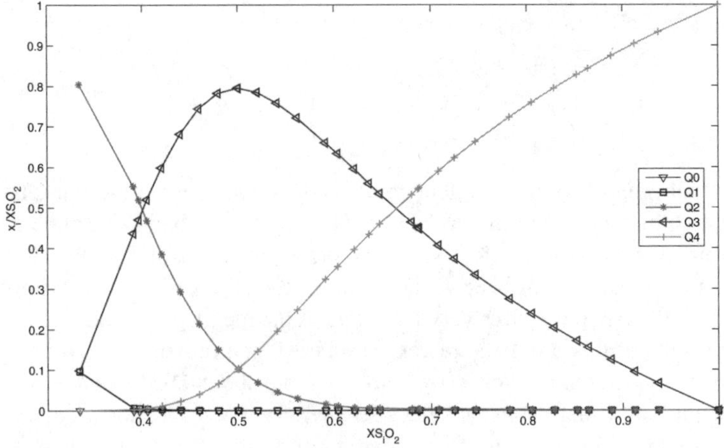

Fig. 9.12. Computed speciation data for the system $K_2O - SiO_2$.

abundance of 0.7 at the glass transition temperature (see the maximum value of Q_3 in Figure 9.11)[8].

The speciation data depend highly on the constant values which should be set in the Equation (9.20). In some chemical systems, the constants are not given and should be considered as parameters. The idea of using LMOPSO helps to find the speciation data for any desired constant values. This is actually one part of a project which will be integrated into a large optimization problem.

The multi-objective optimization problem in Equation (9.20) can be used to find the speciation data for multi-component chemical systems in general. In the literature, there is no reference data for other systems like $K_2O - SiO_2$, as this should be done by some tedious techniques like using spectroscopic techniques. Theoretically these values can be obtained by the LMOPSO method. Figure 9.12 shows the obtained results for this system with the constants ΔH^o and ΔS^o as 13000 and -7.1422. This figure shows that these values look similar to the speciation data of the system $Na_2O - SiO_2$. Here, the maximum abundance of the Q_3 species is approximately 0.8 at the glass transition temperature.

9.5.4 Uncertainty on $XSiO_2$

Due to the uncertainty in the measurements, there is an uncertainty in the data for $XSiO_2$ of about ± 0.01. Uncertainty in optimization can be handled in several ways [22, 6]. Here, we consider them in the constraints as follows:

[8] This kind of evaluation can be performed for the system $Na_2O - SiO_2$.

$$
\begin{aligned}
&x_0+x_1+x_2+x_3+x_4 \leq XSiO_2+0.01 \\
&x_0+x_1+x_2+x_3+x_4 \geq XSiO_2-0.01 \\
&4x_0+3x_1+2x_2+x_3 \leq 2(1-XSiO_2+0.01) \\
&4x_0+3x_1+2x_2+x_3 \geq 2(1-XSiO_2-0.01) \\
&0 \leq x_i \leq 1 \;\; \forall i=0,1,\cdots,4
\end{aligned} \tag{9.22}
$$

This new formulation of the problem contains only inequality constraints. The same optimization method as before has been applied to this problem. It can be observed that the number of basic feasible solutions increases, e.g., for $XSiO_2 = 0.2203$ 16 basic feasible solutions has been obtained (4 for the case without uncertainty). This indicates that feasible region is very large due to the inequality constraints.

The results of this problem are very similar to the problem without uncertainty. The maximum difference is for the x_0 parameter and about 0.025 which is negligible. These results show that the mean values selected for $XSiO_2$ are correct and it can be concluded that $XSiO_2$ has a uniform distribution in the uncertainty range $XSiO_2 \pm 0.01$.

9.6 Conclusion

Linear Multi-objective Particle Swarm Optimization (LMOPSO) is a new approach to solve linearly constrained multi-objective optimization problems by using particle swarm optimization methods. This method can be applied to problems which cannot be solved by numerical methods due to their properties like discrete functions, high number of objectives, and etc. The main idea of this approach is to evolve a set of optimal solutions like in multi-objective optimization while guaranteeing the feasibility of the solutions. Therefore the search strategy will have a great impact on the solutions.

Typically constraints constitute a feasible region which is smaller than the search space. The common way to handle constraints by many of the iterative population based search methods like evolutionary algorithms, is to define a population in the search space and to select the feasible solutions. But when the feasible region is smaller than the search space, most of the population members are infeasible and finding some good feasible solutions would require a high computational effort.

The idea of LMOPSO is to find the feasible region and only explore this region. In other words, the particles are not allowed to be infeasible. For this, first the feasible region must be defined. Then an exploration strategy must be applied to the particles so that they never leave the feasible region. LMOPSO starts with a set of feasible solutions and explores the feasible region by linear combinations of them. The linear combination is integrated into the main formulas of particle swarm optimization which makes to explore the feasible region efficiently. However, the linear combinations may lead some solutions to violate the inequality constraints while they satisfy the equality constraints.

Therefore, in order to guarantee the feasibility of solutions, the feasibility of solutions must be preserved. Here, three methods in particular the Pbest, the

Random, and the Boundary method have been proposed. These methods have been implemented and compared with each other in terms of the diversity and convergence of the obtained solutions on several test problems with different degrees of difficulties. Also, LMOPSO has been studied on a real world problem in identifying the parameters of the chemical system.

Altogether, LMOPSO is a very efficient way to solve linearly constrained multi-objective problems as it guarantees the feasibility of solutions during the iterations while using a very efficient search strategy. However, we have to note that for a high number of constraints and parameters, we would require a high computational time to define the boundaries of the feasible region. This could be considered as one of the disadvantages of this approach.

References

1. J. E. Alvarez-Benitez, R. M. Everson, and J. E. Fieldsend. A MOPSO algorithm based exclusively on pareto dominance concepts. *Lecture Notes in Computer Science (LNCS), EMO 2005*, pages 459–473, 2005.
2. D. Bertsimas and J. N. Tsitsiklis. *Introduction to Linear Optimization*. Athena Scientific, 1997.
3. K. Deb. *Multi-Objective Optimization using Evolutionary Algorithms*. John Wiley & Sons, 2001.
4. S. J. Gurman. Bond ordering in silicate-glasses - a critique and a re- solution. *Journal of Non-Crystalline Solids 125(1-2)*, pages 151–160, 1990.
5. W. E. Halter and B. O. Mysen. Melt speciation in the system Na2O-SiO2. *Chemical Geology 213(1-3)*, pages 115–123, 2004.
6. E. J. Hughes. Evolutionary multi-objective ranking with uncertainty and noise. *Lecture Notes in Computer Science (LNCS), EMO 2001*, pages 329–343, 2001.
7. Fernando Jiménez and José L. Verdegay. Evolutionary techniques for constrained optimization problems. In Hans-Jürgen Zimmermann, editor, *7th European Congress on Intelligent Techniques and Soft Computing (EUFIT'99)*. Verlag Mainz, 1999.
8. J. Kennedy and R. C. Eberhart. *Swarm Intelligence*. Morgan Kaufmann, 2001.
9. Angel Fernando Kuri-Morales and Jesús Gutiérrez-García. Penalty Functions Methods for Constrained Optimization with Genetic Algorithms: A Statistical Analysis. In Carlos A. Coello Coello and et al., editors, *Proceedings of the 2nd Mexican International Conference on Artificial Intelligence (MICAI 2002)*, pages 108–117. Springer-Verlag, 2001. Lecture Notes in Artificial Intelligence Vol. 2313.
10. M. Laumanns, L. Thiele, K. Deb, and E. Zitzler. Archiving with guaranteed convergence and diversity in multi-objective optimization. In *Genetic and Evolutionary Computation Conference (GECCO02)*, pages 439–447, 2002.
11. Z. Michalewicz. A survey of constraint handling techniques in evolutionary computation methods. *Proceedings of the 4th Annual Conference on Evolutionary Programming*, pages 135–155, 1995.
12. Z. Michalewicz and C. Janikow. Handling constraints in genetic algorithms. *Proceedings of the 4th International Conference on Genetic Algorithms*, pages 151–157, 1995.
13. S. Mostaghim. *Multi-objective Evolutionary Algorithms: Data structures, Convergence, and Diversity*. Shaker Verlag, Germany, 2005.

14. S. Mostaghim and J. Teich. The role of e-dominance in multi-objective particle swarm optimization. In *Proceedings CEC'03, the Congress on Evolutionary Computation*, 2003.
15. S. Mostaghim and J. Teich. Strategies for finding good local guides in multi-objective particle swarm optimization. In *IEEE Swarm Intelligence Symposium*, pages 26–33, 2003.
16. Angel-E. Muñoz-Zavala, Arturo Hernández Aguirre, and Enrique R. Villa Diharce. Constrained Optimization via Particle Evolutionary Swarm Optimization Algorithm (PESO). In H. G. Beyer and et al., editors, *Proceedings of the Genetic and Evolutionary Computation Conference (GECCO'2005)*, volume 1, pages 209–216, 2005.
17. Ulrich Paquet and Andries P. Engelbrecht. A new particle swarm optimiser for linearly constrained optimization. In *Proceedings CEC'03, the Congress on Evolutionary Computation*, pages 227–233, 2003.
18. Tapabrata Ray, Tai Kang, and Seow Kian Chye. An Evolutionary Algorithm for Constrained Optimization. In Darrell Whitley and et al., editors, *Proceedings of the Genetic and Evolutionary Computation Conference (GECCO'2000)*, pages 771–777. Morgan Kaufmann, 2000.
19. G. Rudolph and A. Agapie. Convergence Properties of Some Multi-Objective Evolutionary Algorithms. In *Proceedings of the 2000 Congress on Evolutionary Computation*, pages 1010–1016, 2000.
20. Thomas P. Runarsson and Xin Yao. Stochastic Ranking for Constrained Evolutionary Optimization. *IEEE Transactions on Evolutionary Computation*, 4(3):284–294, 2000.
21. Y. Shi and R. C. Eberhart. Parameter selection in particle swarm optimization. *Evolutionary Programming*, pages 591–600, 1998.
22. J. Teich. Pareto-front exploration with uncertain objectives. *Lecture Notes in Computer Science (LNCS), EMO 2001*, pages 314–328, 2001.
23. Gregorio Toscano-Pulido and Carlos A. Coello Coello. A constraint-handling mechanism for particle swarm optimization. In *Proceedings CEC'04, the Congress on Evolutionary Computation*, pages 1396–1403, 2004.
24. E. Zitzler. *Evolutionary Algorithms for Multiobjective Optimization: Methods and Applications*. Shaker Verlag, Germany, Swiss Federal Institute of Technology (ETH) Zurich, 1999.

10

Cooperative Particle Swarm Optimizers: A Powerful and Promising Approach

Mohammed El-Abd and Mohamed Kamel

University of Waterloo, 200 University Avenue West, Waterloo, Ontario, Canada
mhelabd@pami.uwaterloo.ca, mkamel@pami.uwaterloo.ca

Summary. This chapter surveys the different cooperative models that have been implemented using Particle Swarm Optimization (PSO) in many applications. A Taxonomy for classifying these models is proposed. The chapter also identifies the different parameters that can influence the behavior of such models. Experiments run on different benchmark optimization functions show how the performance of a simple cooperative PSO model can change under the influence of these parameters.

10.1 Introduction

Cooperative search algorithms is one of the many areas that have been extensively studied in the past decade to try to solve many large size optimization problems. The basic approach involves having more than one search module running and exchanging information among each other in order to explore the search space more efficiently and reach better solutions. It is sometimes regarded as a parallel algorithms approach since these modules may be running in parallel. On the other hand, some researchers classify it as a hybrid approach since these search modules could employ different heuristics. Different classification schemes were given for both hybrid and parallel approaches [10, 12, 38]. In some of them, the cooperative part is put as a stand alone class. In others, the cooperative class could be inferred. Hence, cooperative search algorithms are considered to lie somewhere in between.

There are many definitions in the literature for cooperative search. In [10] it states that "Teamwork hybridization represents cooperative optimization models, in which we have many parallel cooperating agents, each agent carries out a search in solution space". The authors in [29] consider cooperative search as a category of parallel algorithms. "These algorithms execute in parallel several search programs on the same optimization problem instance". Another definition holding the same idea is found in [28] "Cooperative search algorithms are parallel search methods that combine several individual programs in a single search system". The authors in [4] defines it as "a search performed by agents that exchange information about states, models, entire sub-problems, solutions or other search space characteristics",

M. El-Abd and M. Kamel: *Cooperative Particle Swarm Optimizers: A Powerful and Promising Approach*, Studies in Computational Intelligence (SCI) **31**, 239–259 (2006)
www.springerlink.com

they also indicate that cooperative algorithms could be efficient even if they are sequentially implemented.

Many heuristic search methods have been used in a cooperative search environment including Tabu Search (TS) [11, 14, 13], Genetic Algorithms (GA) [27, 8], Ant Colony Optimization (ACO) [25, 26, 40], and Particle Swarm Optimization (PSO) [30, 37].

This chapter is organized as follows. Section 2 gives an overview of PSO in general. Different cooperative models that were introduced using PSO are surveyed and classified using a proposed taxonomy in section 3. The factors that control the behavior of the different cooperative models are identified in section 4. Section 5 introduces the experimental work. The chapter is concluded in section 6.

10.2 Particle Swarm Optimization

PSO [16, 6] is an optimization method widely used to solve continuous nonlinear functions. It is a stochastic optimization technique that was originally developed to simulate the movement of a flock of birds or a group of fish looking for food. PSO is regarded as a population-based method, where the population is referred to as a swarm. The swarm consists of a number of individuals called particles. Every particle i in the swarm holds the following information: (i) the current position x_i, (ii) the current velocity v_i, (iii) the best position, the one associated with the best fitness value the particle has achieved so far $pbest_i$, and (iv) the global best position, the one associated with the best fitness value found among all of the particles $gbest$.

In every iteration, each particle adjusts its own trajectory in the space in order to move towards its best position and the global best according to the following equations

$$\begin{aligned} v_{ij}(t+1) = w \times v_{ij}(t) + c_1 \times rand() \times (pbest_j - x_{ij}) \\ + c_2 \times rand() \times (gbest_j - x_{ij}), \end{aligned} \tag{10.1}$$

$$x_{ij}(t+1) = x_{ij}(t) + v_{ij}(t+1), \tag{10.2}$$

For $j \in 1..d$ where d is the number of dimensions. $i \in 1..n$ where n is the number of particles, t is the iteration number, w is the inertia weight, *rand()* generates a random number uniformly distributed in the range (0,1), and c_1 and c_2 are the acceleration factors.

Then each particle updates its personal best using the equation

$$pbest_i(t+1) = \begin{cases} pbest_i(t) & \texttt{if} \quad f(pbest_i(t)) \leq f(x_i) \\ x_i(t) & \texttt{if} \quad f(pbest_i(t)) \geq f(x_i) \end{cases} \tag{10.3}$$

And finally, the global best of the swarm is updated using the equation

$$gbest(t+1) = \arg\min_{pbest_i} f(pbest_i(t+1)), \tag{10.4}$$

where *f(.)* is a function that evaluates the fitness value for a given position.

The values of v_i are clamped to the range $(-v_{max}, v_{max})$ to try to prevent the particles from leaving the search space. The value of v_{max} is chosen to be equal to $k \times x_{max}$ where x_{max} specifies the domain in which the initial particles are generated. The value of w drops linearly from 1 at the beginning of the run to near 0 at the end. The value 1 at the beginning of the run helps the particles to achieve a fast movement, and the small value at the end helps the particles to settle at the found minimum. The acceleration factors c_1 and c_2 control how far a particle can move in a single update and are usually set to 2.

The previous model is referred to as the *gbest* (global best) model. Another model is the *lbest* (local best) [7]. In this model, every particle moves towards its own best position and the best position achieved by its neighborhood and not by the whole swarm. Many neighborhood structures have been proposed and tested in [17, 31];

10.3 Cooperative PSO Models

10.3.1 The Taxonomy

In [24], we proposed two different taxonomies for classifying cooperative search algorithms in general. One is based on the level of space decomposition achieved by the cooperative system, and the other one is based on the types of algorithms used in it. In this section, the cooperative PSO models that were introduced in the past few years are surveyed and divided by the type of problems they were applied to. We also show how these models can be classified based on the taxonomy shown in Fig. 10.1.

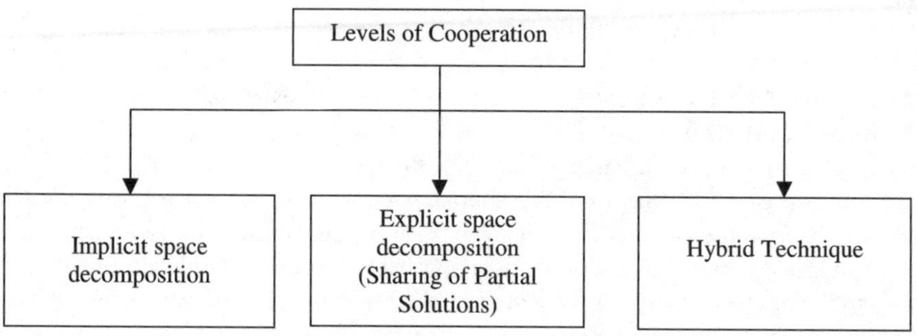

Fig. 10.1. Cooperative PSO taxonomy [1]

The implicit space decomposition involves the decomposition of the search space between different algorithms. This class refers to having different algorithms looking for a solution and sharing useful information between them. There could be many choices for the information to be shared depending on the algorithms being used. The name implicit comes from the fact that the different algorithms explore different areas in the search space due to having different initial solutions, different parameter values, or both.

In the explicit space decomposition class, the search space is explicitly decomposed into sub-spaces. Each algorithm searches for a sub-solution in a different sub-space of the problem. Hence, each algorithm provides a partial solution to the problem, these partial solutions are gathered to provide a complete solution.

The hybrid approach refers to the idea of having a cooperative system that employs both methods of space decomposition.

10.3.2 Single-Objective Optimization

An explicit space decomposition cooperative approach referred to as CPSO_S (Cooperative PSO) was introduced in [35, 37]. The approach relies on splitting the space (solution vector) into sub-spaces (smaller vectors) where each sub-space is optimized using a separate swarm. The overall solution is the vector containing the best particle of each swarm. The algorithm works by sequentially passing over the swarms, each swarm updates its particles and updates its sub-vector in the overall solution, then the algorithm proceeds with the next swarm. This leads to a more thorough investigation of the search space since the solution is evaluated every time a single component changes which results in a more fine-grained type of search. This approach was originally introduced using genetic algorithms [3]. This approach was applied using the original PSO algorithm and it improved the PSO performance in all but one function. Fig. 10.2 illustrates this approach.

A different approach was introduced in [30], referred to as concurrent PSO (CONPSO). This approach adopted implicit space decomposition by having two swarms searching concurrently for a solution with frequent message passing of information. The information exchanged was the global bests of the two swarms. After every exchange point, the two swarms were to track the better global best found. The two swarms were using two different approaches, one adopted the original PSO method, and the other used the Fitness-to-Distance Ratio PSO (FDRPSO) [34]. This approach improved the performance over both methods also minimizing the time requirement of the FDRPSO alone. However, a lot of the aspects that could change the behavior of this approach were not addressed like, for example, how frequent do the swarms communicate. Fig. 10.3 illustrates this approach.

A hybrid decomposition cooperative approach was also introduced in [35, 37], it was referred to as the hybrid CPSO (CPSO_H). It consists of having two search stages working in a serial fashion. Each stage was only run for one iteration then passing the best found solution to the next stage. The first stage applied the CPSO_S technique, and the second stage used the normal PSO algorithm. This approach was applied to take advantage of the ability of PSO to escape pseudo-minimizers while benefiting from the CPSO_S fast convergence property. The information exchange is implemented such that the best solution found by a stage replaces a randomly selected particle of the next stage without changing its global best. This approach is shown in Fig. 10.4.

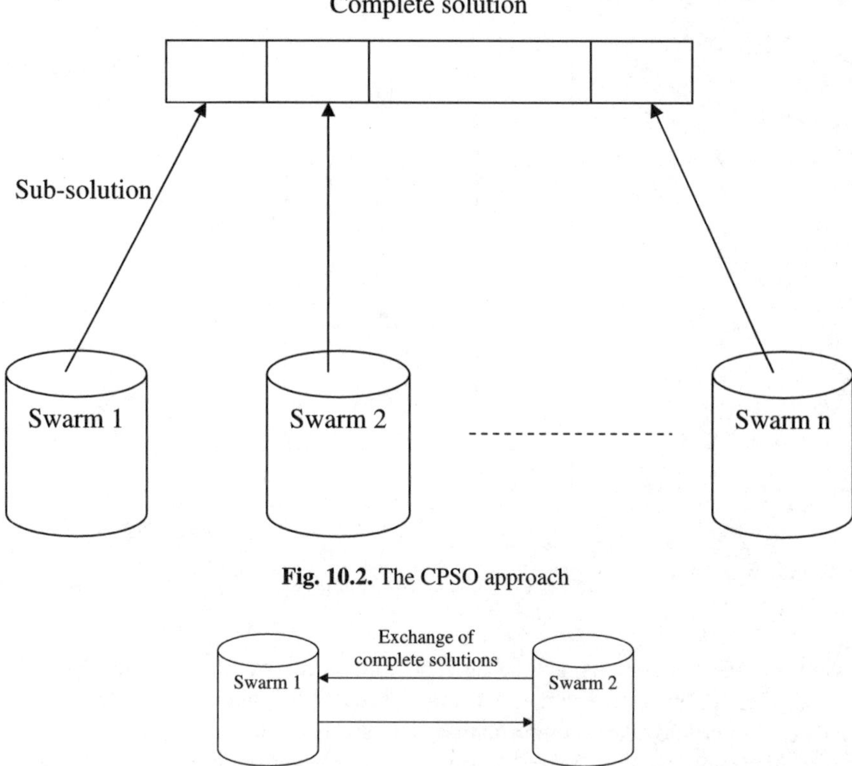

Fig. 10.2. The CPSO approach

Fig. 10.3. The CONPSO approach

In [33], a multi-swarm technique was tested on a dynamic multi-modal benchmark functions. A colony of multiple swarms is used. The swarms exchange information with each other only when they are following attractors that are close to each other. The swarm that has a bad attractor gets all its particles positions and velocities re-initialized. The approach relied on implicit space decomposition and the swarms work in parallel. The communication is carried out synchronously after every iteration to check whether the different attractors are close to each other. However, the actual exchange of information is asynchronous, since it is done when the attractors are close to each other, not at a specific period. The authors showed that the off-line error is reduced while increasing the number of swarms up to a certain limit after which the error starts to increase again.

In all the previous implementations, all the swarms were static. If a particle is assigned to a specific swarm in the beginning of the algorithm, it stays in that swarm till the end. A dynamic multi-swarm approach was presented in [18]. In this approach, each swarm adopted the *lbest* model. After a predefined number of iterations k, each particle gets randomly assigned to a different swarm. Hence, if a particle i gets assigned to a swarm n in the beginning, it will get assigned to another

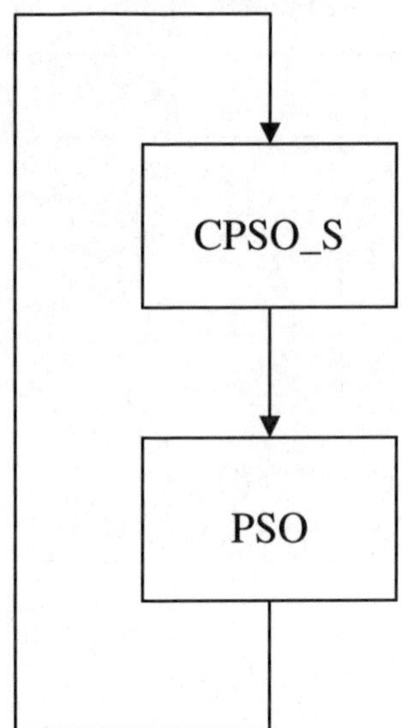

Fig. 10.4. The CPSO_H approach

swarm m after k iterations, and so on. The authors referred to k as the regrouping period. The information exchange in this approach is implicit rather than explicit since every particle takes its information and carries it when assigned to a different swarm. This approach was taken to overcome the problem of immature convergence in PSO by keeping the diversity among the different swarms.

Different parallel PSO models were studied in [20]. The authors experimented with three different parallel versions of PSO.

- The master/slave PSO, which is a simple parallelization approach where the PSO operation is handled by a master processor that delegates some of the work to several slave processors.
- The migration PSO, which is similar to the *coarse-grained* genetic algorithms, also referred to as the island model. In this model, different swarms are run on different processors, and after a finite number of iterations the best solution of each swarm (processor) is migrated to the neighboring swarms (processors).
- The diffusion PSO, which is similar to *fine-grained* genetic algorithms. In this model, each particle is handled by a separate processor. Each particle has only local information about the best position achieved by its neighborhood. The neighborhood topology used was the Von-Neuman model [31].

Based on the complexity analysis, the authors came to the conclusion that the diffusion model can be regarded as a limiting case to the island model. Also, the diffusion model is scalable compared to the master/slave model. Finally, the convergence of the diffusion model is dependent on the neighborhood topology used.

Both the *coarse-grained* and the *fine-grained* types fall under the implicit space decomposition class.

10.3.3 Multi-Objective Optimization

In [9], the cooperating swarms approach was used to solve multi-objective optimization problems. Each swarm is optimizing only one of the objective functions of the problem being solved. The information of this function is exchanged with other swarms via the exchange of the best experience of the swarm. The velocity update equation of a certain swarm i, is updated using the global best position of another swarms j. The method in which j is chosen results in different swarms topologies. The authors implemented with both the single-node (all the swarms on the same CPU) and the parallel (a single swarm per CPU) approaches using the ring topology. The parallel implementation provided better execution times. However, increasing the number of CPUs over six resulted in increased times due to the communication overhead.

The autonomous agent response learning problem was addressed in [5] using a cooperative PSO model. The award function was divided into several award functions, hence, the response extraction process is modelled as a multi-objective optimization problem. The problem was solved using a multi-species PSO. Each objective function was solved using a different swarm. The information exchanged among the different swarms was the best particles. In addition, the velocity update equation of the particles in any given swarm was modified to account of the global best of the neighboring swarm.

A multi-swarm approach for multi-objective optimization was introduced in [15]. This scheme was adopted to overcome the problem of having discontinuous decision variable space. In this approach, each sub-swarm performs a predetermined number of iterations, then the sub-swarms exchange information. This is done by grouping all the leaders in a single set. This set is again divided into groups, and each resulting group will be assigned to a different swarm. The splitting is done with respect to the closeness in the decision variable space. The period after which the swarms communicate was empirically chosen and was kept fixed during the experiments. Changing this period could have an effect on the performance.

10.3.4 Co-evolutionary Constrained Optimization

A co-evolutionary PSO for constrained optimization problems was proposed in [39]. The authors started by transforming the constrained problem into a min-max problem. The min-max problem was solved using a co-evolutionary approach. Two separate swarms were used, one swarm optimizing the min part of the problem and another swarm optimizing the max one. The two swarms interacted together through the fitness evaluation process. The model works in a serial fashion by evolving the first swarm for a determined number of generations, this swarm uses the particles information of the second swarm during the fitness evaluation process. Then, the second swarm evolves using the particles information of the first swarm during the fitness evaluation process, and so on. The swarms stop after obtaining an acceptable solution or after the maximum number of generations has been reached. However, the authors found it difficult to fine tune the solution using a uniform distribution. This problem was addressed in [2] by adopting the Gaussian probability distribution in generating the random numbers for updating the particles velocities. Fig. 10.5 illustrates one iteration of this approach.

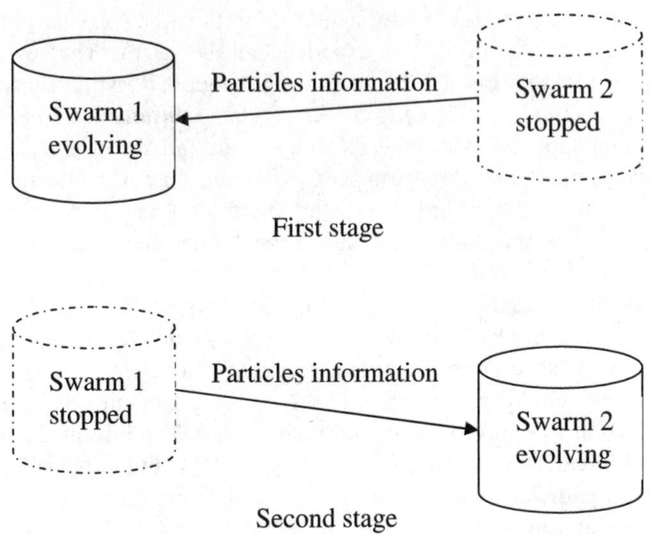

Fig. 10.5. The co-evolving PSO approach

10.3.5 Other Applications

In [36], the CPSO was used to train unit product neural networks. The solution of the problem was the vector containing all the weights of the network. This vector was decomposed among several swarms, each swarm optimizing a specified number of weights. The authors studied the effect of changing the number of swarms, also referred to as the split factor, on the training performance. They concluded that the training performance improves with increasing the split factor until a critical ratio is reached. This ratio was found to be W/5, where W is the total number the optimized weights.

In [32], improvised music (i.e. music that is composed and performed immediately, in real-time and without revision) was played using a multi-swarm approach. Each swarm represents a musical entity, and the particles in the swarm are musical events. The system as a whole is regarded as an improvising ensemble. The velocity update equation is under the influence of four different accelerations:

- Attractive acceleration towards the swarm center of mass (preserving the swarm identity as a whole)

- Attractive acceleration towards the center of mass of another swarm (fragmenting the swarm)
- Attractive acceleration towards a target derived from collaborating humans
- Repulsive acceleration away from neighboring particles or a certain target (preventing the swarm from playing identical notes).

Each particle was a 3-dimensional vector in the music space representing loudness, pulse, and pitch. The ability of each individual to produce a coherent improvisation is ensured by the principles of self-organization. The colony organizes these different improvisations into a large-scale structure.

In [19], the authors proposed a co-evolutionary PSO approach for solving the game of seega. Seega is an Egyptian two-stage board game. In the first stage, the two players take turns in placing their disks on the board until there is only one empty cell. In the second stage, the players take turns in moving their disks; if a disk, that belongs to one player, gets surrounded by disks of the other player, it gets captured and removed from the board. The game was solved by evolving two independent swarms representing the two players. The swarms were used in the score evaluation process. The system consisted of two parts, the game engine and the co-evolutionary part. The second part used the game engine in a master-slave relationship in order to evaluate the particles fitness.

The authors in [1] proposed a co-evolutionary PSO approach for tuning the parameters of a 5 degree-of-freedom arm robot torque controller. Two swarms evolved together, one swarm is optimizing the feedforward controller parameters, and the other swarm is searching for the disturbance values in the worst case. The final solution is generated by both swarms. The two swarms were implemented in a serial fashion similar to the one used in [39]. An accelerated version was used to reduce the computation time by reducing the number of fitness evaluations. Closed-loop simulations showed that the proposed strategy improved the trajectory tracking ability of a perturbed robot arm manipulator.

10.4 Controlling The Cooperative PSO Behavior

Though several implementations were reported for different cooperative PSO models, many parameters that could influence the behavior of these models were not addressed. These parameters were usually chosen empirically and were kept fixed during the reported experiments. These parameters are explained in the following subsections.

10.4.1 Communication Type

It defines the type of communication (*synchronous* or *asynchronous*) adopted between the cooperating swarms. In synchronous communication, the cooperating modules exchange information with each other every predetermined number of iterations. On the other hand, asynchronous communication involves the exchange of information when a certain condition occurs (e.g. when the solution found by a certain module does not improve for a specified number of iterations). In [33], asynchronous communication was adopted and the condition was having two swarms with attractors close to each other.

10.4.2 Communication Interval

This parameter defines the time spent by the swarms performing their own iterations between different communication attempts. This parameter could be defined for both communication types identified before:

- Synchronization period [23], which is the number of iterations that have to be performed before sharing information in a synchronous type of model.
- Grace period, which is the number of iterations that a module has to wait before performing the exchange when the solution quality does not improve in an asynchronous type of model.

The algorithm listed in Fig. 10.6 will be used to test the effect of changing the synchronization period on the performance of a synchronous type model (CONPSO).

1. *Initialize the two swarms*
2. ***For** the specified number of iterations*
    2.1 *Update swarm 1*
    2.2 *Update swarm 2*
    2.3 ***If** a synchronization step is required*
        a. *M = min (Global_Best1, Global_Best2)*
        b. *Set the global best of both swarms to M*
   End for

Fig. 10.6. CONPSO algorithm

10.4.3 Communication Strategy

This parameter defines the communication topology used if there is more than two cooperating swarms [21]. Different communication strategies could be tested:

- Fully-connected topology, sharing the exchanged information among all swarms,
- Ring topology, circular communication of the exchanged information,

Fig 10.7 illustrates these topologies assuming the exchange of the *gbest* information, where GB_i refers to the global best of swarm *i*, and *GB* is the minimum of all global bests.

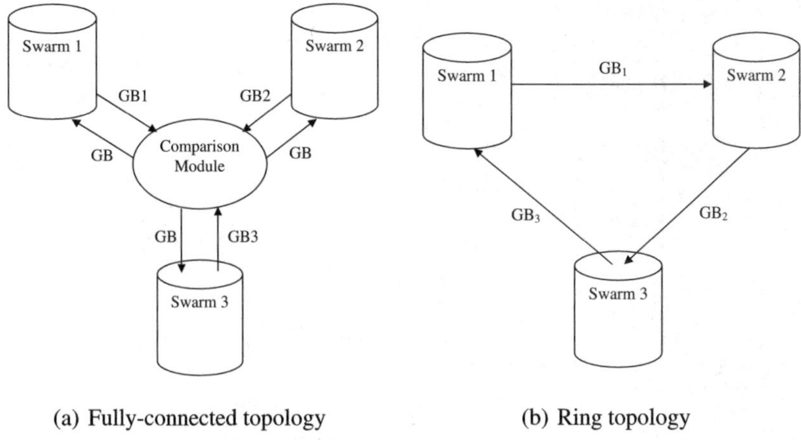

(a) Fully-connected topology (b) Ring topology

Fig. 10.7. Different communication strategies.

10.4.4 Number of swarms

The number of cooperating swarms is a different parameter that affects the performance of the cooperative PSO model. When comparing the performance of the cooperative PSO model with different number of swarms, the number of particles will be kept fixed. This means that increasing the number of cooperating swarms will lead to having less number of particles per swarm. Hence, it is a trade off. Is it better to have small number of swarms with large number of particles per swarm? or is it better to have many swarms with less number of particles in them?

10.4.5 Information Exchanged

It is the kind of information that is shared between the swarms. When having two cooperating swarms sharing their *gbest* value, the flow of information in this case is only from one swarm to the other. The swarm that has the better global best provides new information to the other swarm without gaining anything.

A different approach is to exchange the best particle between the two cooperating swarms instead of just exchanging the global best [22]. There are two main advantages for this approach: (i) when exchanging the best particle between the two swarms, this particle is used to replace the worst particle, (ii) the flow of information is no longer from one swarm to the other, since the best particle in each swarm can replace the worst particle in the other one. This makes it possible for the two swarms to gain new information from each other's experiences. Note that the receiving swarms could also replace a randomly selected particle instead of the worst one.

To boost the performance, one could also adopt the exchange of the best p particles instead of only the best one. Fig. 10.8 shows the algorithm adopted for this approach, where N is the total number of particles in each swarm.

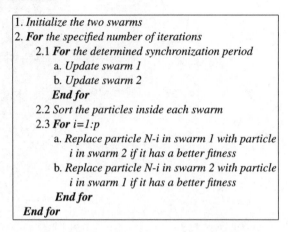

```
1. Initialize the two swarms
2. For the specified number of iterations
   2.1 For the determined synchronization period
       a. Update swarm 1
       b. Update swarm 2
       End for
   2.2 Sort the particles inside each swarm
   2.3 For i=1:p
       a. Replace particle N-i in swarm 1 with particle
          i in swarm 2 if it has a better fitness
       b. Replace particle N-i in swarm 2 with particle
          i in swarm 1 if it has a better fitness
       End for
 End for
```

Fig. 10.8. The exchanging particles algorithm

10.5 Experimental Work

10.5.1 Cooperative PSO model

The experimented model is the same as the one introduced in [30]. We use two cooperating swarms with 10 particles each. Both swarms use the *gbest* model. The two swarms communicate with each other in a synchronous approach by sharing their global best information every predetermined number of iterations. Table 10.1 shows the parameters settings used for applying both the original and the cooperative PSO algorithm.

Table 10.1. Parameters settings

Parameter	Value
c1,c2	2
w	0.9 to 0.1
k	1
Space dimensionality (n)	10
Function evaluations	1000000
Number of runs	50

The experiments are run using four benchmark test functions:
The Spherical function (unimodal)

$$f(x) = \sum_{i=1}^{n} x_i^2 \tag{10.5}$$

where $-100 < x_i < 100$.
The Quadratic function (unimodal)

$$f(x) = \sum_{i=1}^{n} \left(\sum_{j=1}^{i} x_j^2 \right)^2 \tag{10.6}$$

where $-100 < x_i < 100$.
The Rastrigin function (multimodal)

$$f(x) = \sum_{i=1}^{n} (x_i^2 - 10\cos 2\pi x_i + 10) \tag{10.7}$$

where $-5.12 < x_i < 5.12$.
The Ackley function (multimodal)

$$f(x) = 20 + e - 20\exp\left(-0.2\sqrt{\frac{1}{n}\sum_{i=1}^{n} x_i^2}\right) - \exp\left(\frac{1}{n}\sum_{i=1}^{n}\cos 2\pi x_i\right), \tag{10.8}$$

where $-30 < x_i < 30$.

Tables 10.2 and 10.3 show the results obtained using the *gbest* model for a different number of particles. The results generally indicate that increasing the number of particles up to a certain limit leads to improving the solution obtained. After that limit, the quality of the obtained results tend to deteriorate.

Table 10.2. Results for the unimodal functions

Number of	Spherical		Quadratic	
Particles	Avg.	Std.	Avg.	Std.
10	2.7015e-001	3.6979e-001	1.9296e+000	2.4219e+000
20	4.5349e-003	1.0674e-002	4.8139e-002	1.2409e-001
30	2.8494e-005	6.8890e-005	3.1899e-005	9.6830e-005
40	4.2616e-007	2.9139e-006	1.2754e-009	9.0168e-009
50	8.8071e-013	5.7465e-012	2.8867e-017	1.8040e-016
100	9.9650e-031	6.9822e-030	1.2243e-050	7.9382e-050
200	**1.4946e-047**	4.5267e-047	**1.7464e-055**	1.1706e-054
300	2.1495e-041	7.7575e-041	6.7379e-044	1.5792e-043

10.5.2 Synchronization Period

In order to keep the number of function evaluations fixed, each swarm performs 50000 function evaluations when the cooperative model is applied. The model is run several times but each time with a different synchronization period, the results are

Table 10.3. Results for the multimodal functions

Number of Particles	Rastrigin Avg.	Rastrigin Std.	Ackley Avg.	Ackley Std.
10	8.2888e+000	2.8826e+000	1.5315e+000	9.0332e-001
20	8.4325e+000	3.7690e+000	1.4558e+000	9.3398e-001
30	7.1245e+000	2.5923e+000	1.2380e+000	9.2033e-001
40	8.6166e+000	3.4445e+000	1.1236e+000	8.9020e-001
50	8.1787e+000	2.9900e+000	1.1429e+000	8.6812e-001
100	7.9199e+000	2.8637e+000	1.1030e+000	1.0376e+000
200	**7.0481e+000**	2.6026e+000	7.7872e-001	8.5121e-001
300	7.4025e+000	2.6221e+000	**6.4234e-001**	8.2497e-001

Table 10.4. Results for the unimodal functions

Synchronization Period	Spherical Avg.	Spherical Std.	Quadratic Avg.	Quadratic Std.
1	**4.1928e-003**	8.5247e-003	**6.3294e-002**	1.8443e-001
10	6.6530e-003	1.2424e-003	8.6847e-002	1.4278e-001
25	1.0146e-002	1.7151e-002	1.5975e-001	2.4382e-001
50	1.6529e-002	4.1533e-002	3.0114e-001	5.4839e-001
100	5.5588e-002	7.8237e-002	3.5097e-001	4.8750e-001
250	5.6462e-002	7.3723e-002	4.8318e-001	7.2251e-001
500	1.0715e-001	1.0715e-001	7.7956e-001	1.1265e+000
1000	3.3832e-001	3.6499e-001	1.6435e+000	1.9697e+000

Table 10.5. Results for the multimodal functions

Synchronization Period	Rastrigin Avg.	Rastrigin Std.	Ackley Avg.	Ackley Std.
1	7.3981e+000	3.9710e+000	1.4391e+000	7.8571e-001
10	7.3672e+000	3.2048e+000	1.4277e+000	8.9113e-001
25	7.5140e+000	2.8254e+000	1.4648e+000	8.3734e-001
50	7.5515e+000	2.5845e+000	1.3297e+000	1.0296e000
100	7.5891e+000	3.6336e+000	1.3064e+000	9.1858e-001
250	6.9163e+000	2.8151e+000	1.1495e+000	8.1032e-001
500	6.7645e+000	3.0366e+000	1.1102e+000	8.6932e-001
1000	**6.3759e+000**	2.7818e+000	**1.1079e+000**	7.3858e-001

shown in Tables 10.4 and 10.5. The solutions quality shown are the averages obtained over the fifty runs.

The results show that for the unimodal functions, reducing the synchronization period improves the solution quality. Taking a closer look at this approach, one will realize that applying this technique with a synchronization period equal to 1 may be equivalent to performing the original PSO algorithm using 20 particles. Fig. 10.9 clearly asserts this fact. Where the result of the proposed method approaches the result of the original PSO algorithm using 20 particles as the synchronization period decreases.

On the other hand, for the multimodal functions, the results are totally different. These two function are different because they have a rugged surface containing a lot of local minima. Having two independent swarms searching the space is more productive than having one swarm. Increasing the synchronization period produces better results. As the synchronization period increases, the probability of having the two swarms stuck in the same local minimum decreases.

Fig. 10.9 shows the results of both the serial and parallel implementations of the cooperative model. It can be seen that the implementation factor does not have a great effect on the performance of the model.

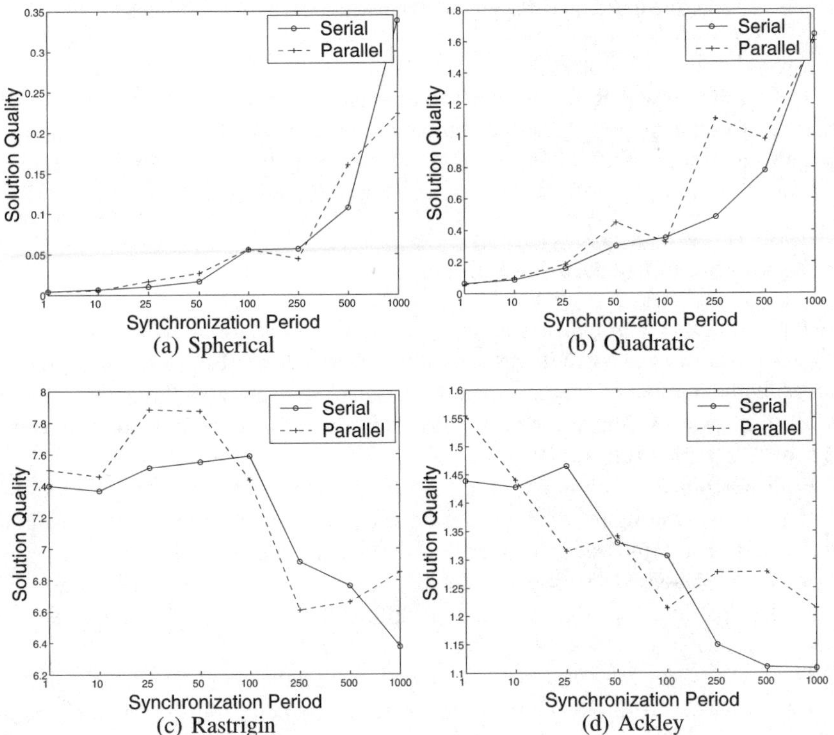

Fig. 10.9. Results of the first set of experiments [1]

10.5.3 Communication Strategy

Two different communication strategies (neighborhood topologies) are tested in this work (i) fully-connected topology, (ii) ring topology. Experiments are run using three swarms, with 10 particles each, at different synchronization periods. Fig. 10.10 shows the results.

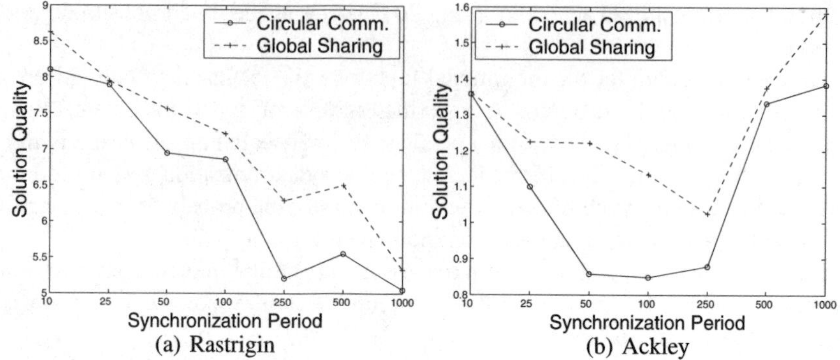

Fig. 10.10. Results of applying different communication strategies

The results show that the circular communication approach gives systematically better results than the global sharing approach. This is due to the diversity maintained using this approach. If all the swarms follow the same global best, this will increase the probability of having all the swarms stuck at the same local minimum.

10.5.4 Number of Cooperating Swarms

After studying two communication approaches. The circular one will be used in the following experiments since it was shown to give better results. To test the effect of increasing the number of swarms, experiments are run using two, three, five then ten cooperating swarms. The number of particles is always kept fixed at 30. The results are shown in Tables 10.6 and 10.7.

Both functions are similar in terms of the property that increasing the number of cooperating swarms puts a limit on the synchronization period used to achieve good results. The results show that more communication is needed when the number of swarms is increased. This is due to the fact that when more swarms are used, having less particles per swarm makes it difficult to escape local minima. Hence, more communication is needed.

The results also show that there should be some balancing between the number of swarms used and the synchronization period depending on the function being optimized. The Ackley function is better solved using 5 cooperating swarms while

[1]With kind permission of Springer Science and Business Media

Table 10.6. Results for the Ackley function

No. of Swarms	Synchronization Period	Best Result Avg.	Std.
2	250	1.1561e+000	8.4678e-001
3	100	8.4882e-001	8.4549e-001
5	50	**7.5795e-001**	7.0312e-001
10	25	9.1770e-001	6.9751e-001

Table 10.7. Results for the Rastrigin function

No. of Swarms	Synchronization Period	Best Result Avg.	Std.
2	1000	**4.8730e+000**	1.7737e+000
3	1000	5.0272e+000	2.2823e+000
5	250	5.3223e+000	1.8664e+000
10	50	5.7603e+000	1.8079e+000

maintaining a short synchronization period. On the other hand, the Rastrigin function is better solved using only two cooperating swarms with a rather long synchronization period. This is due to the different topology of the two functions. The Rastrigin function has a relatively more condensed and sharp edged surface.

10.5.5 Information Exchanged

A cooperative model adopting the *lbest* model is used to compare these two information exchange approaches. This model suffers from a slow flow of information problem [17]. If one particle has useful information, it might take a while for other particles to benefit from it. That is why it is interesting to see if the cooperative approach is going to be efficient using this model. In this case however, in order to change the global best of a swarm it is necessary to physically change the value of the best particle in that swarm.

The experiments are run using the algorithm listed in Fig. 10.8 with $p = 1$. The algorithm is applied to the same cooperative model adopted with the swarms using the *lbest* model and having 15 particles each. The neighborhood sizes selected are 2 for the Rastrigin function and 4 for the Ackley function (since these sizes produced the best results when having one swarm with 15 particles). The results are compared to a single swarm having 30 particles.

When exchanging the best particle instead of the global best, the experiments conducted show that adopting this approach always gives better results. However, having one swarm using the *lbest* model is still better. The reason for that is due to the slow flow of information in the *lbest* model. When introducing new information to that model, it takes a while for the whole swarm to gain benefit from this information. Fig. 10.11 shows the comparison of both information exchange approaches.

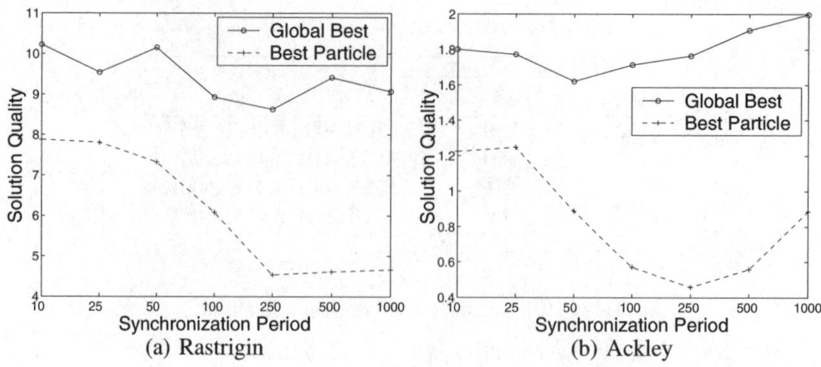

Fig. 10.11. Comparing two information exchange approaches (©2005 IEEE)

To further improve the performance of this approach, the best p particles are exchanged between the two swarms instead of just the best one. This will allow each swarm to gain more information at each exchange point allowing it to overcome the slow flow of information problem. Results of exchanging 1, 3, and 5 particles are illustrated in Fig. 10.12.

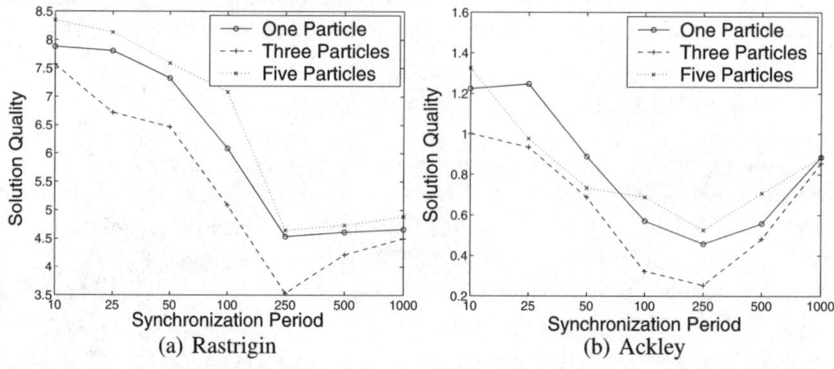

Fig. 10.12. Results of exchanging different number of particles (©2005 IEEE)

The results show that exchanging the best three particles achieved better results than what was given by a single swarm. However, increasing the number of exchanged particles up to five caused the obtained solution quality to deteriorate again.

Exchanging the best p particles helps the two swarms to gain from each other's experiences. Exchanging one particle faces the difficulty of the slow flow of information problem, that is why the solution provided by a single swarm is better. On the other hand, exchanging five particles is too much, both swarms may get stuck in the same region in the search space. Exchanging three particles proved to be the better choice. To achieve better results, it is important to choose the right number of particles to be exchanged. Too little or too much information will not improve the obtained solution quality.

10.6 Conclusion

This chapter surveyed the many implementations reported in the literature for cooperative PSO models. These models were divided by the type of applications they were used in. A proposed taxonomy relying on the level of space decomposition achieved by the cooperative model is used to classify them. The different parameters that control the performance of such models were identified. These parameters are the communication type, the communication interval, the communication strategy, the number of swarms, and the information exchanged. Experiments show that adopting a cooperative models would be more beneficial in optimizing multimodal functions. Increasing the synchronization period of a synchronous cooperative model helped in improving the obtained solution quality. In order to have the right degree of separation between the cooperative swarms, one should carefully choose the number of cooperating swarms and the synchronization period. The proper adjustment of these parameters could be different depending on the function being optimized. Exchanging information both ways between two cooperating swarms is better than just have them follow the same global best value. Increasing the number of particles that are exchanged between the cooperating swarms up to a certain limit helped to improve the obtained solution quality.

References

1. Asmara A., Krohling R. A., and Hoffmann F. Parameter tuning of a computed-torque controller for a 5 degree of freedom robot arm using co-evolutionary particle swarm optimization. In *Proc. IEEE Swarm Intelligence Symposium*, pages 162–168, 2005.
2. Krohling R. A., Hoffmann F., and Coello Ld. S. Co-evolutionary particle swarm optimization to solve min-max problems using gaussian distribution. In *Proc. Congress on Evolutionary Computation*, volume 1, pages 959–964, 2004.
3. Potter M. A. and de Jong K. A. A cooperative coevolutionary approach to function optimization. In *Proc. 3rd Parallel problem Solving from Nature*, pages 249–257, 1994.
4. Blum C. and Roli A. Metaheuristics in combinatorial optimization: Overview and conceptual comparison. *ACM Computing Surveys*, 35(3):268–308, 2003.
5. Chow C. and Tsui H. Autonomous agent response learning by a multi-species particles swarm optimization. In *Proc. Congress on Evolutionary Computation*, pages 778–785, 2004.
6. Eberhart R. C. and Kennedy J. A new optimizer using particle swarm thoery. In *Proc. of the 6th International Symposium on Micro Machine and Human Science*, pages 39–43, 1995.
7. Eberhart R. C., Simpson P., and Dobbins R. *Computational Intelligence*, chapter 6, pages 212–226. PC Tools: Academic, 1996.

8. Cantu-Paz E. A survey pf parallel genetic algorithms. Technical Report IlliGAL 97003, The University of Illinois, 1997.
9. Parsopoulos K. E., Tasoulis D. K., and Vrahatis M. N. Multiobjective optimization using parallel vector evaluated particle swarm optimization. In *Proc. International Conference on Artificial Intelligence and Applications*, volume 2, pages 823–828, 2004.
10. Talbi E. A taxonomy of hybrid metaheuristics. *Journal of Heuristics*, 8(5):541–564, 2002.
11. Crainic T. G. and Grendeau M. Cooperative parallel tabu search for capacitated network design. *Journal of Heuristics*, 8:601–627, 2002.
12. Crainic T. G. and Toulouse M. Parallel strategies for metaheuristics. In Glover F. and Kochenberger G., editors, *State-of-the-Art Handbook in Metaheuristics*. Kluwer Academic Publishers, 2002.
13. Crainic T. G., Toulouse M., and Grendeau M. Parallel asynchronous tabu search for multicommodity location-allocation with balancing requirements. Technical Report 935, Centre de recherche sur les transports, Universite de Montreal, 1993.
14. Crainic T. G., Toulouse M., and Grendeau M. Synchronous tabu search parallelization strrategies for multicommodity location-allocation with balancing requirements. Technical Report 934, Centre de recherche sur les transports, Universite de Montreal, 1993.
15. Toscano G. and Coello A. C. C. Using clustering techniques to improve the performance of a multi-objective particle swarm optimizer. In *Proc. Genetic and Evolutionary Computation Conference*, pages 225–237, 2004.
16. Kennedy J. and Eberhart R. C. Particle swarm optimization. In *Proc. IEEE International Conference on Neural Networks*, volume 4, pages 1942–1948, 1995.
17. Kennedy J. and Mendes R. Population structure and particle swarm performance. In *Proc. IEEE Congress on Coevolutionary Computation*, volume 2, pages 1671–1676, 2002.
18. Liang J. J. and Suganthan P. N. Dynamic multi-swarm particle swarm optimizer. In *Proc. IEEE Swarm Intelligence Symposium*, pages 124–129, 2005.
19. Abdelbar A. M., Ragab S., and Mitri S. Co-evolutionary particle swarm optimization applied to the 7x7 seega game. In *Proc. IEEE International Joint Conference on Neural Networks*, volume 1, pages 243–248, 2004.
20. Belal M. and El-Ghazawi T. Parallel models for particle swarm optimizers. *International Journal for Intelligent Computing and Information Sciences*, 4(1):100–111, 2004.
21. El-Abd M. and Kamel M. Factors governing the behavior of multiple cooperating swarms. In *Proc. Genetic and Evolutionary Computation Conference*, volume 1, pages 269–270, 2005.
22. El-Abd M. and Kamel M. Information exchange in multiple cooperating swarms. In *Proc. IEEE Swarm Intelligence Symposium*, pages 138–142, 2005.
23. El-Abd M. and Kamel M. Multiple cooperating swarms for non-linear function optimization. In *Proc. 4th IEEE International Workshop on Soft Computing as Transdisciplinary Science and Technology, 2nd International Workshop on Swarm Intelligence and Patterns*, pages 999–1008, 2005.
24. El-Abd M. and Kamel M. A taxonomy of cooperative search algorithms. In *Proc. 2nd International Workshop on Hybrid Metaheuristics, LNCS*, volume 3636, pages 32–41, 2005.
25. Middendorf M. and Reischle F. Information exchange in multi colony ant algorithms. In *Proc. 3rd workshop on Biologically Inspired Solutions to Parallel Processing Problems*, pages 645–652, 2000.
26. Middendorf M., Reischle F., and Schmeck H. Multi colony ant algorithms. *Journal of Heuristics*, 8:305–320, 2002.

27. Nowostawski M. and Poli R. Prallel genetic algorithms taxonomy. In *Proc. 3rd international Conference on Knowledge-Based Intelligent Information Engineering Systems*, pages 88–92, 1999.
28. Toulouse M., Crainic T. G., and Sanso B. An experimental study of the systemic behavior of cooperative search algorithms. In Osman I. Voss S., Martello S. and Roucairol C., editors, *Meta-Heuristics: Advances and Trends in Local Search Paradigms*, pages 373–392. Kluwer Academic Publishers, 1999.
29. Toulouse M., Crainic T. G., Sanso B., and Thularisaman K. Self-organization in cooperative tabu search algorithms. In *Proc. IEEE International Conference on Systmes, Man, and Cybernetics*, volume 3, pages 2379–2384, 1998.
30. Baskar S. and Suganthan P. N. A novel concurrent particle swarm optimization. In *Proc. IEEE Congress on Evolutionary Computation*, volume 1, pages 792–796, 2004.
31. Peer E. S., van der Bergh F., and Engelbrecht A. P. Using neighbourhood with guaranteed convergence pso. In *Proc. IEEE Swarm Intelligence Symposium*, pages 235–242, 2003.
32. Blackwell T. Swarm music: Improvised music with multi-swarms. In *Proc. Symposium on Artificial Intelligence and Creativity in Arts and Science*, pages 41–49, 2003.
33. Blackwell T. and Branke J. Multi-swarm optimization in dynamic environments. In Raidl G. R., editor, *Applications in Evolutionary Computing*, pages 488–499. LNCS, Springer-Verlag, 2004.
34. Peram T., Veeramachaneni K., and Mohan C. K. Fitness-distance-ratio based particle swarm optimization. In *Proc. IEEE Swarm Intelligence Symposium*, pages 174–181, 2003.
35. van den Bergh F. and Engelbrecht A. P. Effect of swarm size on cooperative particle swarm optimizaters. In *Proc. Genetic and Evolutionary Computation Conference*, 2001.
36. van den Bergh F. and Engelbrecht A. P. Training product unit neural networks using cooperative particle swarm optimisers. *Proc. IEEE International Joint Conference on Neural Networks*, 1:126–131, 2001.
37. van den Bergh F. and Engelbrecht A. P. A cooperative approach to particle swarm optimization. *IEEE Transactions on Evolutionary Computation*, 8(3):225–239, 2004.
38. Treinekens H. W. and de Bruin A. Towards a taxonomy of parallel branch and bound algorithms. Technical Report EUR-CS-92-01, Department of Computer Science, Erasmus University, Rotterdam, 1992.
39. Shi Y. and Krohling R. A. Co-evolutionary particle swarm optimization to solve min-max problems. In *Proc. Congress on Evolutionary Computation*, volume 2, pages 1682–1687, 2002.
40. Yang Y. and Kamel M. Clustering ensemble using swarm intelligence. In *Proc. IEEE Swarm Intelligence Symposium*, pages 65–71, 2003.

11

Parallel Particle Swarm Optimization Algorithms with Adaptive Simulated Annealing

Shu-Chuan Chu[1], Pei-Wei Tsai[2], and Jeng-Shyang Pan[2,3]

[1] Department of Information Management, Cheng Shiu University, 840 ChengCing Rd., NiaoSong Township, Kaohsiung County 833, Taiwan.
Email:scchu@csu.edu.tw

[2] Department of Electronic Engineering, National Kaohsiung University of Applied Sciences, 415 Chien-Kung Rd, Kaohsiung City 807, Taiwan.
Email:jspan@cc.kuas.edu.tw

[3] Department of Automatic Test and Control, Harbin Institute of Technology, Harbin, Heilongjiang 150001, China

Summary. In this chapter, a series of Particle Swarm Optimization (PSO) algorithms is introduced, which is a subclass of optimization algorithms. In the field of optimization algorithm, most algorithms gain a better solution by evolving themselves. PSO evolves as this way too. In the following sections, we are going to present the original PSO system, PSO with weighted factor system (PSO), Parallel Particle Swarm Optimization (PPSO). Then we will introduce a new combination of Simulated Annealing (SA) with PPSO, which we called Adaptive Simulated Annealing - Parallel Particle Swarm Optimization (ASA-PPSO). Thereafter, we use five test functions, Rosenbrock function, Rastrigrin function, Griewank function, and two other test functions in each of the methods (i.e., PSO, PPSO, and our new method, ASA-PPSO) in order to compare our results with previous research. Experimental results confirm the usefulness of the proposed Parallel Particle Swarm Optimization with adaptive simulated annealing.

11.1 Introduction

Since Genetic Algorithm (GA) [1, 2, 3] being published at the end of the twentieth century, many similar, but different methodologies have been presented, such as, Ant Colony System (ACS) [4, 5, 6] and Particle Swarm Optimization (PSO) [7, 8, 9]. However, PSO is a methodology that is based on social behavior of evolution, which means it is naturally not alike those methodologies that use natural evolution as the weeding-out process [1, 2, 3]. Consequently, PSO usually can find the nearly best solution in much lesser evolution than the others [10, 11].

S.-C. Chu et al.: *Parallel Particle Swarm Optimization Algorithms with Adaptive Simulated Annealing*, Studies in Computational Intelligence (SCI) **31**, 261–279 (2006)
www.springerlink.com

In 1999, Shi [12] presented a new kind of PSO with weighted factor to control the movement velocity; this refines the convergence of the original PSO. Subsequently, the rules for separating particles into several parallel sets and creates three strategies for the communication between each set are set up [13]. The experiments show that the methods make better results than the method of Shi [12].

After describing the PSO-type algorithms, we are going to lead in simulated annealing (SA), which is a different evolutionary algorithm, into PPSO as a catalyzer. By the way of dynamic detection, SA process is appropriate added in to increase the movement of particles and to obtain a larger chance of better solutions when the convergence of particles reaches a bottleneck. For the experiments, we compare the Particle Swarm Optimization (PSO) with weighted factor and Parallel Particle Swarm Optimization (PPSO) with the Adaptive Simulated Annealing - Parallel Particle Swarm Optimization (ASA-PPSO) by five different test functions.

In this chapter, we briefly explain how the original PSO, PSO with weighted factor and parallel PSO work in Section 11.2, and then SA is introduced in Section 11.3. In Section 11.4, the embedding conditions of Adaptive Simulated Annealing - Parallel Particle Swarm Optimization are described; then the processes of our proposed algorithm is defined in Section 11.5. The comparison of several methodologies will be discussed in Section 11.6. At last, the conclusion is presented in Section 11.7.

11.2 The process of Particle Swarm Optimization

The traditional particle swarm optimization (PSO) can be described as follows and shown in Fig. 11.1.

1. Particle initialization,
2. Velocity updating,
3. Particle position updating,
4. Memory updating, and
5. Termination Checking.

These steps are shown as follows:

First of all, we decide how many particles we want to use to solve the problem. Every particle has its own position, velocity and best solution. For example, if we use M particles, the particles, their best solutions and their velocities can be represented as:

$$X = \{x_0, x_1, x_2, \cdots, x_{M-1}\} \tag{11.1}$$

$$B = \{b_0, b_1, b_2, \cdots, b_{M-1}\} \tag{11.2}$$

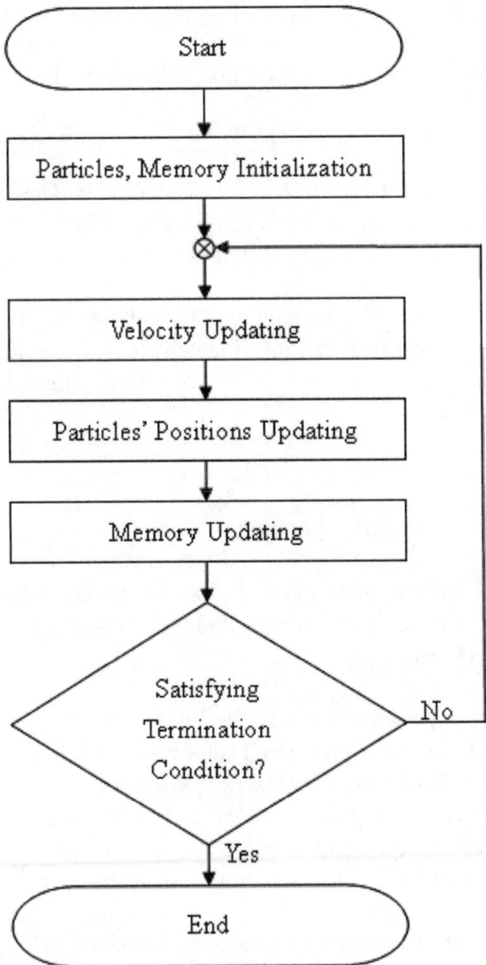

Fig. 11.1. The Process of Particle Swarm Optimization

$$V = \{v_0, v_1, v_2, \cdots, v_{M-1}\} \tag{11.3}$$

In step 2, the process of velocity update is shown in eq. 11.4, where c_1 and c_2 are constants, r_1 and r_2 are random variables in the range from 0 to 1, $b_i(t)$ is the best solution of the ith particle for the iteration number up to the tth iteration and the $G(t)$ is the best solution of all particles:

$$v_i(t+1) = v_i(t) + c_1 \times r_1 \times (b_i(t) - x_i(t)) + c_2 \times r_2 \times (G(t) - x_i(t)) \tag{11.4}$$

To prevent the velocity becomes too large, we set a maximum value to limit the range of velocity as eq. 11.5.

$$-V_{MAX} \leq V \leq V_{MAX} \quad (11.5)$$

In step 3, movement of the particles is processed by eq. 11.6:

$$x_i(t+1) = x_i(t) + v_i(t); \quad 0 \leq i < M \quad (11.6)$$

If we find a better solution than $G(t)$ in $G(t+1)$, $G(t)$ will be replaced by $G(t+1)$ in step 4. Otherwise, there will be no change for $G(t)$. These recursive steps will go on unless we reach the termination condition in step 5.

The difference between the weighted PSO and the original PSO is that the weighted PSO has a weighted factor, which decreases the velocity of particles according to the increase of evolution time. The velocity function of the weighted PSO can be presented as:

$$v_i(t+1) = w(t) \times v_i(t) + c_1 \times r_1 \times (b_i(t) - x_i(t)) + c_2 \times r_2 \times (G(t) - x_i(t)) \quad (11.7)$$

where $w(t)$ is the inertia weight at iteration t.

In the experiment of Shi, he found that if we set c_1 and c_2 equal to 2.0, we will get a better solution than set to the others. Under this condition, we set r_1 and r_2 be random variables between 0 to 1.

We also have the same steps in describing PPSO. But in step1, users must define how many groups will be in the process (It can be designed to be 2^n sets). To decide how many groups should be carefully reference to the size of the problem to solve. In general, we use four groups to evolve, but we use eight groups only if the problem, which we want to solve, is quite large. After several iterations, each group exchanges information in accordance with an explicit rule chosen by users.

The parallel particle swarm optimization (PPSO) method [13] gives every group of particles the chance to have the global best and the local best solutions of other groups. It increases the chance of particles to find a better solution and to leap the local optimal solutions.

Three communication strategies for PPSO are presented. The first one was developed to deal with the loosely correlated parameters in functions. When all the particles evolved a period of time (Here we described as R_1 iterations), communication strategy 1 would exchange the local best solution from one group to others. The way of communication strategy 1 is shown in Fig. 11.2.

The second communication strategy was developed based on self-adjustment in each group. In each group, the best particle would be migrated to its neighborhood to supplant some particles, who present a deteriorated solution than the group's best one. In implementation, the number of groups for the parallel structure was defined as a power of two. Thus, the neighborhoods are defined as one bit difference in

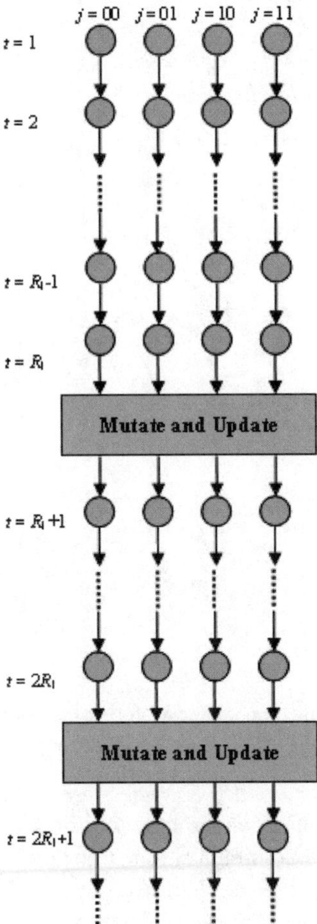

Fig. 11.2. Communication strategy 1 for loosely correlated parameters

the binary view of group number between two groups. The second communication strategy would be applied every R_2 iterations to decide which groups to exchange the local best solution for replacing the poorly performing particles. The way of communication strategy 2 is shown in Fig. 11.3.

The third communication strategy of PPSO is the combination of communication strategy 1 and 2. Under this premise, when using communication strategy 3, the particles must be separate into at least four groups. In communication strategy 3, all particles are separated into two subgroups. Under the subgroups, communication strategy 1 and 2 are imitated. In other words, communication strategy 3 is a general communication strategy for PPSO. The process is shown in Fig. 11.4.

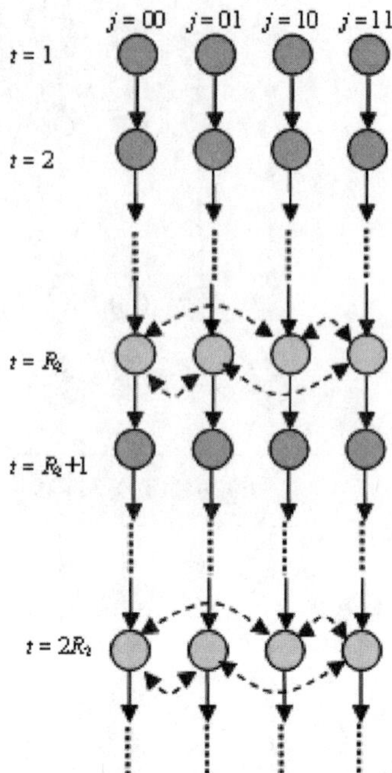

Fig. 11.3. Communication strategy 2 for strongly correlated parameters

Based on the observation, if the parameters are independent or are only loosely correlated, the first communication strategy will obtain good results quite quickly. On the contrary, the second communication strategy obtains higher performance if the parameters are tightly correlated. However, these communication strategies may perform poorly if they have been applied in the wrong situation. Consequently, when the correlation of parameters is unknown, the communication strategy 3 is the better choice to apply.

11.3 Simulated Annealing

Simulated Annealing (SA) [14, 15] is also an evolutionary algorithm for optimization problems. It can be positioned as a landmark in the optimization field, which brings about a radical change in the existing state of optimization algorithms. Prior to SA be proposed, the existing optimization algorithms are almost greedy algorithms, which only accept better performances from preceding iteration to present. In other words, SA brings the idea of leaping across the local optimal into the optimization algorithms.

According to the idea of annealing, SA has a probability function for choosing solutions. Thus, the solution may be worse than present. For this reason, it has the

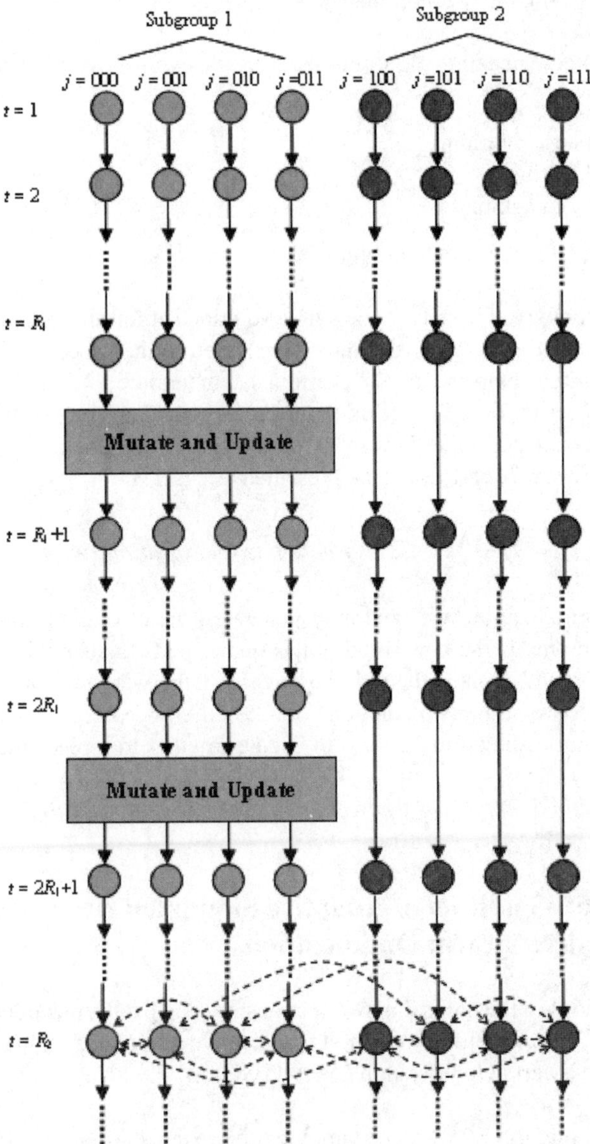

Fig. 11.4. Communication strategy 3 for general purpose

potential for hopping the local optimal.

The Simulated Annealing algorithm (SA) can be expressed as follows:

1. Initialization,
2. Generate a new solution,
3. Solution Updating,
4. Termination Checking.

The whole processes are shown in Fig. 11.5.

This methodology is based on one single solution set for all iterations (from the initial temperature to the terminal temperature). Before the process starts, we have to define the initial temperature (T_S), terminal temperature (T_E) and the decrease scale (ΔT) subject to $T_S < T_E$. ΔT is the decrease scalar of temperature for every iteration, that means, it decides how many iterations will there be in the process. The relationship between T_S and T_E can be presented as eq.11.8.

$$T_E = T_S + \sum_{k=0}^{n} -\Delta T, \quad where\ n\ is\ the\ iteration\ times. \tag{11.8}$$

After the initialization, we randomly generate a new solution and compare it with the existed one. If the new solution has higher performance than the old one, the new solution will replace the old one directly. Otherwise, we have to compute the difference between the two solutions (let the difference be d), and then use the probability function shown in eq. 11.9 to decide whether to replace the solution or not.

$$min(1, exp^{\frac{-d}{T}}), \quad where\ T\ is\ the\ current\ temperature. \tag{11.9}$$

11.4 The infix condition of Adaptive Simulated Annealing Based Parallel Particle Swarm Optimization

After introductions in the preceding two sections, we can list the differences between PPSO and SA below before we get into Adaptive Simulated Annealing Based Parallel Particle Swarm Optimization (ASA-PPSO).

After presenting the PPSO algorithm, we tried to embed SA process into PPSO to improve the performance of the original PPSO. Due to optimization algorithms cannot use the same fitness function for different problems; the adaptive capacity becomes more important. Thus, we notice that the best way to select infix points should be decided dynamically.

In ASA-PPSO, an infix condition was presented. When the convergence of the global best solution is inferior to the past continually over K1 iterations, the infix condition is constituted. To infix the SA process into PPSO, it means to insert five

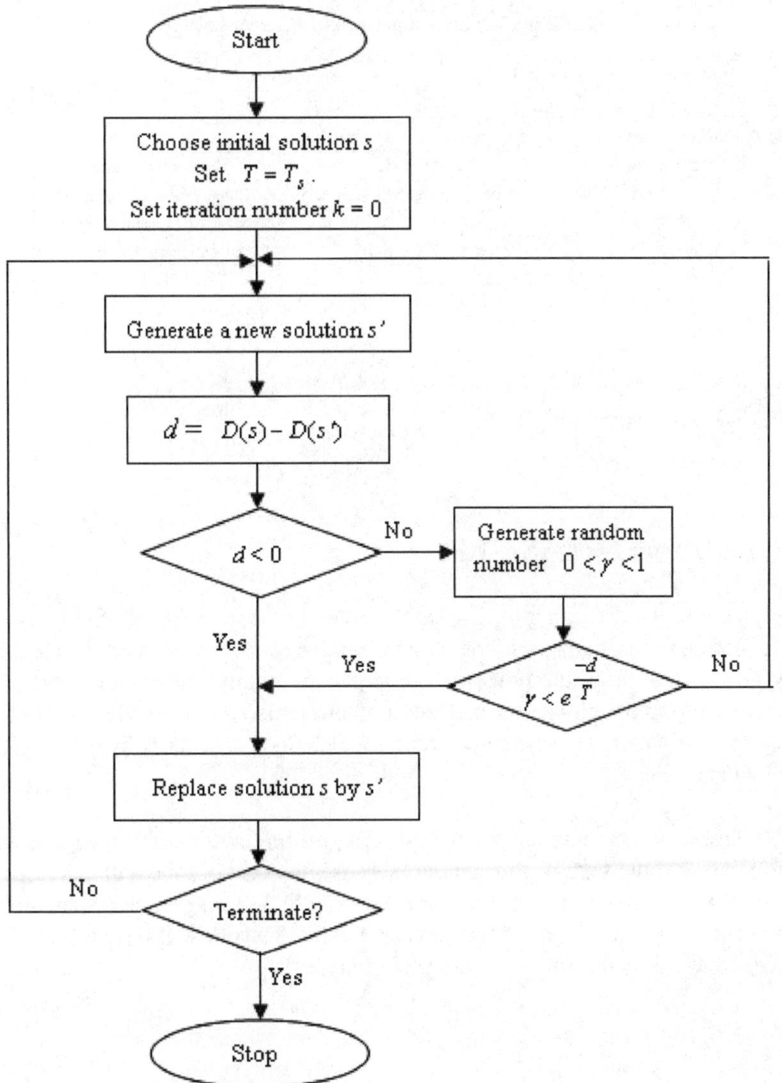

Fig. 11.5. The process of Simulated Annealing (SA)

Table 11.1. The comparison of PSO and SA

Compared Item	Algorithm	
	PPSO	SA
Number of Particles	Several groups of particles per iteration	Single particle per iteration
Convergence Condition	Fitness value	Current temperature
The Way to Leap out of Local Optimal	Lure of the global best solution	Choose the worse solution according to probability function
Algorithm Type	Evolutionary Algorithm	Evolutionary Algorithm
Convergence Time	Fast	Slow

iterations of SA process into every particle for the proposed algorithm. The infix condition is presented in Fig.11.6.

11.5 The Process of ASA-PPSO

In this section, we are going to discuss the details of how ASA - PPSO works. Of course, before the initialization, the user has to decide how many particles and how many groups will be in the process. The larger the number of particles, the larger the chance to find the global optimal solution, but it also increases the process time. Thus, how many particles to use in a process is decided particularly by the problems, which you are facing.

Remember, the main process is the PPSO algorithm, not SA. The decrease scalar of temperature should not be too small; otherwise, the cost of time will increase a lot with only the improvement of the performance a little. We suggest that when infixing SA process, it should be only 5 iterations per infix. Thus, it will only spend a little bit more time, but better improves the performance.

11.6 Experimental Results of PSO, PPSO and ASA-PPSO

The experiments were processed on the machine, which is described as the following: CPU: AMD AthlonXP P1600 (1400.05-MHz), Motherboard: Tyan TigerMP (s2460), RAM: Transcend ECC Registerd DDR 256MB Module (TS32MDR72V6D5)$*$1, HDD: Seagate 10.7GB (ST310212A), OS: Freebsd 4.4-STABLE ♯1, Compiler: gcc version 2.95.3.

In the experiments, we use five test functions, namely Rosenbrock, Rastrigrin, Griewank, and two other test functions, for testing the three methods, weighted PSO, PPSO and ASA-PPSO. Each solution in the functions was set to be a

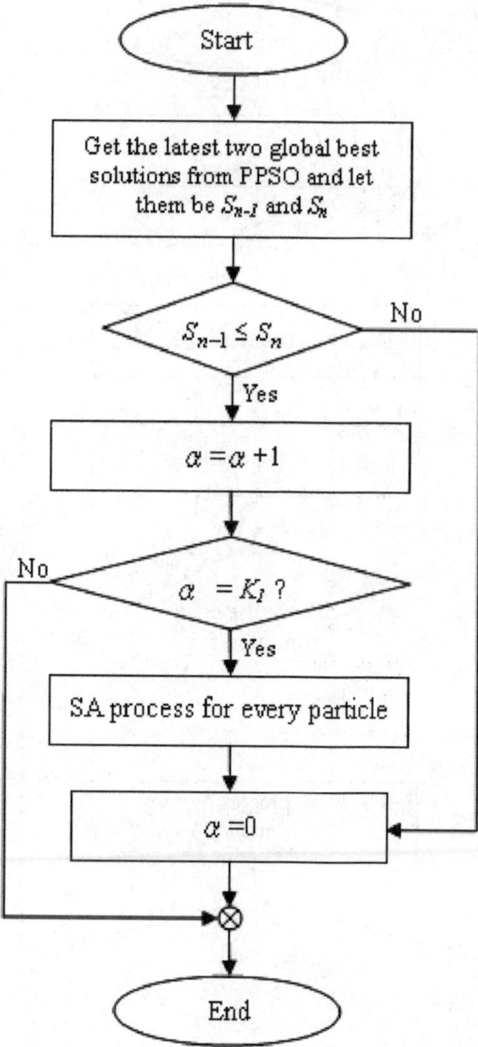

Fig. 11.6. Infix Condition

30 - dimensional real number vector. The solutions of all the test functions were found by 160 particles for each method, and measured by the average of 50 different random seeds with 2000 iterations per seed. The goal of the solutions was to minimize the fitness values of these five functions.

The first experiment was processed with Rosenbrock function, which can be described as:

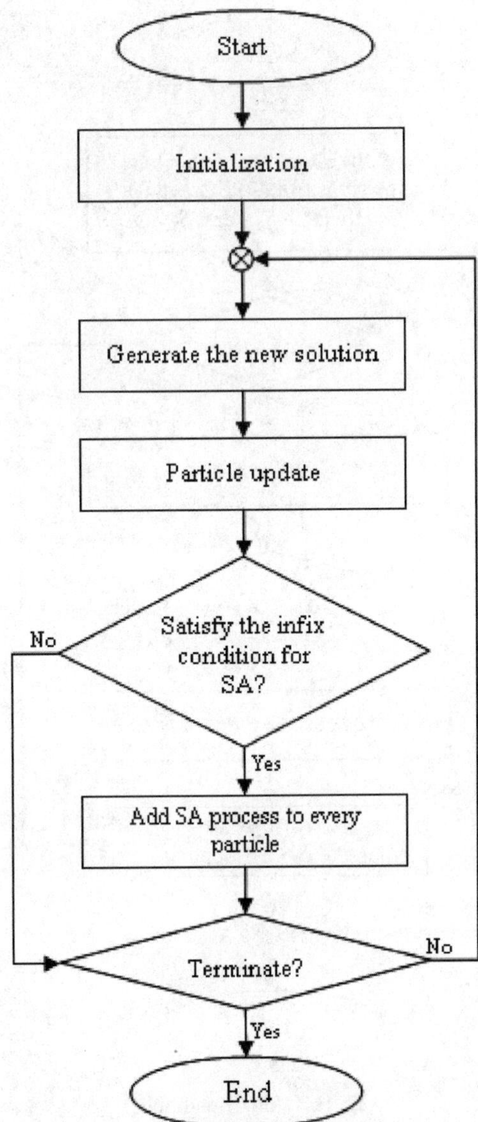

Fig. 11.7. Process of ASA-PPSO

Algorithm 11.1 Adaptive Simulated Annealing Based Parallel Particle Swarm Optimization

In the initial stage, as the same as normal PPSO process, we have to initialize the particles. During the process, we randomly set the global best solution, self's best solutions, velocities and positions of all particles under the initial ranges, which limit the solution spaces.

Subsequently, we move into the evolution stage. In this stage, we calculate the fitness value, which is the solution of the particle, of all particles by the fitness function. If the new fitness value is better than the old one, the replacement will be triggered. After calculating all the fitness values of particles, a sorting process is needed in order to find the particle, which has the best solution, and replace the memory. In the traditional PPSO, these are called one time of iteration, but in ASA-PPSO, there exists one last step, considering whether infix the SA process or not.

After the steps, which we just discussed, now we have the best solution at present. According to the infix condition, we have to compare the latest two best solutions to update the variable "α". When α reaches the trigger point, we add SA process to the specific particles (the particles can be some recommended ones or all of them).

So far, the ASA-PPSO finishes one time of iteration. The recursive process will keep go on until the fitness value fit in with the terminal condition. The ASA-PPSO can be presented as shown in Fig.11.7.

$$f_1(X) = \sum_{i=1}^{n}(100(x_{i+1} - x_i^2)^2 + (x_i - 1)^2). \tag{11.10}$$

The test function of second experiment was Rastrigrin function, which can be described as:

$$f_2(X) = \sum_{i=1}^{n}(x_i^2 - 10cos(2\pi x_i)^2 + 10). \tag{11.11}$$

The third experiment was processed with Griewank function, which can be described as:

$$f_3(X) = \frac{1}{400}\sum_{i=1}^{n} x_i^2 - \prod_{i=1}^{n} cos\left(\frac{x_i}{\sqrt{i}}\right) + 1. \tag{11.12}$$

The test function of fourth experiment was the function of eq.11.13.

$$f_4(X) = \sum_{i=1}^{n} \left| \frac{sin(10 \times x_i \times \pi)}{10 \times x_i \times \pi} \right| \tag{11.13}$$

The last experiment was processed with eq.11.14.

$$f_5(X) = 6 \times n + \sum_{i=1}^{n} \lfloor x_i \rfloor \tag{11.14}$$

According to the experiments, the results indicated that ASA-PPSO obtains the best performance overall, the second one is PPSO, and the last one is weighted PSO.

In the experiments, the setting of parameters is described as follows: The boundary conditions for initialization are shown in Table 11.2. Note that the boundary conditions only limit the solution sets in the initialization stage, not in the evolution stage. However, for test function five, we limit the boundary not only in the initial stage, but also the evolution stage due to we may easily find that if we do not limit the boundary in the evolution stage, the solution sets would move into the certain space with huge minus values.

The constants c_1 and c_2 of the velocity function are set to 2, and the initial weight is set to 0.9, the final weight is set to 0.4. The initial temperature of SA is set to 0.9, the final temperature is set to 0.4, and the decrease scalar of temperature is set to 0.1. The dimensions for every solution sets are set to 30 in real number space.

Table 11.2. The initial rage of parameters

Experimental Function	Range of Initial Position	Maximum Velocity
Rosenbrock	$15 \leq x_i \leq 30$	100
Rastrigrin	$2.56 \leq x_i \leq 5.12$	10
Griewank	$300 \leq x_i \leq 600$	600
Function 4	$-0.5 \leq x_i \leq 0.5$	10
Function 5	$-5.12 \leq x_i \leq 5.12$	0.5

To insure that each algorithm owns the same resources and has the same foundation for comparison, we assign 160 particles for every method. In other words, in the algorithms, which have the parallel structure, such as PPSO and ASA-PPSO, the particles are allotted equally into several groups. For example, if we have 2 groups in PPSO, there will be 80 particles in each group; however, there will be 40 particles in each group for 4 groups.

Here we define the symbols for presenting the results: PSO substitutes the Particle Swarm Optimization with weighted factor algorithm, which was presented by Shi. The parallel particle swarm optimization is referred to as PPSO, and ASA-PPSO means the Adaptive Simulated Annealing Based Parallel Particle Swarm Optimization. For PPSO and ASA-PPSO, we made some conditions: the number of groups was set to 4, in the experiment; we use communication strategy 2, exchange 1 particle each time in PPSO. For ASA-PPSO, we use communication strategy 2, exchange 2 particles each time, and the infix condition.

In Fig.11.8, we present the results of comparing weighted PSO, PPSO and ASA-PPSO with Rosenbrock function. We can find that ASA-PPSO obtains the highest performance, the PPSO gets the second high performance, and the weighted PSO obtains the worst performance. Note that the axis Y is presented in log scalar.

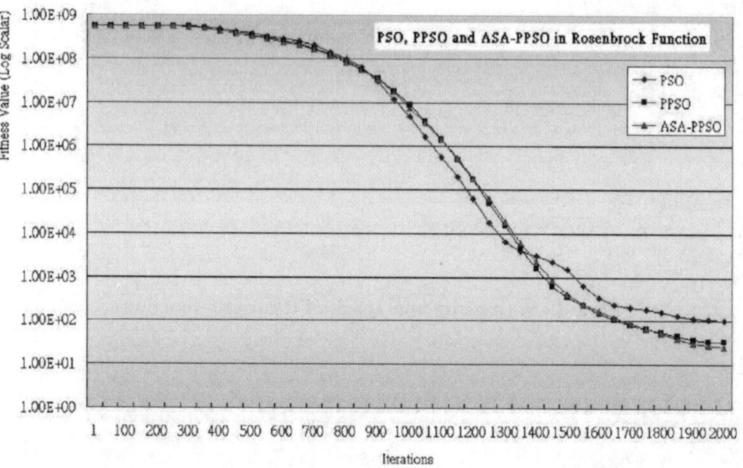

Fig. 11.8. Experimental result of Rosenbrock function

In Fig. 11.9, we present the results of comparing weighted PSO, PPSO and ASA-PPSO with Rastrigrin function.

In Fig. 11.10, we present the results of comparing weighted PSO, PPSO and ASA-PPSO with Griewank function. The axis Y in Fig. 11.10 is still presented in log scalar, although PPSO and ASA-PPSO found the minimum solution "zero", it can not be presented on the figure. That is the result of why the curve of PPSO and ASA-PPSO stop before the whole iterations be finished.

The results of comparing weighted PSO, PPSO and ASA-PPSO with function 4 is shown in Fig. 11.11.

In Fig. 11.12, we present the results of test function five. Since the solution sets are limited in the range [-5.12, 5.12], the results move in a much smaller range than the other test functions. To obviously view the result of the curves, the axis Y is in the normal scalar in Fig. 11.12.

By examining test function one, three and four, we can easily figure out that the weighted PSO cannot regularly leap the local optimal solutions, but the PPSO-type methodologies can. As the result, PPSO and ASA-PPSO all obtain the global

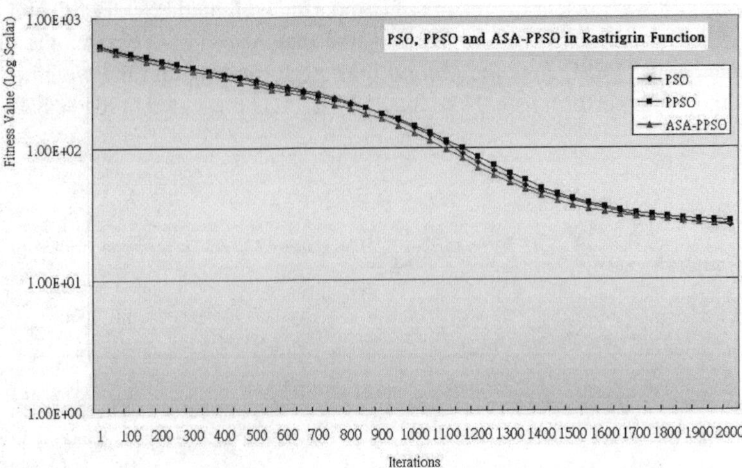

Fig. 11.9. Experimental result of Rastrigrin function

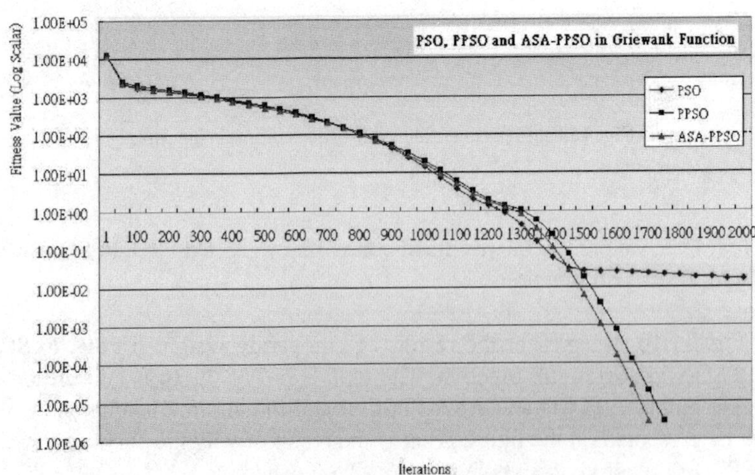

Fig. 11.10. Experimental result of Griewank function

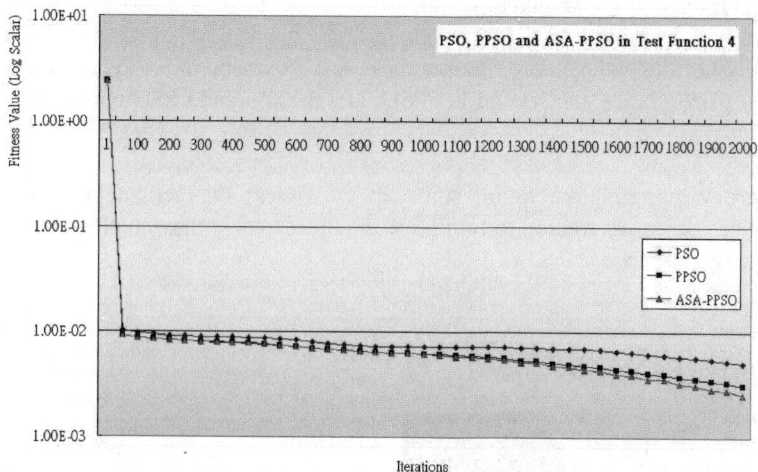

Fig. 11.11. Experimental result of function 4

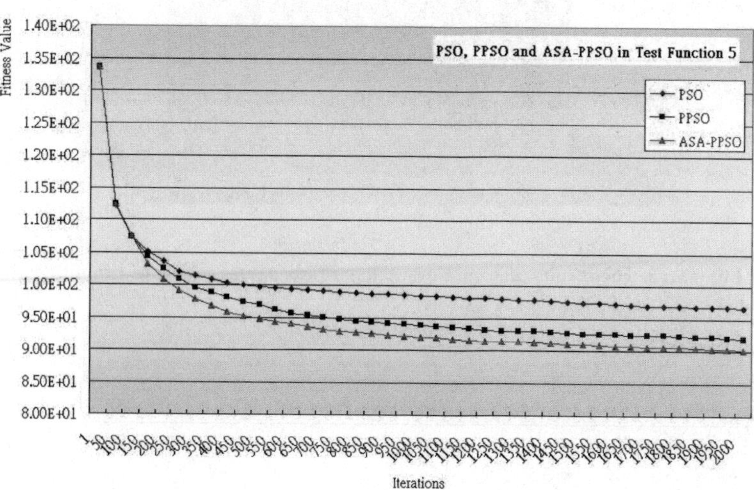

Fig. 11.12. Experimental result of function 5

best solution (for Griewank function in this experiment, the global best solution is zero). In addition, ASA-PPSO obtains the global best solution much earlier than PPSO, thus we can say that for these test functions, ASA-PPSO obtains the highest performance, the second is PPSO, and the weighted PSO obtains the worst performance.

Here we present the figure sums up the fitness of each method shown in Fig.11.13. Since we wanted to minimize the fitness value, the smallest one has the highest performance.

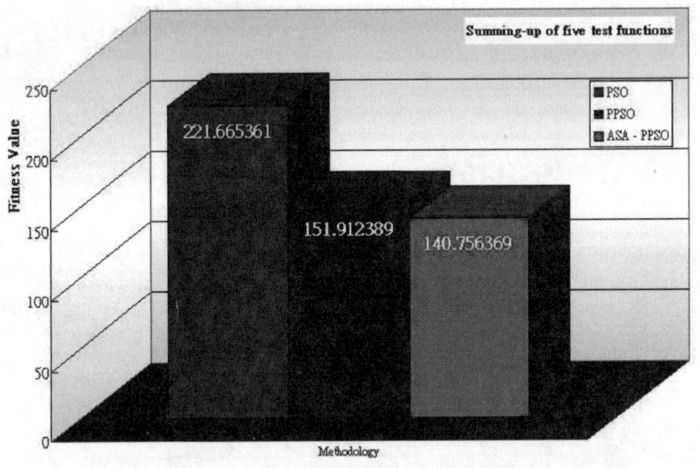

Fig. 11.13. The results of summing-up the fitness values of five test functions

11.7 Conclusion

In this chapter, we gave a briefly introduction of original particle swarm optimization algorithm, particle swarm optimization with weighted factor, parallel particle swarm optimization, and adaptive simulated annealing – parallel particle swarm optimization algorithm.

Through the experiments, the results indicate that ASA-PSO can better improve the performance to find the best solution. Indeed, parallel framework of particle swarm optimization certainly helps the algorithm to be more effective of finding solutions. In the algorithms with evolutionary type, particle swarm optimization is newer than genetic algorithm and simulated annealing, which means it may have more potential to be improved.

References

1. D. E. Goldberg, Genetic Algorithm in Search, Optimization and Machine Learning, Addison-Wesley Publishing Company, 1989.
2. L. Davis, Handbook of genetic algorithms, Van Nostrand Reinhold, New York, 1991.
3. M. Gen, R. Cheng, Genetic algorithm and engineering design, John Wiley and Sons, New York, 1997.
4. M. Dorigo, V. Maniezzo, A. Colorni, The ant system: optimization by a colony of cooperating agents, IEEE Transaction on Systems, Man and Cybernetics-Part B 26 (2) (1996) 29-41.
5. M. Dorigo, L. M. Gambardella, Ant colony system: a cooperative learning ap- proach to the traveling salesman problem, IEEE Transaction on Evolutionary Computation 26 (1) (1997) 53-66.
6. S. C. Chu, J. F. Roddick, J. S. Pan, Ant colony system with communication strategies, Information Sciences 167 (2004) 63-76.
7. R. Eberhart, J. Kennedy, A new optimizer using particle swarm theory, in: Sixth International Symposium on Micro Machine and Human Science, 1995, pp. 39-43.
8. J. Kennedy, R. Eberhart, Particle swarm optimization, in: IEEE International Conference on Neural Networks, 1995, pp. 1942-1948.
9. P. Tarasewich, P. R. McMullen, Swarm intelligence, Communications of the ACM 45 (8) (2002) 63-67.
10. P. Angeline, Evolutionary optimization versus particle swarm optimization: phi- losophy and performance di?erences, in: Proc. Seventh Annual Conference on Evolutionary Programming, 1998, pp. 601-611.
11. R. Eberhart, Y. Shi, Comparison between genetic algorithms and particle swarm optimization, in: Proc. Seventh Annual Conference on Evolutionary Program- ming, 1998, pp. 611-619.
12. Y. Shi, R. Eberhart, Empirical study of particle swarm optimization, in: Congress on Evolutionary Computation, 1999, pp. 1945-1950.
13. J.-F. Chang, S. C. Chu, J. F. Roddick, J. S. Pan, A parallel particle swarm optimization algorithm with communication strategies, Journal of Information Science and Engineering 21 (4) (2005) 809-818.
14. S. Kirkpatrick, J. C. D. Gelatt, M. P. Vecchi, Optimization by simulated annealing, Science 220 (4598) (1983) 671-680.
15. H. C. Huang, J. S. Pan, Z. M. Lu, S. H. Sun, H. M. Hang, Vector quantization based on genetic simulated annealing, Signal Processing 81 (7) (2001) 1513-1523.

12

Swarm Intelligence: Theoretical Proof That Empirical Techniques are Optimal

Dmitri Iourinski[1], Scott A. Starks[2], Vladik Kreinovich[2], and Stephen F. Smith[3]

[1] School of Computing Science, Middlesex University, North London Business Park, Oakleigh Road South, London N11 1QS, UK d.iourinski@mdx.ac.uk
[2] NASA Pan-American Center for Earth and Environmental Studies, University of Texas at El Paso, El Paso, TX 79968, USA sstarks@utep.edu, vladik@utep.edu
[3] Robotics Institute, Carnegie Mellon University, 500 Forbes Ave., Pittsburgh, PA 15213, USA sfs@cs.cmu.edu

Summary. A natural way to distribute tasks between autonomous agents is to use *swarm intelligence* techniques, which simulate the way social insects (such as wasps) distribute tasks between themselves. In this paper, we theoretically prove that the corresponding successful biologically inspired formulas are indeed statistically optimal (in some reasonable sense).

Key words: swarm intelligence, autonomous agents, optimality proof

12.1 Introduction

12.1.1 What Is Swarm Intelligence

In many real-life situations, we have a large number of tasks, and a large number of autonomous agents which can solve these tasks. The problem is how to best match agents and tasks. This problem is typical:

- in manufacturing, where we have several machines capable of performing multiple tasks;
- in robotics, when we need to coordinate the actions of several autonomous robots;
- in computing, when several parallel computers are available, etc.

In general, if we want an optimal matching, then this problem is difficult to solve. For example, it is known that the problem of optimal manufacturing scheduling is NP-hard; see, e.g., [19]. Since we cannot have an optimal solution, we must look for heuristic solutions to such problems.

One of the natural sources of such heuristics is biology, specifically, the biology of insects. Insects are usually small, so it is difficult for an individual insect to perform complex tasks. Instead, they swarm together and perform tasks in collaboration. Since the existing social insects are the result of billions of years of survival-of-the-fittest evolution, we expect that all the features of their collaboration

D. Iourinski et al.: *Swarm Intelligence: Theoretical Proof That Empirical Techniques are Optimal*, Studies in Computational Intelligence (SCI) **31**, 281–295 (2006)
www.springerlink.com

have been perfected to being almost optimal. Thus, it is reasonable to copy the way social insects interact. The resulting multi-agent systems are called *swarm intelligence* [8, 24].

12.1.2 What Formulas Are Used in the Existing Swarm Intelligence Systems

The biological observations led researchers to the following model for the insect collaboration: We have several classes of tasks. Each task T of type t is characterized by its degree of relevance $R_t(T)$; in biology, this degree of relevance is called a *stimulus*.

In principle, each agent can perform each task; in this sense, the agents are *universal*. However, different agents have different abilities with respect to different tasks. If an agent is not very skilled in a certain type of tasks, then this agent picks tasks of this type only when they are extremely important, i.e., when the stimulus is very high. If an agent is reasonable skilled in tasks of certain type, then this agent will also pick such tasks when the corresponding stimulus is much lower. This behavior can be characterized by assigning, to each agent A and to each type of tasks t, a *threshold* $\theta_t(A)$:

- if the stimulus $R_t(T)$ corresponding to a task T is much smaller than the threshold, then the agent will not take this task;
- if the stimulus is much larger than the threshold ($R_t(T) \gg \theta_t(A)$), then the agent will take this task.

In other words, whether the agent takes the task or not depends on the ratio $r \stackrel{\text{def}}{=} R_t(T)/\theta_t(A)$: if $r \ll 1$, the agent does not take the task; if $r \gg 1$, the agent takes the task.

When the ratio is close to 1 (i.e., when the stimulus is of the same order of magnitude as the threshold), then the same insect sometimes takes the task, sometimes does not. The frequency (probability) P with which an insect picks the task increases with the ratio r. From the biological observations, it was determined that the dependence of the probability P on the ratio r has the following form:

$$P(r) = \frac{r^2}{1+r^2}. \tag{12.1}$$

In other words, the probability P of an agent A to pick the task T of type t is equal to:

$$P = \frac{R_t(T)^2}{R_t(T)^2 + \theta_t(A)^2}. \tag{12.2}$$

This formula was proposed in the 1990s in [9, 10, 39]. Since then, it has been used in the existing swarm intelligence systems, and it has led to reasonable results [1, 2, 8, 13, 14, 15, 16, 24, 31, 32].

12.1.3 Formulation of the Problem

The idea that a probability P should depend on the ratio r is very convincing. However, the specific dependence of P on r (as described by the formula (12.1)) is rather ad hoc. Since this formula is successful, it is reasonable to try to find a justification for its use.

In this paper, we provide such a justification.

12.2 Main Idea

Since we want to design an *intelligent* system, we should allow agents to learn, i.e., to use their experience to correct their behavior. In the swarm intelligence model, at any given moment of time, the behavior of an agent A towards tasks of all possible types t is characterized by its thresholds $\theta_t(A)$. Thus, learning means changing the agent's thresholds, from the original values $\theta_t(A)$ to new values $\theta'_t(A)$. As a result, the probability

$$P = P(r) = P\left(\frac{R_t(A)}{\theta_t(A)}\right) \tag{12.3}$$

of an agent A taking the task T changes to a new value

$$P' = P(r') = P\left(\frac{R_t(A)}{\theta'_t(A)}\right). \tag{12.4}$$

The formula describing the transition from the original probabilities (12.3) to the new probabilities (12.4) can be further simplified if we denote the ratio of the old and the new thresholds by

$$\lambda = \frac{\theta_t(A)}{\theta'_t(A)}. \tag{12.5}$$

In terms of λ, we have $r' = \lambda \cdot r$, hence the new probability is equal to

$$P' = P(\lambda \cdot r). \tag{12.6}$$

From the statistical viewpoint (see, e.g., [21, 40, 42]), the optimal way of updating probabilities is by using the Bayes formula. Specifically, if we have n incompatible hypotheses $H_1, \ldots, H_n$ with initial probabilities

$$P_0(H_1), \ldots, P_0(H_n), \tag{12.7}$$

then, after observations E, we update the initial probabilities to the new values:

$$P(H_i \mid E) = \frac{P(E \mid H_i) \cdot P_0(H_i)}{P(E \mid H_1) \cdot P_0(H_1) + \ldots + P(E \mid H_n) \cdot P_0(H_n)}. \tag{12.8}$$

Thus, *an optimal function $P(r)$ can be determined as the one for which the transition from the old probabilities (12.3) to the new probabilities (12.4), (12.6) can be described by the (fractionally linear) Bayes formula (12.8).*

12.3 From the Main Idea to the Exact Formulas

Let us formalize the above condition. In our case, we have two hypotheses: the hypothesis H_1 that it is reasonable for an agent A to take a task of given type t, and the opposite hypothesis H_2 that it is not reasonable for the agent A to take such a task. Initially, the probability of the hypothesis H_1 is equal to P, and the probability of the opposite hypothesis H_2 is equal to $1-P$. According to Bayes formula, after some experience E, the probability P should be updated to the following new value $P' = P(H_1 \,|\, E)$:

$$P' = \frac{P(E\,|\,H_1)\cdot P}{P(E\,|\,H_1)\cdot P + P(E\,|\,H_2)\cdot(1-P)}. \tag{12.9}$$

If we denote $P(E\,|\,H_1)$ by a, $P(E\,|\,H_2)$ by b, and explicitly mention that the probability P depends on the ratio r, then the formula (12.9) takes the following form:

$$P' = \frac{a\cdot P(r)}{a\cdot P(r) + b\cdot(1-P(r))}. \tag{12.10}$$

We want the expression (12.6) to be representable in this form (12.10). So, we arrive at the following definition:

12.4 First Result

Definition 1. *A monotonic function $P(r): [0,\infty) \to [0,1]$ is called* optimal *if, for every $\lambda > 0$, there exist values $a(\lambda)$ and $b(\lambda)$ for which*

$$P(\lambda\cdot r) = \frac{a(\lambda)\cdot P(r)}{a(\lambda)\cdot P(r) + b(\lambda)\cdot(1-P(r))}. \tag{12.11}$$

Comment. In other words, we require that the 2-parametric family of functions $F = \left\{\dfrac{a\cdot P(r)}{a\cdot P(r)+b}\right\}$ corresponding to Bayesian learning be *scale-invariant* under a "re-scaling" $r \to \lambda\cdot r$.

Theorem 1. *Every optimal function $P(r)$ has the form*

$$P(r) = \frac{r^\alpha}{r^\alpha + c} \tag{12.12}$$

for some real numbers α and c.

In other words, for the optimal function $P(r)$, we have

$$P = \frac{R_t(T)^\alpha}{R_t(T)^\alpha + c\cdot\theta_t(A)^\alpha}. \tag{12.13}$$

If we re-scale the threshold by calling $\theta' = c^{1/\alpha}\cdot\theta$ the new threshold, then the formula (12.13) simplifies into

$$P = \frac{R_t(T)^\alpha}{R_t(T)^\alpha + \theta_t(A)^\alpha}. \tag{12.14}$$

Thus, we show that formula (12.14) – which is a minor generalization of the original formula (12.2) – is indeed optimal.

12.5 Proof of Theorem 1

It is known that many formulas in probability theory can be simplified if instead of the probability P, we consider the corresponding odds

$$O = \frac{P}{1-P}. \tag{12.15}$$

(If we know the odds O, then we can reconstruct the probability P as $P = O/(1+O)$.) The right-hand side of the formula (12.11) can be represented in terms of odds $O(r)$, if we divide both the numerator and the denominators by $1 - P(r)$. As a result, we get the following formula:

$$P(\lambda \cdot r) = \frac{a(\lambda) \cdot O(r)}{a(\lambda) \cdot O(r) + b(\lambda)}. \tag{12.16}$$

Based on this formula, we can compute the corresponding odds $O(\lambda \cdot r)$: first, we compute the value

$$1 - P(\lambda \cdot r) = \frac{b(\lambda)}{a(\lambda) \cdot O(r) + b(\lambda)}, \tag{12.17}$$

and then divide (12.16) by (12.17), resulting in:

$$O(\lambda \cdot r) = c(\lambda) \cdot O(r), \tag{12.18}$$

where we denoted $c(\lambda) = a(\lambda)/b(\lambda)$. It is known (see, e.g., [3, 29]) that all monotonic solutions of the functional equation (12.18) are of the form $O(r) = C \cdot r^\alpha$. Therefore, we can reconstruct the probability $P(r)$ as

$$P(r) = \frac{O(r)}{O(r)+1} = \frac{C \cdot r^\alpha}{C \cdot r^\alpha + 1}. \tag{12.19}$$

Dividing both the numerator and the denominator of the right-hand side by C and denoting $c = 1/C$, we get the desired formula (12.12). Q.E.D.

12.6 From Informally "Optimal" to Formally Optimal Selections

12.6.1 In the Previous Section, We Used Informal "Optimality"

In the above text, we argued that if a selection of a probability function is optimal (in some reasonable sense), than it is natural to expect that this selection should be scale-invariant. We used this argument to justify the empirical selection of a probability function – or, to be more precise, the empirical selection of a 2-parametric family $F = \left\{ \dfrac{a \cdot P(r)}{a \cdot P(r) + b} \right\}$.

In this section, we will go one step further, and explain that the empirical selection is indeed optimal – in the precise mathematical sense of this word.

In these terms, the question is how to select, out of all possible families, the family which is optimal in some reasonable sense, i.e., which is optimal in the sense of some optimality criterion.

12.6.2 What is an Optimality Criterion?

When we say that some *optimality criterion* is given, we mean that, given two different families F and F', we can decide whether the first or the second one is better, or whether these families are equivalent w.r.t. the given criterion. In mathematical terms, this means that we have a *pre-ordering relation* $\preceq$ on the set of all possible families.

One way to approach the problem of choosing the "best" family F is to select *one* optimality criterion, and to find a family that is the best with respect to this criterion. The main drawback of this approach is that there can be different optimality criteria, and they can lead to different optimal solutions. It is, therefore, desirable not only to describe a family that is optimal relative to some criterion, but to describe *all* families that can be optimal relative to different natural criteria[4]. In this section, we are planning to implement exactly this more ambitious task.

12.6.3 Examples of Optimality Criteria

Pre-ordering is the general formulation of optimization problems in general, not only of the problem of choosing a family F. In general optimization theory, in which we are comparing arbitrary *alternatives* a', a'', $\ldots$, from a given set A, the most frequent case of such a pre-ordering is when a *numerical criterion* is used, i.e., when a function $J : A \to R$ is given for which $a' \preceq a''$ iff $J(a') \le J(a'')$.

Several natural numerical criteria can be proposed for choosing a function J. For example, we can take, as a criterion, the *average* computation time (average in the sense of some natural probability measure on the set of all problems).

Alternatively, we can fix a class of problems, and take the largest (worst-case) computation time for problems of this class as the desired (numerical) optimality criterion.

Many other criteria of this type can be (and have actually been) proposed. For such "worst-case" optimality criteria, it often happens that there are several different alternatives that perform equally well in the worst case, but whose performance

[4]In this phrase, the word "natural" is used informally. We basically want to say that from the purely mathematical viewpoint, there can be weird ("unnatural") optimality criteria. In our text, we will only consider criteria that satisfy some requirements that we would, from the common sense viewpoint, consider reasonable and natural.

differ drastically in the average cases. In this case, it makes sense, among all the alternatives with the optimal worst-case behavior, to choose the one for which the average behavior is the best possible. This very natural idea leads to the optimality criterion that is not described by one numerical optimality criterion $J(a)$: in this case, we need *two* functions: $J_1(a)$ describes the worst-case behavior, $J_2(a)$ describes the average-case behavior, and $a \preceq b$ iff either $J_1(a) < J_1(b)$, or $J_1(a) = J_1(b)$ and $J_2(a) \leq J_2(b)$.

We could further specify the described optimality criterion and end up with *one* natural criterion. However, as we have already mentioned, the goal of this chapter is not to find *one* family that is optimal relative to some criterion, but to describe *all* families that are optimal relative to some natural optimality criteria. In view of this goal, in the following text, we will not specify the criterion, but, vice versa, we will describe a very general class of *natural* optimality criteria.

So, let us formulate what "natural" means.

12.6.4 What Optimality Criteria are Natural?

It is reasonable to require that the relative quality of two families does not change if we simply change the threshold, i.e., replace the function $P(r)$ with $P(\lambda \cdot r)$, and correspondingly, the family $F = \left\{ \dfrac{a \cdot P(r)}{a \cdot P(r) + b} \right\}$ with the "re-scaled" family $T_\lambda(F) \stackrel{\text{def}}{=} \left\{ \dfrac{a \cdot P(\lambda \cdot r)}{a \cdot P(\lambda \cdot r) + b} \right\}$.

There is one more reasonable requirement for a criterion, that is related with the following idea: If the criterion does not select a single optimal family, i.e., if it considers several different families equally good, then we can always use some other criterion to help select between these "equally good" ones, thus designing a two-step criterion. If this new criterion still does not select a unique family, we can continue this process until we arrive at a combination multi-step criterion for which there is only one optimal family. Therefore, we can always assume that our criterion is *final* in this sense.

Definition 2. *By an* optimality criterion, *we mean a pre-ordering (i.e., a transitive reflexive relation) $\preceq$ on the set A of all possible families. An optimality criterion $\preceq$ is called:*

- *scale-invariant if for all F, F', and $\lambda > 0$, $F \preceq F'$ implies $T_\lambda(F) \preceq T_\lambda(F')$;*
- *final if there exists one and only one family F that is preferable to all the others, i.e., for which $F' \preceq F$ for all $F' \neq F$.*

Theorem 2.

- *If a family F is optimal w.r.t. some scale-invariant final optimality criterion, then this family F is generated by $P(r) = r^\alpha$ for some $\alpha > 0$.*
- *For families corresponding to $P(r) = r^\alpha$, there exists a scale-invariant final optimality criterion for which the only optimal family is this family.*

Comment. In other words, if the optimality criterion satisfies the above-described natural properties, then *the optimal function is* $P(r) = r^\alpha$.

12.6.5 Proof of Theorem 2

We have already shown, in the proof of Theorem 1, that:

- for $P(r) = r^\alpha$, the corresponding family is scale-invariant, and
- vice versa, that if a family is scale-invariant, then it corresponds to $P(r) = r^\alpha$.

1°. To prove the first part of Theorem 2, we thus need to show that for every scale-invariant and final optimality criterion, the corresponding optimal family $F_{\rm opt}$ is scale-invariant, i.e., that $T_\lambda(F_{\rm opt}) = F_{\rm opt}$ for all $\lambda > 0$. Then, the result will follow from Theorem 1.

Indeed, the transformation T_λ is invertible, its inverse transformation is a scaling by $1/\lambda$: $T_\lambda^{-1} = T_{1/\lambda}$. Now, from the optimality of $F_{\rm opt}$, we conclude that for every $F' \in A$, $T_\lambda^{-1}(F') \preceq F_{\rm opt}$. From the invariance of the optimality criterion, we can now conclude that $F' \preceq T_\lambda(F_{\rm opt})$. This is true for all $F' \in A$ and therefore, the family $T(F_{\rm opt})$ is optimal.

But since the criterion is final, there is only one optimal indicator function; hence, $T_\lambda(F_{\rm opt}) = F_{\rm opt}$. So, the optimal family is indeed invariant and hence, due to Theorem 1, it corresponds to $P(r) = r^\alpha$. The first part is proven.

2°. Let us now prove the second part of Proposition 2. Let $P(r) = r^\alpha$, and let F_0 be the corresponding family. We will then define the optimality criterion as follows: $F \preceq F'$ iff F' is equal to this F_0.

Since the family F_0 is scale-invariant, thus the defined optimality criterion is also scale-invariant. It is also clearly final.

The family F_0 is clearly optimal w.r.t. this scale-invariant and final optimality criterion. The theorem is proven.

12.7 Discussion

Traditionally, the choice of a function $P(r)$ is done empirically, by comparing the results of different choices. Two related questions naturally arise:

- first, a *theoretical* question: how can we explain the empirical selection?
- second, a *practical* question: an empirical choice is made by using only finitely many functions; is this choice indeed the best – or there are other, even better functions $P(r)$, which we did not discover because we did not try them?

Our result answers both questions:

- first, we provide a theoretical explanation for the optimality of the empirical choice;

- thus, by proving that these empirical formulas are optimal not only in comparison with other functions that we have tried, but in comparison with all possible functions $P(r)$, we enable the practitioners not to waste time on trying different functions $P(r)$.

12.8 Extending the Optimality Result to a Broader Context

12.8.1 Formulation of a More General Problem

Swarm intelligence techniques are a class of methodology for solving global optimization problems. In this chapter, we have discussed how to optimally select techniques from this class. It is reasonable to consider this problem in a broader setting: how can we optimally select techniques for solving global optimization problems – without necessarily restricting ourselves to swarm intelligence.

The need to make such a selection comes from the fact that, in general, the problem of finding the exact values x that minimize a given objective function $f(x)$ is computationally difficult (NP-hard); see, e.g., [41]. Crudely speaking, NP-hardness means that (provided that P≠NP) it is not possible to have an algorithm that solves *all* optimization problems in reasonable time. In other words, no matter how good is an algorithm for solving global optimization optimization problems, there will always be cases in which better results are possible.

Since we cannot hope for a single algorithm for global optimization, new algorithms are constantly designed, and the existing algorithms are constantly modified. As a result, we have a wide variety of different global optimization techniques and methods; see, e.g., [17, 20, 22, 30, 38]. In particular, there exist numerous heuristic and semi-heuristic techniques which – similar to swarm intelligence techniques – emulate the way optimization is done in nature; e.g., genetic algorithms simulate the biological evolution which, in general, leads to the birth and survival individuals and species which are best fit for a given environment; see, e.g., [28].

Because of this variety of different global optimization techniques, every time we have a new optimization problem, we must select the best technique for solving this problem. This selection problem is made even more complex by the fact that most techniques for solving global optimization problems have *parameters* that need to be adjusted to the problem or to the class of problems. For example, in gradient methods, we can select different step sizes.

When we have a *single* parameter (or few parameters) to choose, it is possible to empirically try many values and come up with an (almost) optimal value. Thus, in such situations, we can come up with optimal version of the corresponding technique. In other approaches, e.g., in swarm intelligence, instead of selecting the value of single *number*-valued parameter, we have select the auxiliary *function*. It is not practically possible to test all possible functions, so it is not easy to come up with an optimal version of the corresponding technique.

In this chapter, we described an indirect way of finding the optimal version of swarm intelligence techniques. It is desirable to consider a more general problem of selecting the best auxiliary function within a given global optimization technique – so that we would be able to either analytically solve the problem of finding the optimal auxiliary function, or at least reduce this problem to an easier-to-solve problem of finding a few numerical parameters.

12.8.2 Case Study: Optimal Selection of Penalty (Barrier) Functions

A well-known Lagrange multiplier method minimizes a function $f(x)$ under a constraint $g(x) = 0$ by reducing it to the un-constrained problem of optimizing a new objective function $f(x) + \lambda \cdot g(x)$. One of the known approaches to solving a similar problem with a constraint $g(x) \geq 0$ is the *penalty* (*barrier*) method in which we reduce the original problem to the un-constrained problem of optimizing a new objective function $f(x) + \lambda \cdot g(x) + \mu \cdot P(g(x))$, for an appropriate (non-linear) penalty function $P(y)$. Traditionally, the most widely used penalty functions are $P(y) = y \cdot \ln(y)$ and $P(y) = y^2$. How can we select an optimal penalty function? Or, to be more precise, how can we select the optimal family $\{\lambda \cdot y + \mu \cdot P(y)\}$?

The first natural requirements is that the optimal penalty function $P(y)$ should be smooth. Smoothness is needed because smooth functions are easier to optimize, and we therefore want our techniques to preserve smoothness.

In solving a similar problem from swarm intelligence, we used the argument that the optimal expression should not change if we simply change the threshold and thus, re-scale the parameter r. For penalty functions, similarly, the measuring unit for measuring the quantity y is often a matter of arbitrary choice: we can use meters or feet to measure length, we can use pounds or kilograms to measure weight, etc. If a selection of the penalty function $P(y)$ is "optimal" (in some intuitive sense), then the results of using this penalty functions should not change if we simply change the measuring unit for measuring y – i.e., replace each value y with a new value $C \cdot y$, where C is the ratio of the corresponding units. Indeed, otherwise, if the "quality" of the resulting penalty method changes with this "re-scaling", we could change the unit and get a better penalty function $P(y)$ – which contradicts to our assumption that the selection of $P(y)$ is already optimal.

So, the "optimal" choices $P(y)$ can be determined from the requirement that the family $\{\lambda \cdot y + \mu \cdot P(y)\}$ be invariant under the corresponding re-scaling.

Definition 3. *A 2-parametric family of functions* $F = \{\lambda \cdot y + \mu \cdot P(y)\}$ *is called* scale-invariant *if for every* $C > 0$*, it coincides with the family* $T_C(F) \stackrel{\text{def}}{=} \{\lambda \cdot C \cdot y + \mu \cdot P(C \cdot y)\}$.

At first glance, scale-invariance is a reasonable but weak property. It turns out, however, that this seemingly weak property actually almost uniquely determines the optimal selection of penalty functions; see, e.g., [29].

Proposition 1. *If a family* $\{\lambda \cdot y + \mu \cdot P(y)\}$ *is scale-invariant, then this family corresponds to* $P(y) = y^\alpha$ *or to* $P(y) = y \cdot \ln(y)$.

12.8.3 Proof of Proposition 1

Since the family is scale-invariant, for every C, the re-scaled function $P(C\cdot y)$ must belong to the same family, i.e., there must exist $\lambda(C)$ and $\mu(C)$ for which

$$P(C\cdot y)=\lambda(C)\cdot y+\mu(C)\cdot P(y) \tag{12.20}$$

for all C and y.

Differentiating both sides of (12.20) by C and setting $C=1$, we conclude that

$$y\cdot\frac{dP}{dy}=\lambda_0\cdot y+\mu_0\cdot P(y), \tag{12.21}$$

where $\lambda_0\stackrel{\text{def}}{=}\frac{d\lambda(C)}{dC}_{|C=1}$ and $\mu_0\stackrel{\text{def}}{=}\frac{d\mu(C)}{dC}_{|C=1}$. One can check that the only solutions to these equation are $P(y)=C_1\cdot y+C_2\cdot y^{\mu_0}$ (when $\mu_0\neq 1$) and $P(y)=C_1\cdot y+C_2\cdot y\cdot\ln(y)$ (when $\mu_0=1$). Thus, the only scale-invariant families $\{\lambda\cdot y+\mu\cdot P(y)\}$ are families corresponding to $P(y)=y\cdot\ln(y)$ and $P(y)=y^{\alpha}$ for some real number α.

Thus, under any scale-invariant optimality criterion, the optimal penalty function must indeed take one of the desired forms. Q.E.D.

Comments.

- We can also show that for every scale-invariant final optimality criterion, the optimal family corresponds to $P(y)=y\cdot\ln(y)$ and $P(y)=y^{\alpha}$.
- This example also shows that we can go beyond theoretical justification of empirically best heuristic, towards finding new optimal heuristics: indeed, for penalty functions, instead of two-parameter families $\{\lambda\cdot y+\mu\cdot P(y)\}$, we can consider multiple-parameter families

  $$\{\lambda\cdot y+\mu_1\cdot P_1(y)+\ldots+\mu_m\cdot P_m(y)\}$$

  for several functions $P_1(y),\ldots,P_m(y)$. In this case, the optimal functions have also been theoretically found: they are of the type

  $$P_i(y)=y^{\alpha_i}\cdot(\ln(y))^{p_i}$$

  for real (or complex) values α_i and non-negative integer values of p_i [29].

12.8.4 Other Examples

Similar symmetry-based techniques provide an explanation of several other empirically optimal techniques.

How to bisect a box. For example, many optimization algorithms are based on the branch-and-bound idea, where we subdivide the original domain into several smaller subdomains – and thus, reduce the original difficult-to-solve problem of optimizaing the objective function $f(x)$ over the entire domain to several easier-to-solve problems of optimizing $f(x)$ over smaller domains (usually, boxes). Some of these boxes must be further subdivided, etc. Two natural questions arise:

- which box should we select for bisection?
- which variable shall we use to bisect the selected box?

To answer both questions, several heuristic techniques have been proposed, and there has been an extensive empirical comparative analysis of these techniques. It turns out that for both questions, the symmetry-based approach enables us to theoretically justify the empirical selection:

- Until recently, for subdivision, a box B was selected for which the computed lower bound $\underline{f}(B)$ was the smallest possible. Recently (see, e.g, [11, 12]), it was shown that the optimization algorithms converge much faster if we select, instead, a box B with the largest possible value of the ratio

  $$I_0 = \frac{\widetilde{f} - \underline{f}(B)}{\overline{f}(B) - \underline{f}(B)},$$

  where $\widetilde{f}$ is a current upper bound on the actual global minimum. In [25], we give a symmetry-based theoretical justification for this empirical criterion. Namely, we consider all possible indicator functions $I(\underline{f}(B), \overline{f}(B), \widetilde{f})$, and we show that:
  - first, that the empirically best criterion I_0 is the only one that is *invariant* w.r.t. some reasonable symmetries – namely, shift and scaling; and
  - second, that this criterion is *optimal* in some (symmetry-related) reasonable sense.
- We can bisect a given box in n different ways, depending on which of n sides we decided to halve. So, the natural question appears: which side should we cut? i.e., where to bisect a given box? Historically the first idea was to cut the *longest* side (for which $x_i^U - x_i^L \to \max$). It was shown (in [33, 34]) that much better results are achieved if we choose a side i for which $|d_i|(x_i^U - x_i^L) \to \max$, where d_i is the known approximation for the partial derivative $\dfrac{\partial f}{\partial x_i}$. In [23], we consider arbitrary selection criteria, i.e., functions

  $$S(f, d_1, \ldots, d_n, x_1^L, x_1^U, \ldots, x_n^L, x_n^U),$$

  which map available information into an index $S \in \{1, 2, \ldots, n\}$, and we show that the empirically best box-splitting strategy is the only scale-invariant one – and is, thus, optimal under any scale-invariant final optimality criterion.

How to enlarge a box. Sometimes, it is beneficial to (slightly) enlarge the original (non-degenerate) box $[x^L, x^U]$ and thus improve the performance of the algorithm; the empirically efficient "epsilon-inflation" technique [35, 36]

$$[x_i^L, x_i^U] \to [(1+\varepsilon)x_i^L - \varepsilon \cdot x_i^U, (1+\varepsilon)x_i^U - \varepsilon \cdot x_i^L]$$

was proven to be the only shift- and scale-invariant technique and thus, the only one optimal under an arbitrary shift-invariant and scale-invariant optimality criterion [26] (see also [37]).

Convex-based techniques. Several algorithms for solving convex global optimization problems are based on the fact that for convex functions there exist efficient algorithms for finding the global minimum. There are numerous effective global optimization techniques that reduce the general global optimization problems to convex ones; see, e.g., [17, 38]. Empirically, among these techniques, the best are αBB method [4, 5, 17, 27] and its modifications recently proposed in [6, 7]. It turns out [18] that this empirical optimality can also be explained via shift- and scale-invariance.

Simulated annealing and genetic algorithms. By using shift-invariance, we can also explain why the probability proportional to $\exp(-\gamma \cdot f(x))$ is optimal in simulated annealing [29].

By using scale- and shift-invariance, we explain why exponential and power re-scalings of the objective function are optimal in genetic algorithms [29].

Acknowledgments

This work was supported in part by NASA under cooperative agreement NCC5-209 and grant NCC 2-1232, by the Future Aerospace Science and Technology Program (FAST) Center for Structural Integrity of Aerospace Systems, effort sponsored by the Air Force Office of Scientific Research, Air Force Materiel Command, USAF, under grants numbers F49620-95-1-0518 and F49620-00-1-0365, by grant No. W-00016 from the U.S.-Czech Science and Technology Joint Fund, by NSF grants CDA-9522207, ERA-0112968 and 9710940 Mexico/Conacyt, and by IEEE/ACM SC2001 Minority Serving Institutions Participation Grant.

Smith's work was sponsored in part by the Department of Defense Advanced Research Projects Agency and the U.S. Air Force Rome Research Laboratory under contract F30602-00-2-0503, by the National Aeronautics and Space Administration under Contract NCC2-1243 and by the CMU Robotics Institute.

The authors are thankful to the editors and to the anonymous referees for valuable suggestions.

References

1. Abraham A, Grosan C, Ramos V, Eds. (2006) Swarm Intelligence and Data Mining. Springer-Verlag, Berlin – Heidelberg

2. Abraham A, Grosan C, Ramos V, Eds. (2006) Stigmergic Optimization. Springer-Verlag, Berlin – Heidelberg
3. Aczel J (1966) Lectures on functional equations and their applications. Academic Press, New York, London
4. Adjiman CS, Androulakis I, Floudas CA (1998) A global optimization method, αBB, for general twice-differentiable constrained NLP II. Implementation and computational results. Computers and Chemical Engineering, 22:1159–1179
5. Adjiman CS, Dallwig S, Androulakis I, Floudas CA (1998) A global optimization method, αBB, for general twice-differentiable constrained NLP I. Theoretical aspects. Computers and Chemical Engineering, 22:1137–1158
6. Akrotirianakis IG, Floudas CA (2004) Computational experience with a new class of convex underestimators: box-constrained NLP problems. Journal of Global Optimization, 29:249–264
7. Akrotirianakis IG, Floudas CA (2006) A new class of improved convex underestimators for twice continuously differentiable constrained NLPs. Journal of Global Optimization, to appear.
8. Bonabeau E, Dorigo M, Theraulaz G (1999) Swarm Intelligence: From Natural to Artificial Systems. Oxford University Press, Oxford
9. Bonabeau E, Sobkowski A, Theraulaz G, Jean-Louis Deneubourg J-L (1997) Adaptive Task Allocation Inspired by a Model of Division of Labor in Social Insects. In: Lundh D, Olsson B (eds), Bio Computation and Emegent Computing, World Scientific, Singapore, 36–45
10. Bonabeau E, Théraulaz G, Denebourg J-L (1996) Quantitative study of the fixed response threshold model for the regulation of division of labour in insect societies. Proc. Roy. Soc. B 263:1565–1569
11. Casado LG, García I (1998) New load balancing criterion for parallel interval global optimization algorithm, In: Proc. of the 16th IASTED International Conference, Garmisch-Partenkirchen, Germany, February 1998, 321–323
12. Casado LG, García I, Csendes T (2000) A new multisection technique in interval methods for global optimization. Computing, 65:263–269
13. Cicirello VA, Smith SF (2001) Ant colony control for autonomous decentralized shop floor routing. In: Proc. 5th Int'l Symposium on Autonomous Decentralized Systems ISADS'2001, IEEE Computer Society Press, March 2001.
14. Cicirello, VA, Smith SF (2001) Insect societies and manufacturing. In: Proc. of IJCAI'01 Workshop on AI and Manufacturing: New AI Paradigms and Manufacturing, August 2001.
15. Cicirello VA, Smith SF (2001) Wasp nests for self-configurable factories. In: Agents'2001, Proc. 5th Int'l Conference on Autonomous Agents, ACM Press, May-June 2001.
16. Cicirello VA, Smith SF (2001) Improved routing wasps for distributed factory control. In: Proc. of IJCAI'01 Workshop on AI and Manufacturing: New AI Paradigms and Manufacturing, August 2001.
17. Floudas CA (2000) Deterministic Global Optimization: Theory, Methods, and Applications. Kluwer, Dordrecht
18. Floudas CA, Kreinovich V (2006) Towards Optimal Techniques for Solving Global Optimization Problems: Symmetry-Based Approach. In: Torn A, Zilinskas, J (eds.), Models and Algorithms for Global Optimization, Springer, Dordrecht (to appear)
19. Garey M, Johnson, D (1979) Computers and intractability: a guide to the theory of NP-completeness. Freeman, San Francisco
20. Horst R, Pardalos PM, eds. (1995) Handbook of Global Optimization, Kluwer, Dordrecht

21. Jaynes ET, Bretthorst GL, ed. (2003) Probability Theory: The Logic of Science. Cambridge University Press, Cambridge, UK
22. Kearfott RB (1996) Rigorous Global Search: Continuous Problems. Kluwer, Dordrecht
23. Kearfott RB, Kreinovich V (1998) Where to Bisect a Box? A Theoretical Explanation of the Experimental Results. In: Alefeld G, Trejo RA (eds.), Interval Computations and its Applications to Reasoning Under Uncertainty, Knowledge Representation, and Control Theory. Proceedings of MEXICON'98, Workshop on Interval Computations, 4th World Congress on Expert Systems, México City, México
24. Kennedy J, Eberhart R, Shi Y (2001) Swarm Intelligence. Morgan Kaufmann, San Francisco, California
25. Kreinovich V, Csendes T (2001) Theoretical Justification of a Heuristic Subbox Selection Criterion. Central European Journal of Operations Research CEJOR, 9:255–265
26. Kreinovich V, Starks SA, Mayer G (1997) On a Theoretical Justification of The Choice of Epsilon-Inflation in PASCAL-XSC. Reliable Computing, 3:437–452
27. Maranas CD, Floudas CA (1994) Global minimal potential energy conformations for small molecules. Journal of Global Optimization, 4:135–170
28. Michalewicz Z (1996) Genetic Algorithms + Data Structures = Evolution Programs. Springer, Berlin
29. Nguyen HT, Kreinovich V (1997) Applications of continuous mathematics in computer science. Kluwer, Dordrecht
30. Pinter JD (1996) Global Optimization in Action. Kluwer, Dordrecht
31. Ramos V, Merelo JJ (2002) Self-Organized Stigmergic Document Maps: Environment as a Mechanism for Context Learning. In: Alba E, Herrera F, Merelo JJ, et al. (Eds.), Proc. of 1st Spanish Conference on Evolutionary and Bio-Inspired Algorithms AEB'02, Centro Univ. de Mèrida, Mérida, Spain, Feb. 6–8, 2002, pp. 284–293
32. Ramos V, Muge F, Pina P (2002) Self-Organized Data and Image Retrieval as a Consequence of Inter-Dynamic Synergistic Relationships in Artificial Ant Colonies. In: Ruiz-del-Solar A, Abraham A, Köppen M (Eds.), Frontiers in Artificial Intelligence and Applications, Soft Computing Systems – Design, Management and Applications, 2nd Int. Conf. on Hybrid Intelligent Systems, Santiago, Chile, Dec. 2002, IOS Press, 87:500–509
33. Ratz D (1992) Automatische Ergebnisverifikation bei globalen Optimierungsproblemen. Ph.D. dissertation, Universität Karlsruhe
34. Ratz D (1994) Box-Splitting Strategies for the Interval Gauss–Seidel Step in a Global Optimization Method. Computing, 53:337–354
35. Rump SM (1980) Kleine Fehlerschranken bei Matrixproblemen, Ph.D. dissertation, Universität Karlsruhe
36. Rump SM (1992) On the solution of interval linear systems. Computing, 47:337–353
37. Rump SM (1998) A Note on Epsilon-Inflation. Reliable Computing, 4(4):371–375
38. Tawarmalani M, Sahinidis NV (2002) Convexification and Global Optimization in Continuous and Mixed-Integer Nonlinear Programming: Theory, Algorithms, Software, and Applications. Kluwer, Dordrecht
39. Theraulaz G, Bonabeau E, Deneubourg JL (1998) Reponse threshold reinforcement and division of labor in insect societies, Proceeeings of the Royal Society of London B 263(1393):327–335
40. Vapnik VN (2000) The Nature of Statistical Learning Theory. Springer-Verlag, New York
41. Vavasis SA (1991) Nonlinear Optimization: Complexity Issues. Oxford University Press, New York
42. Wadsworth HM (ed) (1990) Handbook of statistical methods for engineers and scientists. McGraw-Hill Publishing Co., New York

Index

Printing: Krips bv, Meppel
Binding: Stürtz, Würzburg